Materials Issues in Art and Archaeology X

MATERIALS RESEARCH SOCIETY
SYMPOSIUM PROCEEDINGS VOLUME 1656

Materials Issues in Art and Archaeology X

Symposium held December 1-6, 2013, Boston, Massachusetts, U.S.A.

EDITORS

Pamela B. Vandiver
University of Arizona
Tucson, Arizona, U.S.A.

Weidong Li
Shanghai Institute of Ceramics, CAS
Shanghai, China

Philippe Sciau
Universite de Toulouse
Toulouse, France

Christopher Maines
National Gallery of Art
Washington, DC, U.S.A.

Materials Research Society
Warrendale, Pennsylvania

CAMBRIDGE
UNIVERSITY PRESS

University Printing House, Cambridge CB2 8BS, United Kingdom

One Liberty Plaza, 20th Floor, New York, NY 10006, USA

477 Williamstown Road, Port Melbourne, VIC 3207, Australia

314-321, 3rd Floor, Plot 3, Splendor Forum, Jasola District Centre, New Delhi - 110025, India

79 Anson Road, #06-04/06, Singapore 079906

Cambridge University Press is part of the University of Cambridge.

It furthers the University's mission by disseminating knowledge in the pursuit of education, learning and research at the highest international levels of excellence.

www.cambridge.org
Information on this title: www.cambridge.org/9781605116334

Materials Research Society
506 Keystone Drive, Warrendale, PA 15086
http://www.mrs.org

First published 2017

CODEN: MRSPDH

A catalogue record for this publication is available from the British Library

ISBN 978-1-605-11633-4 Hardback

CONTENTS

*Invited Paper

THE NARRATIVE INTERPRETATION OF PROCESS RECONSTRUCTION AND MATERIALS CHARACTERIZATION IN TECHNICAL ART HISTORY AND ARCHAEOLOGICAL SCIENCE

TECHNICAL ART HISTORY: PIGMENT IDENTIFICATION, REACTIVITY AND TRANSFORMATION

*Invited Paper

ALTERATION, TECHNOLOGY AND INTERPRETATION OF ARCHAEOLOGICAL CERAMICS, GLAZES AND GLASSES

*Invited Paper

CHARACTERIZATION OF METAL AND STONE CHARACTERIZATION, AND METAL NANOPARTICLE CORROSION

*Invited Paper

METHOD DEVELOPMENT IN IMAGE PROCESSING, ANALYSIS AND PROOF OF CONCEPT

*Invited Paper

PREFACE

Materials Issues in Art and Archaeology (MIAA) X, the tenth in a regularly occurring symposium series, was held at the Materials Research Society Fall Meeting, Dec. 1-6, 2013, in Boston, Massachusetts. Forty papers were presented of which only twenty-seven are published in this volume. The MIAA symposium began in 1987, at the instigation of Cliff Draper, as a single plenary lecture by Pieter Meyers, then at the Metropolitan Museum of Art. The first symposium was published as MRS Symposium Proceedings volume 123. Since then, eight other symposia have been held and published as MRS Symp. Proc. vols. 185, 267, 352, 462, 712, 852, 1044 and 1319. By the rapid publication standards of the MRS, our volumes have been slow to come to fruition due to the standard of having more peer review attention for each manuscript. Such care has helped coalesce identity of and support for this hybrid field. This is the only symposium in the U.S. where scientists in the field can meet together and share their work with one another and the wider community of materials researchers doing cutting-edge science.

Although the focus of topics has changed and evolved in the previous twenty-six years, four areas have assumed significance: archaeological science, heritage conservation science, technical art history, and reverse-engineering ancient and traditional technologies. Another area has involved instrumentation development, testing, and adaptation for non-destructive analysis and/or analysis of micro-samples, and many of these studies have been collaborations been with scientists we have met at the MRS meetings. Another area significant for conservation science has been preventive conservation and stabilization that has involved sensor development, monitoring, modeling and long-term best-practices for preserving art objects and archaeological artifacts. Integral to this process has been assessment of mechanical properties, and the construction of three-dimensional plots of strength, relative humidity and temperature of cultural materials as well as those used for conservation treatments, especially of fragile materials, such as lacquer, paint films, paper, ivory, plastics and others. We have endeavored to present some sessions on outreach to non-scientists and under-represented minorities, and some on the use of our studies of beautiful and culturally important objects to attract middle and high school students and challenge them to understand the methods and results of scientific research into materials. At many meetings we have toured the conservation and scientific research laboratories at museums in Boston, and at several others we have arranged workshops, for instance, in faience-making and various glass-blowing, core-forming and millefiori techniques at the Massachusetts Institute of Technology Department of Materials Science and Engineering glass laboratory, thanks to laboratory director, Peter Houk. Also held at MIT was a workshop on colored coatings on metals and alloys presented by jewelry professors from the Museum of Fine Arts, Boston's Museum School, a bronze casting demonstration in the MIT DMSE foundry, and an investigation of the visualization and tactile projects at the MIT Media Lab.

As our interdisciplinary and multi-disciplinary field has grown and coalesced, we have tried to attract one or more international organizers at each symposium, who in turn draw presenters from their respective regions. Now the European Materials Research Society, the Chinese Materials Research Society and Sociedad Mexicana de Materiales annual meetings hold symposia on conservation and archaeological science, technical studies of objects, buildings and sites, and the variation of materials and their properties. In addition, the National Science Foundation has initiated a SCI-ART initiative with three rounds of funding, such that proposals now compete in the Chemistry and Materials Research directorates. The Gordon Research Conference program has hosted summer meetings on advanced studies and application of nuclear methods. The American Ceramic Society has formed an Art, Archaeology and Conservation Science Division, that meets on alternate years at the Materials Science and Technology national conventions.

Educational offerings also have increased in number and scope. At the first MIAA symposium, a program in Heritage Conservation Science was being established at Johns Hopkins University that is now defunct, but presently six universities recognize the field as a worthwhile topic of doctoral research, including the University of Arizona, the University of Delaware and Northwestern University. The MIAA symposia have contributed to increasing awareness and opportunities in the field by leading or helping with the production of four focused *MRS Bulletin* issues, entitled "Microscopy in Archaeology" edited by Terry Childs, "Art and Technology" edited by P.B. Vandiver and James Druzik, "The Science of Art," edited by James Mavor, and "Preserving Art Through the Ages" edited by P.B. Vandiver. More diverse students are pursuing STEM degrees, many embracing new interconnections among humanities, art and science. Underserved minorities and especially women are registered in higher proportions in MSE graduate programs with such a focus. Case studies published from MIAA symposia excite children and adults about art and science, cultural heritage and the engineering involved in its preservation.

Much of the world's cultural heritage is in peril, and many of the underlying technologies, art objects and culturally significant artifacts, buildings and sites are not being preserved. Organizers of the MIAA symposia seek contributions of cutting-edge, interdisciplinary research that provides new insights into our common cultural heritage and that leads to its long-term preservation. The preservation of cultural heritage includes developing a critical understanding of how people developed, used and transferred technologies to solve problems of survival, organization and the making of objects that represented what was important to them. Preservation also includes case studies of material and object degradation, stabilization, documentation, monitoring and conservation.

Materials research, when applied to cultural heritage, allows us to analyze and reconstruct the compositional and microstructural variability of significant objects and processes, to measure and gain understanding of special properties and performance characteristics. We discover artists' processes, appreciate their special knowledge and sometimes find innovations, mistakes, corrections or the rate-limiting steps that bounded their practices. Researchers focus on (1) the integration of high-resolution imaging using a variety of high-precision, non-destructive techniques to learn about and understand complex objects, (2) the physical and chemical characterization of art objects and archaeological artifacts and their ranges of variability, (3) analysis and reconstruction of the technologies of selection, preparation, production, testing and performance by which materials are transformed into useful, significant and beautiful objects, (4) studies of the properties and performance of ancient objects and the processes underlying their deterioration, and (5) the development of sensors, proxies and other tools and methods for evaluation of long-term stability and testing of new methods and materials for conservation treatment.

We hope to continue to add to the tradition of creating new and significant knowledge to the investigation and preservation of art objects, archaeological artifacts, historic buildings, natural and archaeological sites and environments and to the knowledge of intangible cultural heritage, such as craft practices and knowledge.

Pamela B. Vandiver
Weidong Li
Philippe Sciau
Christopher Maines

May 2014

Figure 1. Jose Luis Ruvalcaba Sil conducting in-situ, non-invasive Raman and XRF analysis of Mexican colonial cathedral mural paintings.

Figure 2. Egyptian core-formed vessel, left with height of 8.4cm and computed radiograph of the structure, right, from B. McCarthy et al.

Figure 3. Two slip-glazed and inscribed shards, 7th century BCE, from Corinth, Greece, showing layered applications, incising (left) and variations in gloss banding (right), width 2 cm, from J. Stephens et al.

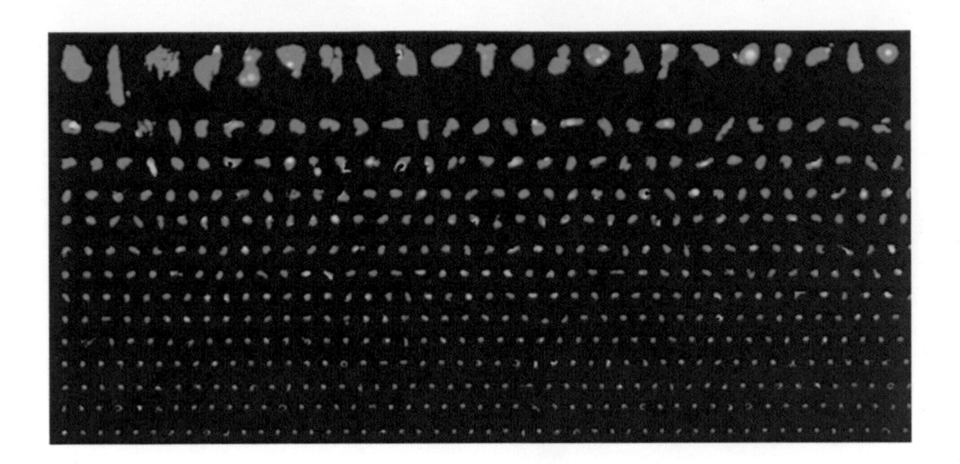

Figure 4. Detail of image analysis ranking size and showing shape of porosity in an earthenware body, fired to 950°C., from C. Reedy et al.

The Conservation Science of Stone Consolidation, Site Material Condition and a New Pollution Source in the Museum Environment

Mater. Res. Soc. Symp. Proc. Vol. 1656 © 2014 Materials Research Society
DOI: 10.1557/opl.2014.814

Nucleation, Growth and Evolution of Hydroxyapatite Films on Calcite

Sonia Naidu[1], Jeremy M. Blair[2], George W. Scherer[3]
[1,2] Department of Chemical and Biological Engineering, Princeton University, Eng. Quad. E-226, Princeton NJ 08544, USA
[3] Department of Civil and Environmental Engineering, Princeton University, Eng. Quad. E-319, Princeton NJ 08544, USA

ABSTRACT

Marble, a non-porous stone composed of calcite, is subject to acid rain dissolution due to its relatively high dissolution rate. With the goal of preventing such damage, we have investigated the deposition of films of relatively insoluble hydroxyapatite (HAP) on marble. This paper investigates the factors that affect the nucleation and growth kinetics of HAP on marble. A mild, wet chemical synthesis route, in which diammonium hydrogen phosphate (DAP) salt was reacted with marble, alone and with cationic and anionic precursors under different reaction conditions, was used to produce inorganic HAP films on the mineral surface. Film nucleation, growth and metastable phase evolution were studied, using techniques such as scanning electron microscopy (SEM) and grazing incidence X-ray diffraction (GID). The onset of nucleation, and the growth rate of the film, increased with cationic (calcium) and anionic (carbonate) precursor additions. The calcium and phosphate precursors also influenced metastable phase formation, introducing a new phase.

INTRODUCTION

The motivation behind this work is to develop a surface protective treatment that is resistant to acid attack, is structurally compatible with calcite, involves relatively simple chemistry, and does not noticeably alter the aesthetics of the stone. Hydroxyapatite (HAP) was chosen due its low solubility product, K_{sp} ($\sim 10^{-59}$) [1] and dissolution rate, R_{diss} ($\sim 10^{-14}$ moles•cm^{-2}•s^{-1} at pH 5.6) [2]. HAP also has a similar crystal structure and a lattice match to calcite [3]. This structural compatibility is expected to favor nucleation of HAP and permit the formation of a coherent, epitaxial layer of HAP on the surface of calcite. A mild treatment based on reacting a solution of diammonium hydrogen phosphate (DAP) with calcite under ambient temperature and pressure to produce the mineral HAP was developed by Kamiya et al. [4]:

$$10CaCO_3 + 5(NH_4)_2HPO_4 \rightarrow Ca_{10}(PO_4,CO_3)_6(OH,CO_3)_2 + 5(NH_4)_2CO_3 + 3CO_2 + 2H_2O \quad (1)$$

This chemistry has been successfully applied to restore the mechanical properties of thermally damaged limestone [5-8] and marble [9, 10]. An investigation of the acid resistance of DAP-treated marble showed that HAP films were partially protective, but incomplete growth and residual porosity degraded the coating's ability to protect the stone from acid dissolution [11].

The objectives of this paper are to investigate the factors that affect nucleation and growth of calcium phosphate (CaP) films on calcite. To study growth kinetics, film coverage was evaluated for different reaction conditions. Phase evolution was studied by distinguishing between phases using differences in crystal morphology and identifying corresponding phases using grazing incidence X-ray diffraction (GID). Further details will be presented in a forthcoming paper [12].

METHODOLOGY

Carrara marble (CM) is a pure white variety quarried in Carrara, Italy with connected porosity of 0.01% - 0.22% [13]. CM samples were saw-cut cubes with dimensions of 0.5 x 0.5 x 0.5 cm^3. The DAP (purity > 99%) and calcium chloride ($CaCl_2 \cdot 2H_2O$, assay > 99.0%) were purchased from Sigma-Aldrich and ammonium carbonate ($(NH_4)_2CO_3$, assay > 99.8%) from Mallinckrodt Chemicals, their concentrations made up with deionized (DI) water. Ethanol solvent (200 Proof) was manufactured by Decon Labs, Inc.

CM cubes were divided into three groups of six cubes each and subjected to the treatment conditions described in Table I. After being reacted, the cubes were soaked in DI water for 60 seconds and left to dry in the fume hood overnight.

Table I. Treatment definitions for samples used to measure CaP growth rate

Sample	Treatment	Reaction time, t (hours)
CM 1	1M DAP	3, 4, 6, 12, 18, 24
CM 2	1M DAP + 1mM $CaCl_2$	1, 2, 3, 6, 9, 12, 24
CM 3	1M DAP + 1mM $CaCl_2$ + 150 μM $(NH_4)_2CO_3$	1, 3, 6, 9, 12, 24

Scanning electron microscopy (SEM) was used to observe the morphology of growth. To determine film thickness once the CaP crystals started to coalesce and spread across the calcite surface, treated samples were coated with a low viscosity epoxy (Resin G1, Gatan) and polished to reveal their cross-sections. The cross-sections were profiled to obtain film thickness and elemental composition. Compositions were analyzed using Energy dispersion X-ray spectroscopy (EDX). SEM images of CM 1, 2 and 3 were used to detect differences in crystal morphology versus reaction time, as crystals of different polymorphs, such as octacalcium phosphate (OCP) and HAP, crystallize in different structures [14]. For samples with different film morphologies, GID at angle $\theta = 0.5°$ was conducted to identify the CaP phases present.

RESULTS

Figure 1 and Figure 2 reveal that the first CaP crystals to nucleate appear to have random orientations. No orderly crystal overgrowth orientation to suggest epitaxy was observed on any of the three samples. However, the larger crystals that later formed on CM 2 and CM 3 appeared highly oriented and intersected at angles of 90°, suggesting possible epitaxy between those crystals and the early stage film. Cross-sectional EDX line profiles on CM 2 (not shown) reveal that the calcium count drops between substrate and CaP film while the phosphorous count is observed throughout the film layer [12]. The film appears dense in the profile.

Figure 1 consists of SEM images of CM 1 at 4, 12 and 24 hours of reaction. At 3 to 4 hours, film has just begun to nucleate as small flakes of CaP crystals (size ≈ 1.5 μm). At 12 hours, the film partially coats the surface of the substrate while the crystal structure still resembles small flakes. After 24 hours, the film coats a greater portion of the surface.

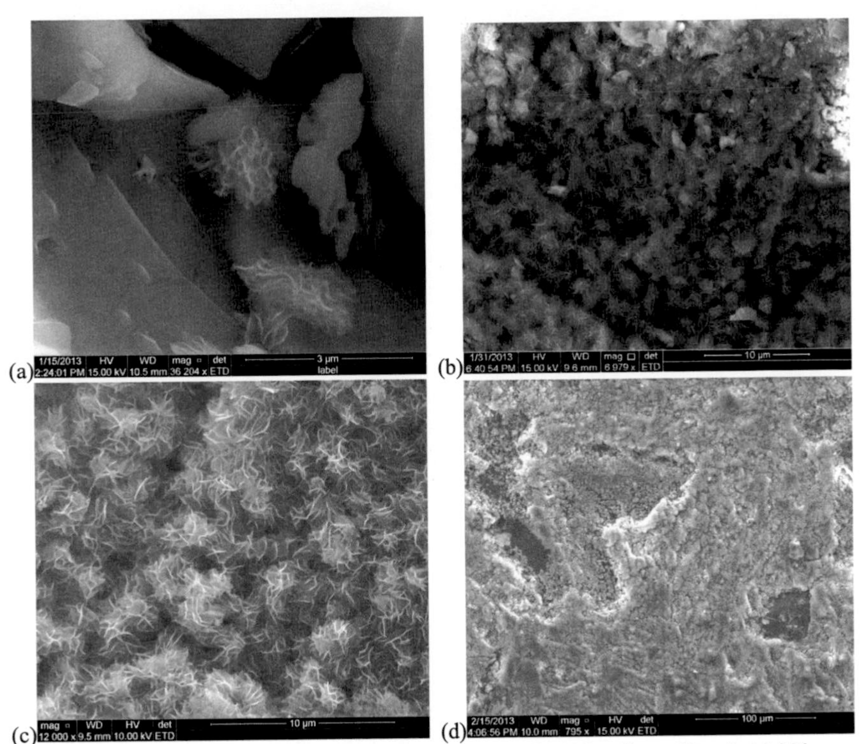

Figure 1. SEM images of CM 1 after (a) 4 hrs (bar = 3 μm), (b) 12 hrs (bar = 10 μm), (c) 24 hrs (bar = 10 μm) and (d) low mag. 24 hrs (bar = 100 μm).

As shown in Figure 2, CaP started to nucleate on CM 2 within 1 hour and by 2 hours, consists of clearly defined small crystal flakes (size ≈ 2 μm), akin in size and structure to the flakes observed on CM 1. At 9 hours, the film consists of large disc-like flakes (width ≈ 5 μm, thickness < 200 nm) and a network of the small flakes (width ≈ 2 μm). At 24 hours, the CaP film still consists of large flakes, and a continuous network of small flakes. Crystal growth on CM 3 is identical to CM 2. The main difference is the higher quantity of crystals on CM 3 than on CM 2 up to 12 hours, at which point there is no longer a visible difference. Film cracking and subsequent spalling were also observed on CM 3.

GID spectra for CM 1 at 4, 6 and 24 hours reveal HAP. GID spectra for CM 2 at 1, 9 and 24 hours reveal peaks for OCP and HAP. A small EDX peak for chloride was obtained on the CM 2 samples, but no peaks for chloroapatite or a chloride salt were detected in GID.

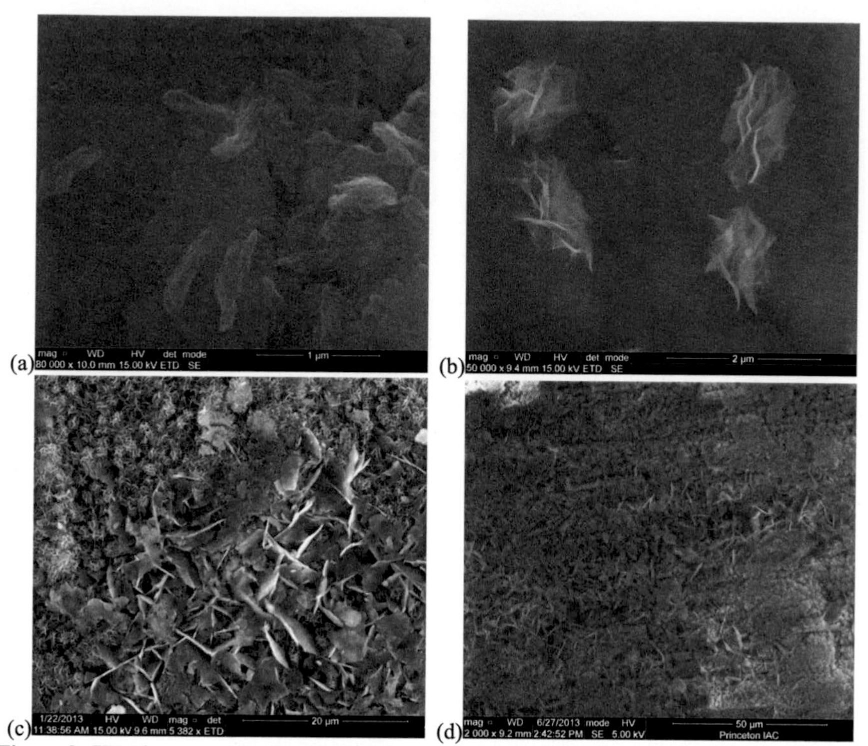

Figure 2. SEM images of CM 2 after (a) 1 hr (bar = 1 µm), (b) 2 hrs (bar = 2 µm), (c) 9 hrs (bar = 20 µm) and (d) 24 hrs (bar = 50 µm).

DISCUSSION

The flowery habit of the HAP crystals makes it difficult to determine whether the initial nucleation was epitaxial. However, the subsequent ordered growth of OCP flakes (apparently nucleated on top of the HAP) implies that epitaxy may have occurred.

The time of nucleation and final surface coverage of the film are affected by addition of both calcium and carbonate ions. Since the presence of even minute amounts of carbonate in solution tends to result in carbonate substitution in the HAP lattice [15], it would be detrimental to the durability of the film, as the solubility of apatite has been shown to decrease with increasing carbonation [16, 17]. On the other hand, additions of carbonate and calcium inhibit dissolution of the substrate. Despite higher film coverage on early stage CM 3, cracking (which presumably occurred during subsequent drying of the film) was already observed after 6 hours of growth and eventually extended throughout the film, creating areas where spalling occurred, which reduced the coverage. This accounts for CM 3's poor acid resistance [11] and this treatment is therefore not recommended. In fact, equilibrium calculations [12] indicate that the addition of calcium alone is sufficient to inhibit dissolution of calcite in water that is in equilibrium with atmospheric CO_2, so carbonate additions are unnecessary. Although the EDX cross-sectional profiles reveal dense-appearing films, the performance in acid attack experiments [11] suggests that the films contain a network of pores that allows solutions to access the underlying substrate. CM 2 was the best of the three treatments due to its fast nucleation, almost complete film coverage after only 12 hours, film stability, lack of cracking, and lastly its superior performance in acid, likely the result of the aforementioned properties.

When CM was subjected to 1M DAP (CM 1), only HAP was detected, while in 1M DAP + 1mM $CaCl_2$ (CM 2) and 1M DAP + 1mM $CaCl_2$ + 150 μM $(NH_4)_2CO_3$ (CM 3), the sequence of evolution was from HAP to both OCP and HAP. As GID spectra detected only HAP in CM 1, but both OCP and HAP in CM 2, and given that there are small flakes in CM 1 and both small and large flakes in CM 2, it can be concluded that the small flakes are HAP and the large flakes are OCP. Based on these experiments, 24 hours reaction time is recommended, as film coverage had already stabilized within this time period, while shorter reaction times would not allow for sufficient film coverage or metastable conversion. More details regarding the growth kinetics and microstructure of the film are presented in a forthcoming paper [12].

CONCLUSIONS

The nucleation and growth of hydroxyapatite crystals on calcite does not appear to be epitaxial. Film growth and uniformity are enhanced by the addition of calcium chloride and ammonium carbonate precursors to the reaction in millimolar and micromolar quantities, respectively. In particular, millimolar calcium chloride additions increase film growth and uniformity without causing cracking or other undesirable effects. Addition of calcium chloride also introduces the metastable octacalcium phosphate to the film composition; this is not detrimental, as OCP has low solubility. Addition of carbonate ions is not necessary to suppress dissolution of the calcite [12], and is associated with cracking of the film, so such additions are not recommended. Current research is aimed at reducing the porosity of the film through modification of the growth habit with organic additives.

ACKNOWLEDGEMENTS

The authors wish to thank NCPTT (grant MT-2210-12-NC-08) for financial support.

REFERENCES

[1] H. McDowell, T.M. Gregory, W.E. Brown, Solubility of Ca5(PO4)3OH in the System Ca(OH)2-H3PO4.H2O at 5, 15, 25 and 37°C, Journal of Research of the National Bureau of Standards: Sect A Phys Chem 81 A (1977) 273-281.
[2] N. Harouiya, C. Chaïrat, S.J. Köhler, R. Gout, E.H. Oelkers, The Dissolution Kinetics and Apparent Solubility of Natural Apatite in Closed Reactors at Temperatures from 5 to 50°C and pH from 1 to 6, Chemical Geology 244 (2007) 554-568.
[3] S. Naidu, E. Sassoni, G.W. Scherer, New Treatment for Corrosion-Resistant Coatings for Marble and Consolidation of Limestone, Jardins de Pierres Section française de l'Institut international de conservation, Champs-sur-Marne, France, 2011, pp. 289-294

[4] M. Kamiya, J. Hatta, E. Shimada, Y. Ikuma, M. Yoshimura, H. Monma, AFM Analysis of Initial Stage of Reaction between Calcite and Phosphate, Materials Science and Engineering B 111 (2004) 226-231.
[5] E. Sassoni, S. Naidu, G.W. Scherer, The Use of Hydroxyapatite as a New Inorganic Consolidant for Damaged Carbonate Stones, Journal of Cultural Heritage 12 (2011) 346-355.
[6] E. Sassoni, E. Franzoni, B. Pigino, G.W. Scherer, S. Naidu, Effectiveness of Hydroxyapatite as a Consolidating Treatment for Lithotypes with Varying Carbonate Content and Porosity, 5th Int. Cong. Sci. Technol. Safeguard of Cultural Heritage in the Mediterranean Basin, Istanbul, 2012, pp. 338-343.
[7] M. Matteini, S. Rescic, F. Fratini, G. Botticelli, Ammonium Phosphates as Consolidating Agents for Carbonatic Stone Materials Used in Architecture and Cultural Heritage: Preliminary Research, International Journal of Architectural Heritage 5 (2011) 717-736.
[8] Fuwei Yang, Bingjian Zhang, Yan Liu, Guofeng Wei, Hui Zhang, W. Chen, Z. Xu, Biomimic Conservation of Weathered Calcareous Stones by Apatite, New Journal of Chemistry 35 (2011) 887–892.
[9] E. Sassoni, E. Franzoni, Evaluation of Hydroxyapatite Effects in Marble Consolidation and Behaviour towards Thermal Weathering, in: M. Boriani, R. Gabaglio, D. Gulotta (Eds.), Proceedings of Built Heritage – Monitoring Conservation Management, Milan (Italy), 2013, pp. 1287-1295.
[10] E. Sassoni, E. Franzoni, Sugaring Marble in the Monumental Cemetery in Bologna (Italy): Characterization of Naturally and Artificially Weathered Samples and First Results of Consolidation by Hydroxyapatite, Applied Physics A: Materials Science & Processing (submitted).
[11] S. Naidu, J. Blair, G.W. Scherer, The Mechanism of Acid Attack on Carrara Marble and the Efficacy of a Hydroxyapatite-based Treatment for Reducing Attack, (to be published).
[12] S. Naidu, G.W. Scherer, Nucleation, Growth and Evolution of Calcium Phosphate Films on Calcite, (submitted to the Journal of Colloid and Interface Science).
[13] F.W. Tegethoff, J. Rohleder, E. Kroker, Ch. 2 in Calcium Carbonate: From the Cretaceous Period into the 21st Century, Birkhäuser, 2001.
[14] E. Vlieg, Understanding Crystal Growth in Vacuum and Beyond, Surface Science 500 (2002) 458-474.
[15] D.G. Nelson, J.D. Featherstone, Preparation, Analysis, and Characterization of Carbonated Apatites, Calcif Tissue Int 34 Suppl 2 (1982) S69-81.
[16] H. Pan, B.W. Darvell, Effect of Carbonate on Hydroxyapatite Solubility, Crystal Growth & Design 10 (2010) 845-850.
[17] R.A. Jahnke, The Synthesis and Solubility of Carbonate Fluorapatite, American Journal of Science 284 (1984) 58-78.

Mater. Res. Soc. Symp. Proc. Vol. 1656 © 2014 Materials Research Society
DOI: 10.1557/opl.2014.712

Novel hydroxyapatite-based consolidant and the acceleration of hydrolysis of silicate-based consolidants

Sonia Naidu[1], Chun Liu[2], George W. Scherer[3]

[1] Department of Chemical and Biological Engineering, Eng. Quad. E-226, Princeton University, Princeton, NJ 08544, USA
[2] Department of Chemistry, University of California, Berkeley, CA 94720, USA
[3] Department of Civil and Environmental Engineering, Eng. Quad. E-319, Princeton University, Princeton, NJ 08544, USA

ABSTRACT

This paper discusses the effectiveness of hydroxyapatite (HAP) as an inorganic consolidant for physically weathered Indiana Limestone, and as a coupling agent between limestone and a silicate consolidant. A double application is investigated, in which samples are coated with HAP followed by a commercially available silicate-based consolidant (Conservare® OH-100). To artificially weather limestone, a thermal degradation technique was utilized. Diammonium hydrogen phosphate (DAP) salt was reacted with limestone, alone and with cationic precursors, to produce HAP films. The dynamic elastic modulus, water sorptivity and tensile strength of the treated stones were evaluated. HAP was found to be an effective consolidant for weathered Indiana Limestone, and its performance was enhanced by addition of millimolar quantities of calcium chloride. However, HAP was not useful as a coupling agent; a double treatment with DAP is more effective than sequential treatment with DAP and Conservare®.

INTRODUCTION

A consolidant is intended to restore the mechanical integrity of deteriorated stones by binding the grain boundaries and fracture surfaces, either physically or chemically [1]. The ideal consolidant should retain the stone's water transport properties and aesthetics, and the treatment should be reversible, or at least not hamper further treatment. Silicate-based consolidants are very effective on silicate stones, such as quartzitic sandstone, due to their chemical compatibility [2,3]. One such consolidant is oligomeric tetra-ethoxy-ortho-silicate (TEOS), $-(Si(OC_2H_5)_4)_n-$ [4] (commercially available in the U.S. as Conservare™ OH-100) [5]. After application, atmospheric water gradually replaces the $-OC_2H_5$ groups with -OH groups. This process is known as hydrolysis, and is very slow, taking six to eight weeks to complete [2]. Once TEOS has hydrolyzed, the oligomers condense with each other to form a silica gel. The gel can then form covalent bonds with the silanol groups on the surface of sandstone [6]. Unfortunately, TEOS is not as effective on carbonate stones, as it can only bond mechanically, owing to the absence of –OH groups in calcite that would allow chemical interaction to occur [3].

To address the lack of an effective consolidant for limestone, hydroxyapatite (HAP) was tested, using the precursor diammonium hydrogen phosphate (DAP) and based on the following reaction [7]:

$$10CaCO_3 + 5(NH_4)_2HPO_4 \rightarrow Ca_{10}(PO_4,CO_3)_6(OH,CO_3)_2 + 5(NH_4)_2CO_3 + 3CO_2 + 2H_2O \quad (1)$$

As indicated in the formula, HAP is usually found carbonated due to its interaction with carbon dioxide from the atmosphere. As illustrated by Sassoni et al. [7], HAP restores the mechanical integrity of deteriorated Indiana limestone. In addition, the benefits of the treatment are: a) retention of water transport properties, b) no chromatic alteration, and c) the inorganic and non-toxic nature of treatment. Consolidation with HAP has been independently investigated by several groups [7, 8, 9, 10].

The motivation for the present study is to evaluate the use of HAP as a coupling agent between calcite and a silicate consolidant. We therefore test the effectiveness of treatment with DAP followed by Conservare, in comparison with DAP or Conservare alone.

METHODOLOGY

Indiana limestone (IL) is composed mainly of calcite ($CaCO_3$, > 97%) and minute amounts of $MgCO_3$, Al_2O_3 and SiO_2. IL has a porosity of approximately 14% [11]. IL samples used in this study were core-drilled cylinders with a diameter of 2 cm and a height of 5 cm. DAP (purity >99%) and calcium chloride ($CaCl_2.2H_2O$, assay >99.0%) were purchased from Sigma-Aldrich. TEOS was obtained from ProSoCo, Inc. as Conservare® OH-100 and diluted in a 1:3 v/v ratio with ethanol (EtOH) before application. This was done as undiluted conservare was found to form cracks inside the stone [6]. Ethanol solvent (200 Proof) was manufactured by Decon Labs, Inc.

Subjecting limestone samples to heat induces accelerated degradation of the stone [7,12]. The higher the temperature, the greater the damage. It was found that subjecting IL samples to 300°C for 1 hour decreases the stone's dynamic elastic modulus by 40%, comparable to the ~36% decrease in E_{dyn} due to 150 years of natural weathering in the field [13]. Hence, using a furnace, IL samples were heated at 300°C for one hour to induce artificial degradation. Thermally weathered samples were divided into six pairs of cylinders, each pair subjected to different treatment conditions, as described in Table I. In each case, there were two cylinders (*e.g.*, IL 1 consisted of IL 1a and IL 1b) subjected to identical conditions.

Table I: Description of treatment conditions

Sample	Treatment conditions
IL 1	1M DAP 48 hrs (single treatment)
IL 2	1M DAP 48 hrs (double treatment)
IL 3	1M DAP + 1mM $CaCl_2$ 48 hrs (single treatment)
IL 4	1M DAP + 1mM $CaCl_2$ 48 hrs (double treatment)
IL 5	1M DAP 48 hrs → Conservare® OH-100 12 hrs → EtOH-H_2O 24 hrs
IL 6	1M DAP 48 hrs → Conservare® OH-100 12 hrs

DAP solutions (1 molar) were applied to both samples in a beaker by filling up to 1/3 the height of the samples. Some of the solutions also contained 1 mM $CaCl_2$ to provide calcium ions for the formation of HAP; without such additions, the calcium would have to be leached from the limestone. After the solution had completely risen to the stone's surface, the remainder was poured in up to 5 mm from the top surface of the stone. After 48 hours, samples were removed, water-saturated for 3 days to remove unreacted DAP, then dried under a fan at ambient conditions until constant weight (approximately 7 days). Subsequent treatment, where indicated, was then applied in the same manner and the process repeated. After samples IL 5 (IL 5a and IL 5b) were dry, they were submerged in Conservare® OH-100 for 12 hours, then removed and dried and finally immersed in a solution of EtOH-H_2O in 1:5 v/v ratio for 24 hours. This was done to accelerate the hydrolysis of TEOS, as ethanol allows transport of water into the now hydrophobic stone to promote hydrolysis. A low ethanol-high water mixture was used to limit the concentration of volatile organic compounds.

To investigate consolidant deposition, binding and penetration depth, specimens from the center of each cylinder were obtained by hammer fracturing, then viewed under a scanning electron microscope (SEM; FEI Quanta 200 ESEM). The instrument's energy dispersive x-ray spectroscopic (EDX) capability was utilized to confirm mineralogical composition.

The dynamic elastic modulus is given by $E_{dyn} = \rho v^2$, where ρ is sample density and v is the ultrasonic pulse velocity [14]. A PUNDIT device (CNS Farnell) with two 54 kHz transducers was used to measure time taken for the sound pulse to travel through the material. E_{dyn} was recorded for all samples when dry - before and after heat treatment, and after each treatment.

Sorptivity is a measure of the rate of water uptake by capillarity. It was used to determine changes in water transport properties of samples. Cylindrical samples were suspended from a balance and a water dish raised until contact with the bottom of the sample was established. Mass increase was then continuously recorded by a computer data acquisition system (DASYLab) and the sorptivity, S calculated from the initial linear portion of the mass change versus time plot as $S = \left(\frac{\Delta mass}{S_A \cdot t} \right)$, where S_A = cylinder surface area and t = time.

RESULTS

SEM images of the limestone sample subjected to treatment IL 4 confirm that calcium phosphate coats the calcite grains, as was observed in ref. 7. EDX spectra taken on the surface of a grain indicate the presence of carbon, oxygen, phosphorous and calcium. However, no nitrogen peaks were detected, thus confirming that there was no residual DAP or ammonium salts in the film. Ammonia odor was detected during the reaction, indicating that the ammonium carbonate produced had evaporated. The sample subjected to treatment IL 5 also showed film formation at pores and grain surfaces, and EDX spectra confirm the presence of oxygen, carbon, calcium and phosphorous. In addition, a silicon peak is obtained, confirming deposition of the TEOS consolidant.

Figure 1 shows that the heating the IL samples at 300°C for 1 hour caused a reduction in dynamic modulus of ~40%. The single DAP treatments brought the modulus back up to the original value, and the double treatments increased the modulus above the original undamaged values. The change in E_{dyn} is most dramatic after the second 1M DAP + 1mM $CaCl_2$ treatment (IL 4). All treated samples displayed a significant increase in E_{dyn}. This increase was most dramatic for sample 1L 4, which went from E_{dyn} (damaged) = 22.0 GPa to E_{dyn} (treated) = 46.3 GPa (Figure 2).

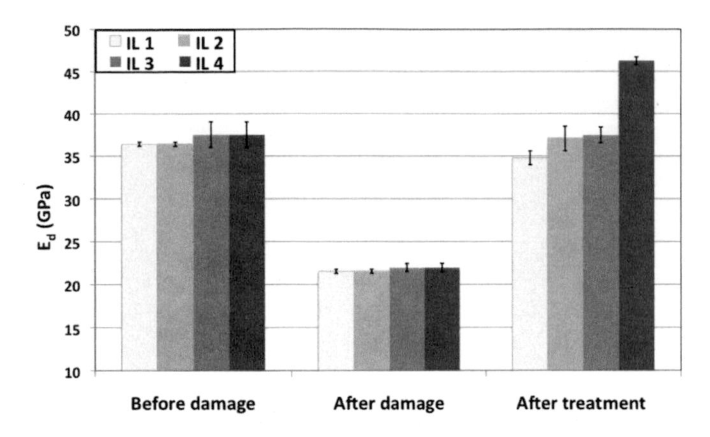

Figure 1: Dynamic modulus of DAP-treated IL samples. Error bars represent the values for the pairs of samples.

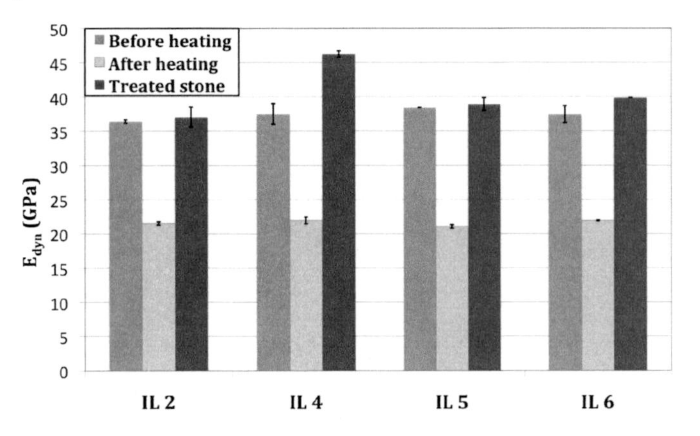

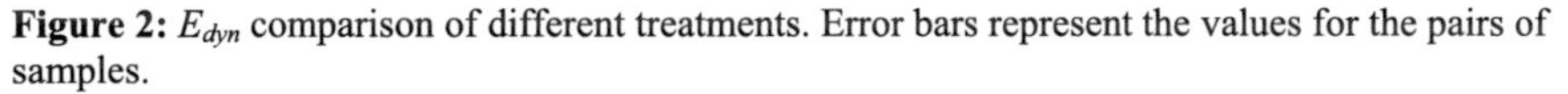

Figure 2: E_{dyn} comparison of different treatments. Error bars represent the values for the pairs of samples.

Water uptake data for untreated and treated IL samples showed a marked reduction in sorptivity in treated samples, except IL 1, which experienced only a slight reduction. A decrease in sorptivity also occurred between the single and double DAP and DAP-$CaCl_2$ applications, with $S_{untreated} = 0.04359$ g•cm^{-2}•min$^{-1/2}$ while $S_{IL1} = 0.03317$ g•cm^{-2}•min$^{-1/2}$, $S_{IL2} = 0.02646$ g•cm^{-2}•min$^{-1/2}$, $S_{IL3} = 0.02828$ g•cm^{-2}•min$^{-1/2}$ and $S_{IL4} = 0.02236$ g•cm^{-2}•min$^{-1/2}$. Limestone treated with the silicate consolidant had sorptivity an order of magnitude lower than those treated with phosphate. Comparing sorptivity between the DAP- Conservare® OH-100 rinsed (IL 5) and unrinsed (IL 6) samples revealed a small increase with ethanol-water rinsing, from S_{IL6} = 0.00224 g•cm^{-2}•min$^{-1/2}$ to S_{IL5} = 0.00300 g•cm^{-2}•min$^{-1/2}$.

DISCUSSION

SEM and EDX analysis of limestone treated in 1M DAP for two days indicated that a film of calcium phosphate had formed around the calcite grains. The film was not cracked and, given that the only species detected were Ca^{2+} and PO_4^{3-}, the presence of damaging salts can be excluded. Stiffness was restored to its original value after treating in DAP for two days as the solution formed a film that entered the pores and bridged the grains. The second application of DAP increased E_{dyn} above the undamaged value, indicating a benefit of a double application of consolidant. The benefit of adding $CaCl_2$ to the DAP solution was apparent from the 100% E_{dyn} restoration after the first DAP-$CaCl_2$ treatment and a further 23.4% increase in E_{dyn} after the second treatment. This is attributed to the enhanced reaction between DAP and Ca^{2+} from the calcium salt, which increased film formation. In the case of DAP alone, the reaction is dependent on the dissolution of limestone to supply Ca^{2+} ions, which is slow at the mildly basic conditions of the solution.

A preceding study showed that limestone treated in 1M DAP for 2 days experienced a slight reduction in total open porosity, with the fraction of coarser pores ($r_p > 1$ μm) reduced and the fraction of finer pores ($r_p < 0.1$ μm) increased [7]. This is consistent with the observed decrease in sorptivity results. The samples treated in DAP followed by Conservare® OH-100 displayed negligible sorptivity, which can be explained by the expected hydrophobicity of the IL samples. The EtOH:H_2O rinse application produced a negligible increase in the sorptivity and failed to restore water uptake. As it has already been established that rinsing improves stone hydrophilicity after treatment with Conservare® OH-100 alone [15], this lack of improvement is most likely due to pore blocking by the large amount of material deposited from both DAP and TEOS, so that the EtOH:H_2O mixture was unable to penetrate the pores and hydrolyze the silicate. Dynamic modulus was restored with both the DAP-Conservare® OH-100 rinsed and unrinsed treatments, with the rinse not affecting IL stiffness beyond the error margin for moduli readings.

CONCLUSIONS

All of the DAP-based treatments applied were effective consolidants for weathered Indiana limestone, based on improvement in mechanical properties, retained water uptake and absence of harmful by-products. The DAP treatment's performance was significantly improved by the addition of millimolar quantities of $CaCl_2$, which provides calcium ions for HAP formation. The double treatment of DAP and DAP-$CaCl_2$ provided further improvement in mechanical properties over the single treatment, but negatively affected the sorptivity due to the additional amount of consolidant in the pores. If a double treatment is applied, the DAP double treatment is superior to a double treatment consisting of DAP followed by Conservare® OH-100, as the latter had extremely poor water sorptivity and no significant improvement in mechanical properties over the single DAP treatment.

ACKNOWLEDGMENTS

The authors wish to thank NCPTT (grant MT-2210-12-NC-08) for financial support.

REFERENCES

[1] E. Doehne and C.A. Price, Stone Conservation: An Overview of Current Research, 2nd ed. (Getty Conservation Inst., Los Angeles, 2010)

[2] C.A. Grissom and N.R. Weiss (1981): "Alkoxysilanes in the conservation of art and architecture: 1861-1981", Art and Archaeology Technical Abstracts 18 (1): 150-202

[3] G. Wheeler, Alkoxysilanes and the Consolidation of Stone (Getty Conservation Institute, Los Angeles, 2005) 196 pp.

[4] C.J. Brinker and G.W. Scherer, Sol-Gel Science (Academic Press, New York, 1990) Ch. 3

[5] PROSOCO, 3741 Greenway Circle Lawrence, KS 66046
http://www.prosoco.com/Products/

[6] G.W. Scherer and G.S. Wheeler (2009): "Silicate Consolidants for Stone", Key Engineering Materials 391: 1-25 (online at http://www.scientific.net © (2009) Trans Tech Publications, Switzerland) ISBN: 978-0-87849-365-4

[7] E. Sassoni, S. Naidu and G.W. Scherer (2011): "The use of hydroxyapatite as a new inorganic consolidant for damaged carbonate stones", J. Cultural Heritage 12: 346–355

[8] M. Matteini, S. Rescic, F. Fratini and G. Botticelli (2011): "Ammonium Phosphates as Consolidating Agents for Carbonatic Stone Materials Used in Architecture and Cultural Heritage: Preliminary Research", Int. J. Architectural Heritage: Conservation, Analysis, and Restoration 5(6): 717-736

[9] F. Yang, B. Zhang, Y. Liu, G. Wei, H. Zhang, W. Chen and Z. Xu (2011): "Biomimic conservation of weathered calcareous stones by apatite", New J. Chem. 35: 887–892

[10] S. Naidu (2014): Novel Hydroxyapatite Coatings for the Conservation of Marble and Limestone, Ph.D. Thesis, Dept. Chemical and Biological Eng., Princeton University

[11] T. Perry, N.M. Smith and W.J. Wayne, Salem limestone and associated formations in south-central Indiana, Indiana Department of Conservation, Geological Survey, Field Conference Guidebook 7, 1954

[12] E. Franzoni, E. Sassoni, G.W. Scherer and S. Naidu (2012): "Artificial weathering of stone by heating", J. Cultural Heritage 14(3): 85-93

[13] T. Esaki and K. Jiang (1999): "Comprehensive study of the weathered condition of welded tuff from a historic bridge in Kagoshima, Japan", Eng. Geol. 55: 121–130

[14] H. Kolsky, Stress waves in solids, Dover, Minneola, New York, 1963

[15] S. Naidu, C. Liu, and G.W. Scherer (2014): "New techniques in limestone consolidation: Hydroxyapatite-based consolidant and the acceleration of hydrolysis of silicate-based consolidants", J. Cultural Heritage, http://dx.doi.org/10.1016/j.culher.2014.01.001

Mater. Res. Soc. Symp. Proc. Vol. 1656 © 2014 Materials Research Society
DOI: 10.1557/opl.2014.824

Properties and Characterization of Building Materials from the Laosicheng Ruins in Southern China*

Ya Xiao[1,2], Ning Wang[2], Haibin Gu[1], Weimin Guo[1], Feng Gao[2,3], Ning Niu[4], Shaojun Liu[2,5**]

[1]Cultural Relics and Archaeology Institute of Hunan Province, Changsha, 410083, China
[2]State Key Laboratory for Powder Metallurgy, Central South University, Changsha, 410083, China
[3]Chinese Academy of Cultural Heritage, Beijing, 100029, China
[4]Henan Research Institute of Ancient Architecture Protection, Zhengzhou, Henan, CHINA
[5]School of Chemistry and Chemical Engineering, Central South University, Changsha 410083, China

ABSTRACT

As one of the most typical ancient cultural relics in southern China's minority regions near Changsha in Hunan province, the magnificent Laosicheng ruins excavated recently have been included in the UNESCO World Cultural Heritage Tentative List. Urgent conservation of excavated Laosicheng ruins brings about the need for a study of the formulation and properties of construction materials used, including earth, stone, mortar, and brick. In the present study, comprehensive analyses were carried out to determine their raw material compositions, mineralogical, and microstructural properties using sheet polarized optical microscopy, scanning electron microscopy with energy dispersive spectrometer, thermogravimetric/differential scanning calorimetry, X-ray powder diffraction, and Fourier transform infrared spectroscopy. Special attention was paid to mortars, which were the most widely used in building the Laosicheng. Results show that mortar used as external render of the city wall is mainly built up from inorganic $CaCO_3$ and $MgCO_3$ based hybrid materials produced by the carbonation of $Ca(OH)_2$ and $Mg(OH)_2$ with a small amount of sticky rice. In contrast, mortar used to bond stones of the city walls is a traditional mortar that does not contain sticky rice. This study is a part of a huge interdisciplinary project aimed to clarify the role of organics in ancient China's organic-inorganic hybrid mortar, which can be considered as one of the greatest invention in construction material history. The results provide valuable basic data and restoration strategies that can be used in the conservation of the ruins as well.

INTRODUCTION

The Laosicheng ruins, currently the largest, best preserved, and oldest ancient military castles in Southwest China, are located in the Yongshun County in the Hunan province, China. The original city was built in 690AD with a well-conceived layout. The Laosicheng used to be

* Note: The characters "cheng" in Laosicheng mean city in Chinese.

** Corresponding author. Tel: +86 731 88876315; Email address: liumatthew@csu.edu.cn (S. J. Liu) the economic, political, military and cultural center of Tujia ethnic group, a traditional minority in Southwest China, during the reign of Tusi system. The Laosicheng ruins are about 25 square kilometers. The central city about 19 hectares includes the palace areas, government offices, residential areas, tombs, and religious worship areas.

As one of the most typical ancient cultural relics in southern China's minority regions, the magnificent ruins excavated recently have been included in the UNESCO World Cultural Heritage Tentative List. Fig.1 shows the geographic distribution diagram of central city of the Laoshicheng ruins. However, urgent conservation of the excavated Laosicheng ruins brings about the need for a comprehensive study of the formulation and properties of construction materials used, including mortars, bricks, masonry stones, and foundation soils, especially because new restoration mortars often fail to ensure a physical, chemical and mechanical compatibility with old mortar and architectural surfaces.

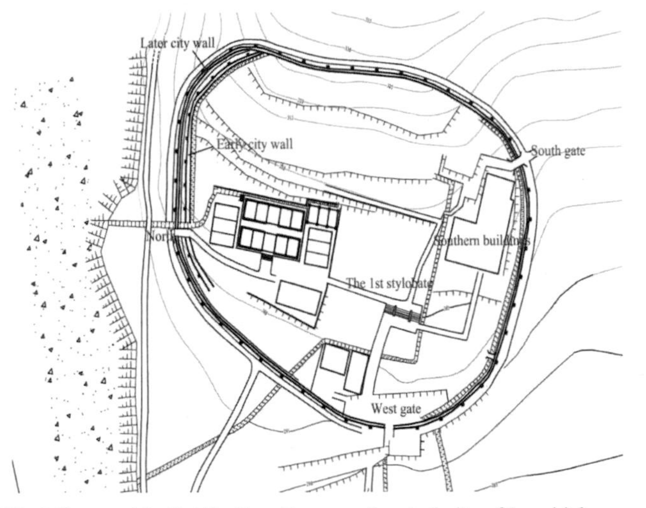

Fig.1 Geographic distribution diagram of central city of Laoshicheng ruins.

As a key material in building ancient cities, mortars are most widely used in the Laosicheng. We would like to mention that lime was widely used in construction in ancient Greece [1-2]. In ancient Roman, hydraulic materials called “Roman mortar” such as ground volcanic ash, ceramic chips, and ground brick were introduced [3-6]. In China, there is a long history of the use of lime. However, the absence of natural hydraulic materials like volcanic ash resulted in the development of distinctive organic-inorganic hybrid mortars by adding natural organic compounds like sticky rice soup, juice of vegetable leaves, egg white, tung oil, fish oil, or animal blood into mortars [7]. Especially, it has been revealed in archeology that sticky rice-lime mortar was widely used in important buildings in ancient China [8]. It is also stressed that high temperature and humidity in summer form an alternating wet and dry microenvironment in the

Laoshicheng ruins due to its unique geographical location, which is well known to be favorable to weathering and erosion of outdoor inorganic monuments [9, 10]. However, the durability of most of mortars is still remarkable after they have been exposed for a long period of time.

EXPERIMENTS

Table 1 Sampling points and sample descriptions

Sample No.	Named Samples	Sampling points	Sample description
LSC-C1	Foundation Soils	West city wall in Palace area	Compacted Miscellaneous grey yellow soil with a small amount tile fragments
LSC-C2	Foundation Soils	South Gate in Palace area	Compacted Miscellaneous light gray black soil
LSC-C3	Bonding mortar	Gaps between cornerstone in West city wall in Palace area	Gray, hard, containing with a small amount tile fragments
LSC-C4	Plastering mortar	Surface of West city wall in Palace area	Hard, warped, thin
LSC-C10	Masonry stones	South city wall Gate in Palace area	Grey, Dolomite
LSC-C11	Plastering mortar	Earth wall outside West city wall in Palace area	Hard, white, relative thick, Containing fibers
LSC-C12	Plastering mortar	Earth wall outside West city wall in Palace area	Hard, thin, outer layer with brownish red color

In order to ensure that the collected samples are representative, 38 typical samples were collected from the whole ruins. Among them, 18 samples were collected from the city wall, including two bonding mortar samples, ten plastering mortar samples, two soil samples, and one decaying product sample. We would like to mention that the city wall is an original well-persevered relic.

In this study, soils, stones, and mortars used in the Laosicheng ruins were characterized in a combination of X-ray fluorescence spectrometer (XRF), polarized light microscopy, scanning electron microscopy with X-ray energy dispersive spectrometer (SEM-EDS), X-ray diffraction spectrometer (XRD), Fourier transform infrared spectroscopy (FT-IR), and thermo gravimetric differential scanning calorimetry (TG-DSC). While extensive building materials had been used to build the ruins, special attention is paid to mortars, which are crucial to the well-conserved foundations and walls of the Laosicheng ruins. Due to the large scale of the Laoshicheng ruins, large numbers of samples have been collected. Description of typical samples and sampling points are summarized in Table 1. Mortar sampling points in the Laosicheng ruins are further shown in Fig.1 (a)-(c). As indicated in Fig.1 (d), it is interesting to notice that most of the external wall surfaces covered by mortars in the Laoshicheng ruins show brownish red, which is rarely observed in ancient China wall ruins. Before optical microscope and SEM-EDX analysis, mortar and masonry stone samples were polished and cleaned carefully. For XRD, FTIR, and TG-DSC measurements, samples were pulverized and dried at 60°C for 48 h in an oven.

Fig.2 (a) West city walls in the Palace area; (b) South city walls in the Palace area; (c) External walls outside the West city walls in the Palace area; (d) Base of External walls outside the West city walls in the Palace area with brownish-red layer.

XRF was used to analyze the chemical compositions of foundation soils. Thin section analysis was done to characterize masonry stones of the city walls by Nikon E400 pol petrographic microscope (PM) using polarized light technique. Microstructure and chemical composition were characterized by SEM-EDS. Gold was sputtered on the sample surface. XRD (Philips X-Pert Pro) was used to determined the mineralogical components. Data were collected at 40 kV and a current of 250 mA, with a step of 0.02°/Second. Identification of crystalline compounds was performed by MDI Jade and then verified manually with the JCPDS database. FTIR (Nicolet380) was used to determine the composition. As a reference, sticky rice and calcite were carried out by FTIR as well. Thermal transformation of samples was recorded with TG-DSC analyzer (SETARAMEV018/24). Measuring temperature is in a range of 20-900°C with a heating rate of 10°C/min in N_2 to determine the components of mortar samples.

RESULTS AND DISCUSSION

Foundation Soil and Masonry Stone

Table 2 XRF results of the Foundation soils

Sample No.	Oxide content (wt.%)											
	SiO_2	Al_2O_3	Fe_2O_3	MgO	CaO	K_2O	P_2O_5	Na_2O	MnO	TiO_2	SO_3	Others
LSC-C1	65.850	17.552	5.556	3.865	2.558	2.058	1.090	0.401	0.114	0.735	0.061	0.160
LSC-C2	61.637	21.349	7.627	3.440	1.799	1.930	0.544	0.274	0.181	0.798	0.042	0.379

Fig. 3 Petrographic microscope images of masonry stones collected from West city walls: (a) Plane polarized light; (b) Cross polarized light

As a large scale ancient architectural site, the Laosicheng ruins exhibit relatively integrated foundations, drainage ditches, streets, arches, and walls. A variety of locally sourced building materials were extensively applied, including stones, soils, and sands. Fig. 5(a) shows the mineralogical compositions of foundation soils by XRD. As shown, mineralogical compositions of foundation soils from the west city wall and the south city wall in the palace area are very similar. In addition to a small amount of dolomite and hematite, main compositions are detected as quartz and muscovite, implying that foundation soils are original local soils without further treatment. Table 2 summarizes chemical compositions of foundation soils by XRF. In a consistent with XRD results, foundation soils in west city walls and the south city walls in the palace area show similar chemical compositions. In addition to a certain amount of Fe, Mg, Ca, and K, and a very small amount of P, Na, Mn, Ti and S, main elements of the foundation soils are detected as Si and Al. To understand the petrographical and mineralogical characterization of stones constructing the city walls, thin section analysis by petrographic microscope was carried out. As shown in Fig. 3, stones were identified as fine-crystalline dolomite (~100wt%) with a very small amount of quartz. Typical band structure of concentric rings implies that stone is of oolitic structure. In contrast, dolomite component can be divided into two types. One functions as cement with a subhedral crystal and grain size about 0.06mm ~ 0.5mm. The other, distributed in oolitics and debris, is structure-dependent crystal.

Mortars

In addition to soils and stones, mortars were extensively used in constructing the Laosicheng ruins. As the most basic and common bonding materials, mortars were used either to bond stones of walls or to plaster walls as a surface rendering layer. We would like to mention that the Laosicheng ruins are located in a deep valley surrounded by a river, having a typical humid subtropical monsoon climate and abundant rainfall.

For a long time, ancient Chinese have realized that sticky rice mortars are of significantly enhanced strength, toughness, and resistance to seepage [11]. This resulted in a wide application of sticky rice mortars in ancient China's important architectures, such as city walls, palaces, tombs, and even water conservancy facilities [12]. It is well known that chromogenic reaction by iodine is a very simple and effective method to preliminarily detect the existence of starch since blue complex forms when iodine is mixed with starch. In contrast, amylopectin, a main component in sticky rice, shows purple when it reacts with iodine. Fig. 4 (a)-(d) show the results of chromogenic reaction experiment by mixing iodine with mortar samples from the Laoshicheng ruins. The same experiments were done on pure $CaCO_3$ and sticky rice powder samples as well. As shown in Fig. 4(a), pure sticky rice powders show purple when iodine solution was dipped into. In contrast, orange color of iodine solution appears in pure $CaCO_3$ powders. Although the color of plastering mortar samples does not show the same color as that of sticky rice powder, it is obvious that the change of color is similar. This implies possible existence of sticky rice in plastering mortars from the West city wall in the Palace area. However, very similar change of bonding mortars to calcite powders indicates that bonding mortars also from the West city wall in the Palace area do not contain sticky rice, as shown in Fig. 4(c).

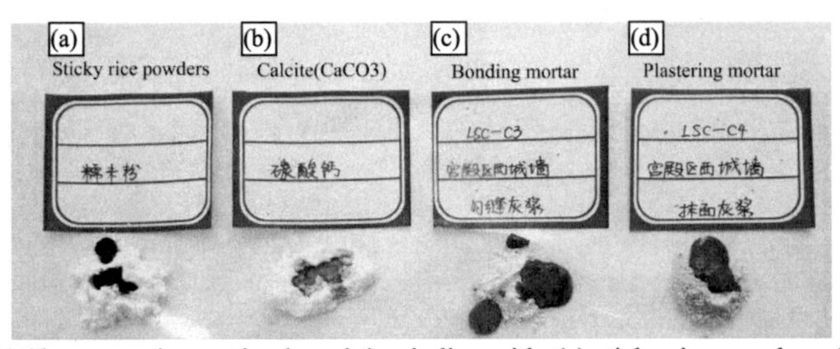

Fig.4 Chromogenic reaction by mixing iodine with: (a) sticky rice powders; (b) calcite powders; (c) bonding mortars from gaps between cornerstone in the West city wall in the Palace area; (d) plastering mortars from surface of the West city wall in the Palace area.

Fig. 5 (b) shows the mineralogical compositions bonding mortars from gaps between cornerstone and plastering mortars form the surface in the West city wall in the Palace area characterized by XRD, respectively. As shown, significant differences can be distinguished from XRD patterns of bonding mortars and plastering mortars from the same city walls. Main mineral compositions of bonding mortars can be identified as calcium carbonate with calcite structure, magnesium hydroxide with brucite structure, and silica with quartz structure. In contrast, several extra diffraction peaks are observed in plastering mortars, which correspond to characteristic diffraction peaks of portlandite (P) and slaked lime ($Ca(OH)_2$), respectively.

SEM images of bonding mortars and plastering mortars from the West city wall in the Palace area are shown in Fig. 6 (a) and (b), respectively. Significantly different microstructures can be observed between bonding mortars and plastering mortars as well. Microstructure of plastering mortars is more compact and denser than that of bonding mortar. Especially, obvious cracks can be observed in bonding mortars. Further EDX analysis results are summarized in Table 3. Chemical components of bonding mortars and plastering mortars are very similar. Both mortars mainly contain O, Ca, C, and Mg. Although magnesium carbonate was scarcely used in conventional fabrication technique of ancient China mortars, relatively high Mg content in mortars of the Laosicheng ruins shows that magnesium carbonate was added during the fabrication of mortars in the Laosicheng. We would like to mention that it was reported that mortars containing magnesium carbonate can show higher strength and better durability than calcite ones [13]. Additionally, a small amount of silicon appears in bonding mortars. It is reasonable to speculate it may stem from the surrounding soils since mortars have been covered by soils until they were excavated recently.

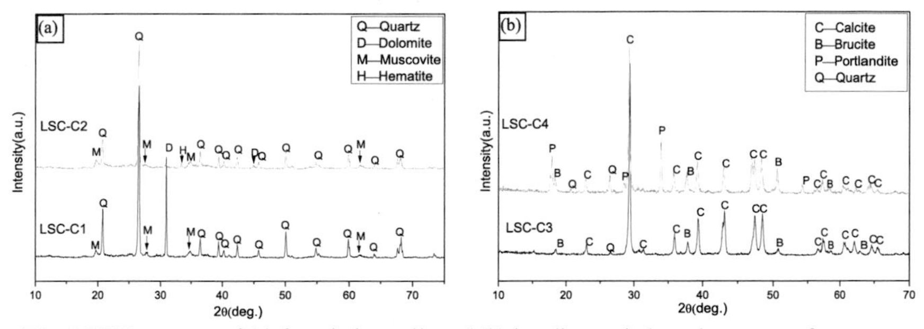

Fig. 5 XRD patterns of (a) foundation soils and (b) bonding and plastering mortar from gaps between cornerstone and surface in West city wall, respectively.

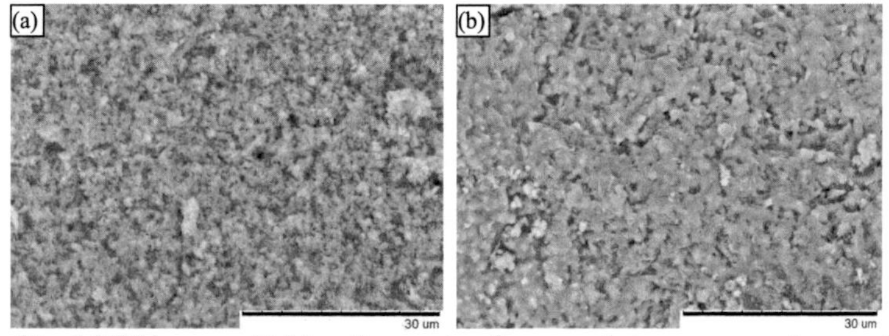

Fig. 6 SEM images of (a) bonding mortars from gaps between corner stones and (b) plastering mortars from surface of city walls. Note: Both samples were collected from the West city wall in the Palace area.

According to SEM and XRD results, differences in the microstructure and mineralogical compositions between the bonding mortars and plastering mortars are obvious. Chromogenic reaction experiment above also shows the existence of sticky rice in plastering mortars, which is not contained in bonding mortars. Thermal gravimetric analysis was further used to carry out comparative analysis of bonding mortars and plastering mortars. Fig. 7(a) and (b) show DSC-TGA curves of bonding mortars and plastering, respectively. Both samples were collected from the West city wall in the Palace area. As shown in Fig. 7(a), three endothermic peaks at 110°C, 400°C and 733°C can be observed, which are believed to be attributed to the evaporation of water in mortar, the decomposition of $Mg(OH)_2$ ($Mg(OH)_2 \rightarrow MgO + H_2O$), and the decomposition of $CaCO_3$ ($CaCO_3 \rightarrow CaO + CO_2$), respectively. In contrast, as shown in Fig. 7(b), two additional peaks appeared at 249°C and 550°C, which corresponding to the decomposition of organic additive and $Ca(OH)_2$ ($Ca(OH)_2 \rightarrow CaO + H_2O$), respectively [14]. These results further clearly indicate the presence of some kind of organic compounds in plastering mortars.

Table 3 EDX results of mortars and masonry

Sample No.	O	Ca	C	Mg	Si
	Contents of major elements in Mass percentage (%)				
LSC-C3	43.3	35.4	11.2	5.8	4.3
LSC-C4	46.5	40.7	5.2	7.6	---
LSC-C10	46.6	30.6	12.2	10.7	---
LSC-C11	45.4	26.3	8.5	14.5	5.3
LSC-C12	49.3	35.1	9.2	4.2	1.8

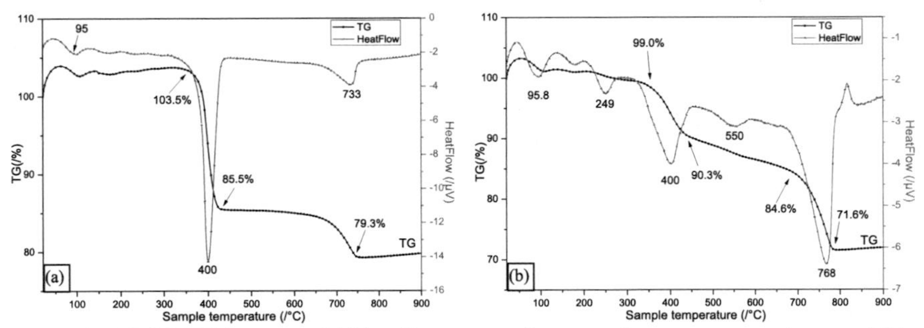

Fig. 7 DSC-TGA curves of (a) bonding mortars from gaps between corner stones and (b) plastering mortars from surface of city walls. Note: Both samples were collected from the West city wall in the Palace area.

As references, Fig.8 (a) shows FTIR spectra of calcite and sticky rice flour. In contrast, Fig.6 (b) shows FTIR spectra for bonding and plastering mortars. Absorbance bands at 713cm^{-1}, 875cm^{-1}, 1430cm^{-1}, 1797cm^{-1}, 2510cm^{-1}, 2873cm^{-1}, 2985cm^{-1}, and 3452cm^{-1} in FTIR spectra of bonding mortars are in a good match with those of calcite [15]. Absorbance bands near 794cm^{-1} and 3692cm^{-1} are ascribed to vibration absorption of quartz and -OH bonding in brucite, respectively. However, absorbance bands near 1658cm^{-1} and 1025cm^{-1} corresponding to absorbance of C-O and -OH group of glucose anhydride ring can be clearly observed, respectively. This further confirms that plastering mortars contain organic additives, such as sticky rice. These observations lead us to make a conclusion that plastering mortars used to plaster the city wall of the Laosicheng ruins are an inorganic-organic composite construction material mainly made up from calcium carbonate, magnesium carbonate, and sticky rice.

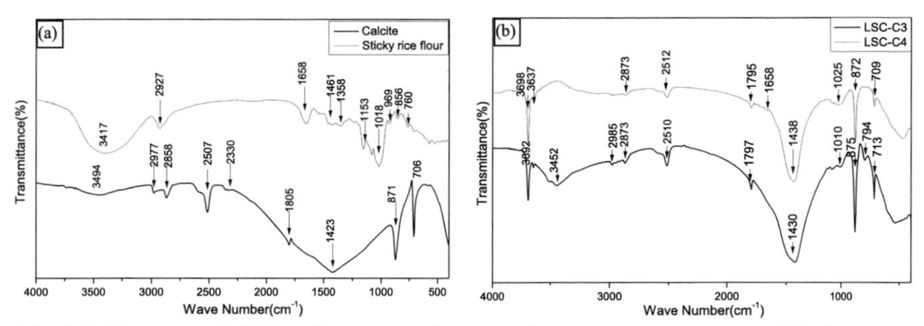

Fig. 8 FTIR curves of: (a) bonding mortars from gaps between corner stones and (b) plastering mortars from surface of city walls.

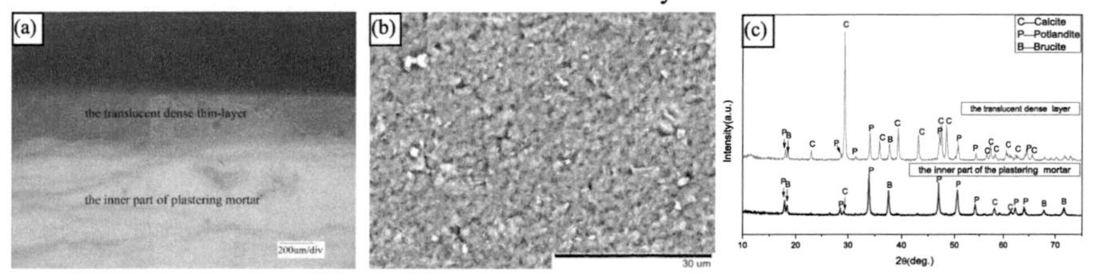

Fig. 9 (a) Optical microscope images; (b) SEM images; (c) XRD patterns of translucent dense thin-layer on plastering mortars.

Intriguingly, a very distinctive thin-layer that is highly translucent and dense was observed on the surface of plastering mortars. Fig.9 (a), (b), and (c) show the optical microscope, SEM-EDX, and XRD results, respectively. As shown in Fig.9 (a), this thin-layer is uniform with an average thickness about 250 μm. SEM image in Fig.9 (b) further shows that microstructure of thin-layer is more compact and particle size is finer compared to the inner part of plastering mortars. Fig.9 (c) indicates that calcite content in plastering mortars was significantly less than that in surface layer. Obviously, further investigation is necessary, but it is speculated that the formation of translucent layer may be related to biochemical mineralization of $Ca(OH)_2$ added by microorganism [16] in the surrounding environment due to high humidity of the Laosicheng. This dense thin-layer can effectively block CO_2 in the air penetrating into the plastering mortars and delay the carbonization process of the inner part. Specifically, it can prevent water and soluble salts penetrating into inner layers and result in enhanced conservation of mortars.

Furthermore, fibers identified as cotton were found in plastering mortar of the city walls by optical microscope as well. As shown in Fig. 10(a), these fibers, which are very similar to commercial cotton fibers, are of width in a range of ~15 and ~26 μm. These fibers can be helpful to improve the performance of mortars and have traditionally been used to improve the tensile strength of mortars as additives [17].

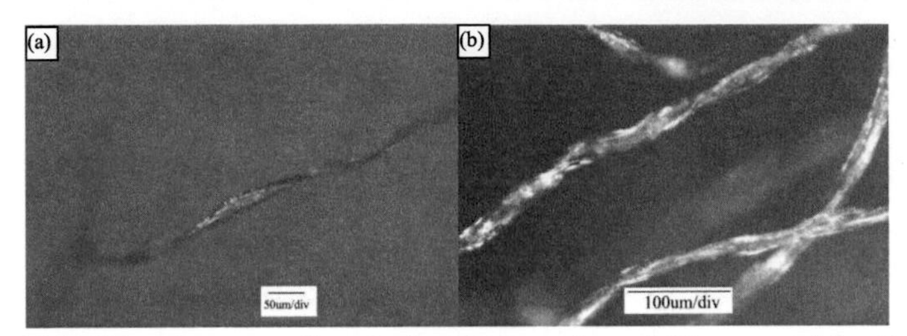

Fig.10 (a) Fibers in mortars from the Laosicheng ruins and (b) commercial cotton fibers

Coatings

It is also interesting to observe a layer of brownish-red coatings on the West city walls treated by plastering mortars. It should be mentioned that the color of soils in the Laoshicheng ruins is very similar to that of coatings. Therefore, it is initially speculated that their presence origins from local soils. Fig. 11 (a)-(b) show cross-sectional view and surface view images by optical microscope of brownish-red coatings, respectively. As indicated in Fig. 11(a), two significantly different layers can be observed. Surface of coatings is relatively uniform with ~200μm thickness. White layer below coatings is detected as mortars. Fig. 11 (c) shows XRD patterns of brownish-red coatings. Although main components can be distinguished as calcite, quartz and vaterite, it is obviously noticed that background diffraction peaks are broad. This indicates that coatings contain materials with poor crystallinity.

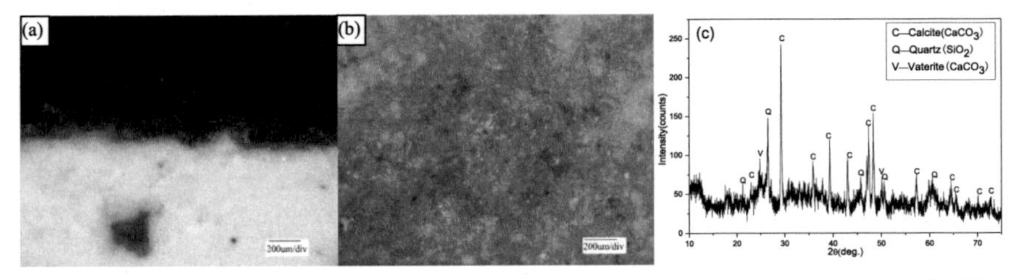

Fig. 11 (a) Cross-sectional view; (b) surface view images by optical microscope; (c) XRD patterns of brownish-red paint on the plastering mortars.

SEM combined with EDX was further used to characterize coatings. Fig. 12(a) and (b) show SEM images and EDX results of plastering mortars below coatings marked by yellow circle, respectively. In contrast, SEM image and EDX results of coatings from the West city walls in the Palace area are shown in Fig. 12(c) and (d), respectively. Compared to mortars below brownish-red coatings, it is clear that carbon, silicon, and aluminum content in coatings are higher. Furthermore, it was easily observed that a large amount of impurity particles are embedded in coatings. Fig. 12(c) and (d) are SEM images and EDX results of impurity particles marked by yellow circle in brownish-red coatings, showing that main chemical composition of impurities are C and O, respectively. Since traditional red mineral pigments (Fig. 12(c)) generally contain iron red, lack of iron evidenced by EDX and XRD shows the brownish-red coating is some kind of plant pigment. While tung oil is popular in the Yongshun County as paint, we speculate the brownish-red coatings on plastering mortars were brushed by a mixture of tung oil and local plant pigment. However, further investigation is necessary to identify the components of coatings to carry out effective conservation strategies of the Laosicheng ruins.

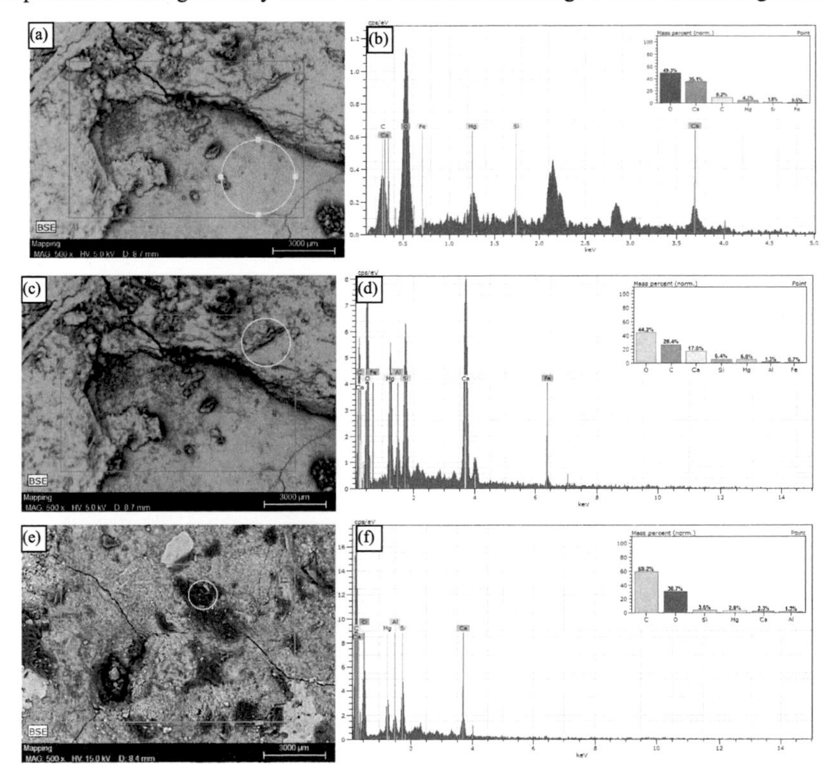

Fig. 12 SEM and EDX results of: (a) and (b) plastering mortars; (c) and (d) brownish-red coatings; (e) and (f) impurity particles embedded in coatings, respectively. Note: samples were collected from the West city walls in the Palace area.

CONCLUSIONS

Comprehensive analyses have been carried out to determine basic physical properties, raw material compositions, mineralogical and microstructural properties of construction materials used to construct the Laosicheng ruins. Special attention was paid to the mortars that were the most widely used in building the Laosicheng. Results show that mortar used as external render of the city wall is mainly built up from inorganic $CaCO_3$ and $MgCO_3$ based hybrid materials developed from the carbonation of $Ca(OH)_2$ and $Mg(OH)_2$ with a small amount of sticky rice. In contrast, mortar used to bond the stones of the city wall is a kind of traditional mortar that does not contain sticky rice. The presented results provide valuable data that can be used in the conservation work of the ruins. However, it is clear that further investigation is necessary and imminent to carry out effective preservation strategies of the Laosicheng ruins.

ACKNOWLEDGMENTS

This work was financially supported by China Science & Technology Polar Program (No. 2012BAK14B05 and 2009BAK53B06) and Compass Special Program of the State Administration of Cultural Heritage of China (No. [2011]1806). Authors also thank Mr. Tao Zhang at Cultural Relics and Archaeology Institute of Hunan Province for drawing the geographic distribution diagram of central city of Laoshicheng ruins.

REFERENCE

1. S. Kramar, V. Zalar, M. Urosevic, W. Körner, A. Mauko, B. Mirtič, Mater. Charact. **62**, 1042(2011).
2. M. Riccardi, P. Duminuco, C. Tomasi, P. Ferloni, Thermochimica Acta., **321(1)**, 207(1998).
3. D. Benedetti, S. Valetti, E. Bontempi, C. Piccioli, L. Depero, Appl. Phys. A, **79(2)**, 341(2004).
4. D. Silva, H. Wenk, P. Monteiro, Thermochimica Acta., **438(1-2)**, 35 (2005).
5. A. Velosa, J. Coroado, M. Veiga, F. Rocha, Mater. Charact., **58**, 1208 (2007).
6. J. Weber, N. Gadermayr, R. Kozłowski, D. Mucha, D. Hughes, D. Jaglin, Mater. Charact. **58**, 1217 (2007).
7. Y. Song, Tian Gong Kai Wu, Commercial Press: Shanghai; 1958, p.197.
8. F. Tie, Sci. Conserv.Archaeol. 16(1), 47 (2004).
9. R. Espinosa-Marzal, G. Scherer, Accounts Chem. Res., **43**, 897 (2010).
10. G. Scherer, R. Flatt, G. Wheeler, MRS Bulletin., **26(1)**, 44 (2001).
11. F. Yang, B. Zhang, Q. Ma, Accounts Chem. Res., **43**, 936 (2010).
12. Y. Zeng, B. Zhang, X. Liang, Sci. Conserv. Archaeol. 20(2), 1 (2008).
13. C. Atzeni, L. Massidda, U. Sanna, Sci. Tech. cultural heritage. **5**, 29 (1996).
14. Y. Zeng, B. Zhang, X. Liang, Thermochimica Acta., **473(1-2)**, 1(2008).
15. F. Yang, B. Zhang, C. Pan, Y. Ceng, Sci. in China E: Tech. Sci., **52(6)**, 1641 (2009).
16. M. Cusack, A. Freer, Chem. Rev., **108**, 4433 (2008).
17. J. Elsen, Cement Concrete Res., **36(8)**, 1416 (2006).

Mater. Res. Soc. Symp. Proc. Vol. 1656 © 2014 Materials Research Society
DOI: 10.1557/opl.2014.827

Dispersions of Surface Modified Calcium Hydroxide Nanoparticles with Enhanced Kinetic Stability: Properties and Applications to Desalination and Consolidation of the Yungang Grottoes

Ya Xiao[1,2], Feng Gao[1,3], Yun Fang[4], Youdan Tan[5], Kaiyu Liu[5], Shaojun Liu[1,5,*]

[1] State Key Laboratory for Powder Metallurgy, Central South University, Changsha, China, 410083
[2]Cultural Relics and Archaeology Institute of Hunan Province, Changsha, China, 410083
[3]Chinese Academy of Cultural Heritage, Beijing 100029, China
[4]China University of Geosciences (Wuhan), Wuhan, 430074, China
[5]School of Chemistry and Chemical Engineering, Central South University, Changsha 410083, China

ABSTRACT

Calcium hydroxide ($Ca(OH)_2$) is one of the most interesting materials used to consolidate stone sculptures, monuments, mortars or wall paintings. In this study, we reported on the synthesis and characterization of surface modified $Ca(OH)_2$ nanoparticles as a dispersion with enhanced kinetic stability and the applications for the conservation of sandstone monuments. Uniform hexagonal $Ca(OH)_2$ nanoparticles (~35nm) were obtained by mixing NaOH and NaCl aqueous solutions at 100~175°C using homogeneous-phase reactions. It was further demonstrated that 3-(Methacryloyloxypropane oxygen) trimethoxysilane surfactant agent can significantly reduce agglomeration and simultaneously improve specific surface area of as-synthesized $Ca(OH)_2$ nanoparticles. Brunauer-Emmett-Teller (BET) measurement showed that specific surface area of modified $Ca(OH)_2$ nanoparticles reaches up to ~48.78m^2/g, about 2.5 and 3.4 times higher than that of unmodified and commercial ones, respectively. The kinetic stability of $Ca(OH)_2$ despersion can be further enhanced and its viscosity can be decreased by optimizing the ratio of ethanol and n-propanol. Especially, a technique, which combined the Ferroni-Dini method and dispersion of $Ca(OH)_2$ nanoparticles with enhanced kinetic stability, was proposed to effectively desalinate and consolidate the decayed stone, as evidenced by significant decreases of the porosity and concentration of detrimental Cl^- and SO_4^{2-} ions in the severely decayed sandstone samples from the Yungang grottoes.

INTRODUCTION

Several nanostructured materials and their applications in the field of cultural heritage conservation have been reported [1-3]. Calcium hydroxide ($Ca(OH)_2$) is one of the most interesting materials used to perform consolidation of stone sculptures, monuments, mortars or wall paintings [4]. Calcium hydroxide reacts with carbon dioxide in the air, producing calcium carbonate that is highly compatible with the inorganic original substrate. However, direct use of aqueous $Ca(OH)_2$ solutions is limited by its low solubility in water [5]. Moreover, commercially

* Email: liumatthew@csu.edu.cn (S.J. Liu)

available $Ca(OH)_2$ particles have a broad size distribution and large range of mean dimensions. This usually results in a formation of white coating on the surface of artifacts because of small surface pore size. In contrast, a nanometric size of $Ca(OH)_2$ particles ensures their good penetration into porous matrix of the substrate and enhanced carbonation process. Its low solubility and physicochemical properties compatible with stone surfaces favor its use in inorganic material-based conservation treatments.

Usually, $Ca(OH)_2$ particles with uniform size and shape can be obtained through homogeneous phase reactions in nonaqueous solution. Influence of precipitation temperature and initial reactant concentrations on the size and shape of precipitated crystals has been extensively reported [6-8]. It was shown that high temperatures are necessary to obtain very fine particles [9, 10]. Furthermore, kinetically stable dispersions can be obtained in environmentally friendly short-chain aliphatic alcohols [11]. Especially, the dispersing solvent can be selected as pure or in a mixture to achieve ideal penetration depth inside the artifact and the ideal rheological properties according to the features of porous materials. Baglioni et.al [12] synthesized nanoparticles in nonaqueous solvents for applications in protection of stucco coatings in the archeological ruins of Calakmul, in Campeche, México. However, it is well-known that relatively large Van der Waals force and the Coulomb force due to high specific surface area and surface energy of nanoparticles result in the occurrence of the agglomeration of $Ca(OH)_2$ nanoparticles. In addition to the retardation of carbonation process of $Ca(OH)_2$ nanoparticles in air, agglomeration can significantly reduce their penetration depth inside the stone. Moreover, their applications in sandstone monuments, which are considered as the most typical outdoor immovable heritage in China, were hardly reported.

The Ferroni-Dini method, also called the ''barium'' method, is the first method that provided reliable results by removing soluble salts threatening the artifacts and reinforcing porous structure simultaneously [11]. The Ferroni-Dini method consists of two steps: the first is the application to the surface layer of stone samples using a saturated solution of ammonium carbonate, $(NH_4)_2CO_3$; the second is the treatment with a barium hydroxide solution, $Ba(OH)_2$. For the reaction mechanism, see the section on "Application to deteriorated stone."

As outstanding examples of Chinese outdoor sandstone artifacts, the Yungang Grottoes in China have been included in the UNESCO world cultural heritage list. We would like to stress that cements in the Yungang sandstone contain a large amount of $CaCO_3$ [13]. It was clearly indicated that the interaction between water and rocks is one of the most important damaging factors leading to the dissolution of carbonate cements, detrital feldspar hydrolysis, and change in the water content of mineral salts [14]. However, it should be mentioned that outdoor cultural heritage often suffer several decaying processes due to environmental pollution, especially in rapidly developing China [15]. For example, chemical corrosion of the cements results in the loss of cohesion between particles in stone. Main decay products of the Yungang grottoes have been identified as gypsum ($CaSO_4 \cdot 2H_2O$) and thenardite ($Na_2SO_4 \cdot 10H_2O$). Their hydration and dehydration in dry and wet cycling environment are considered as one of the most important factors in causing the weathering of the Yungang Grottoes by intraporous crystallization [16, 17]. It is mandatory to remove soluble $CaSO_4 \cdot 2H_2O$ and $Na_2SO_4 \cdot 10H_2O$ salts before consolidation of severely decayed sandstones was carried out for long-term conservation to the Yungang Grottoes.

This study is aimed at understanding the influence of the surfactants on size, shape, specific surface, and dispersibility of synthesized $Ca(OH)_2$ nanoparticles by homogeneous phase reactions. Special attention was paid to the change in kinetic stability and rheological properties of $Ca(OH)_2$ nanoparticle dispersions in short-chain aliphatic alcohols that resulted from addition of surfactants. To demonstrate the efficacy of the technique combining the Ferroni-Dini method with $Ca(OH)_2$ nanoparticle dispersions in sandstone monument conservation, severely decayed sandstone samples from the Yungang grottoes were used to evaluate its effectiveness in desalination and consolidation.

EXPERIMENT

Calcium chloride, sodium hydroxide, ethanol, n-propanol, and isopropanol analysis products were used without further purification. Surfactants, polyvinyl alcohol (PV), sodium dodecylbenzenesulfonate (SDBS), and 3-(Methacryloyloxypropane oxygen) trimethoxysilane (KH-570) were applied to modify the surface of synthesized $Ca(OH)_2$ nanoparticles.

Particle size, shape, and agglomeration of $Ca(OH)_2$ nanoparticles constitute several key parameters for their application in stone conservation. $Ca(OH)_2$ nanoparticles were synthesized by similar method reported by Salvadori as follows: 100 mL of 0.8 M NaOH solution and 100 mL of 0.4 M $CaCl_2 \cdot H_2O$ were separately heated to different temperatures (100, 125, 150, and 175 °C), and were mixed under strong stirring. The reaction mixture was cooled under a nitrogen atmosphere, and the resulting suspension was washed five times with water to remove NaCl. To study the effects of different experimental conditions on the resulting particles, several syntheses were performed, changing one condition each time while the other parameters remained constant. The parameters investigated were as follows: (a) aging time of the solution: t=20 min; (b) reaction temperature: T=100, 125, 150, and 175 °C; (c) molar ratio of $NaOH/CaCl_2$ in the range of N=2.0, 1.8, 1.6, and 1.4. The supernatant solution was discarded, and the remaining suspension was washed five times by water to reduce NaCl concentration below 10^{-6} M.

Specific surface area of $Ca(OH)_2$ nanoparticles was measured by Brunauer-Emmett-Teller (BET) method. Change of SO_4^{2-} and Cl^- concentration in the solution during desalination process is monitored by ion chromatography (IP). Experimental methods are as follows: weighed stone sample was placed in a clean glass tube and then mixed with deionized water. After the sample was stirred and soaked for 12 hrs, it was filtered. A 2ml filtered solution was diluted by deionized water in flask. Then, 60uL solution was extracted for IP testing.

Microstructure was characterized by scanning electron microscope (SEM) coupled with X-ray energy dispersive spectrometer (EDS) (JEOL JSM 6400). Gold was sputtered on sample surface. Determination of mineralogical components was performed using a Philips X-Pert Pro X-Ray Diffractometer equipped with Cu K_α radiation. Data were collected at 40 kV with a scanning step of 0.02°/second with 2θ from 10° to 85°. Molecular structure of $Ca(OH)_2$ nanoparticles before and after surface modification can be characterized by Fourier transform infrared spectroscopy (FTIR).

Weathered stone dropped from the edge of cliff near the Yungang grottos were collected and tested. Stone samples were blocks of 4.5×4.5×1cm. All stone samples were cleaned by deionized water and acetone, and then dried. Before the consolidation of weathered stone by $Ca(OH)_2$ nanoparticle dispersions, the Ferroni-Dini method was applied to desalinate stone samples. Samples were soaked in saturated ammonium sulfate solution by atmospheric pressure impregnation method for 24 hrs. Then, they were removed from the solution and covered completely by clean cotton wool, and subsequently were sealed by plastic wrap at room temperature for 12hrs. These samples were further soaked in saturated $Ba(OH)_2$ solution by atmospheric pressure impregnation method for 24 hrs. Dispersions of 1g/L, 3g/L, 5g/L $Ca(OH)_2$

nanoparticles modified by KH-570 surfactant, and unmodified $Ca(OH)_2$ nanoparticles were applied to consolidate severely weathered stone by vacuum impregnation technique. We would like to state that vacuum impregnation cannot be used in the grotto since it is impractical onsite or on a large stone sculpture. The depth (~10-40 mm) of penetration into the stone by various dispersions was evaluated by SEM coupled with EDS since fine bright points corresponding to $Ca(OH)_2$ nanoparticle can be observed clearly in SEM figures, as shown in Fig. 7 (d).

Capillary water absorption measurement was performed by gravimetric absorption technique [18]. Dried and weighed stone were placed on a filter paper pad while they were partially immersed in distilled water. After 24 hrs, samples were extracted and weighed again to determine the amount of water absorbed by capillary force. Water vapor permeability test was carried out as follows: after the stone was put over a plastic bottle filled with distilled water, the bottle was sealed and placed into a chamber with the constant temperature and humidity. The weight change of distilled water was recorded every 24 hrs to reflect the breathability of stone.

Ethanol, n-propanol, and isopropanol were used as dispersants. Physical properties, including density, boiling point, and saturated vapor pressure, viscosity of ethanol, n-propanol, and isopropanol dispersants are listed in Table 1. As shown, ethanol has lower boiling point, larger vapor pressure, and lower viscosity. Since low boiling point and high vapor pressure of the material usually correspond to high evaporation coefficient, ethanol and isopropanol show higher volatility. Compared to conventional viscometer, a kinetic viscosity tester was used to measure the viscosity of dispersions since the liquid in a glass capillary can simulate the penetration process of dispersions inside multi-porous stone. Measurements were carried out at a constant temperature (25°C) by measuring a volume of solution flowing through a capillary glass. Results show that propanol and isopropanol have similar kinetic viscosity, and their values are greater than that of ethanol. Kinetic stability of dispersions was measured by gravity sedimentation technique at 25°C. After a 5g/L $Ca(OH)_2$ nanoparticle dispersion was prepared, it was ultrasonically dispersed for 30min, and then allowed to stand for half a minute. The greater sediment volume indicates higher kinetic stability of the dispersion.

Spectrophotometry was used to measure the chromatic parameters on the surface of stone before and after consolidation. Total color difference ΔE^* is provided as a result of the formula [19]: $\Delta E^*=((\Delta L^*)^2+(\Delta a^*)^2+(\Delta b^*)^2)^{1/2}$, where L^*, a^*, and b^* are luminosity, the red-green parameter, and the blue-yellow parameter, respectively. Consolidant absorption rate measured by Karsten flask method is calculated by the formula: $M\%=(m_1-m_0)/m_0 \cdot 100\%$, where m_0 (g) and m_1(g) are the mass of samples before and after treatment, respectively.

Table 1. Physical Properties of ethanol, n-propanol, isopropanol dispersants

Dispersant	Density (g/cm^3)	Boiling Points (°C)	Saturated vapor pressure (kPa)	Viscosity (mm^2/s)
ethanol	0.7894	78.40	5.33	1.143
n-propanol	0.8036	97.19	2.00	1.810
isopropanol	0.7863	82.40	4.32	1.738

RESULTS AND DISCUSSION

Synthesis of $Ca(OH)_2$ nanoparticles

It was shown that temperature above 100°C promotes the formation of nanosized particles in nonaqueous media since the solubility of $Ca(OH)_2$ increases with decreasing temperature [20].

Also, some studies reported significant effect of organic solvents on shape and size of particles obtained by precipitation [8]. In this study, we first investigated the size and shape change of $Ca(OH)_2$ nanoparticles setting the NaOH and $CaCl_2{\cdot}2H_2O$ molar ratio (N) as N=1.4, 1.6, 1.8, and 2.0. The other parameters are constant: NaOH solution concentration C=1.5mol/L, oil bath temperature T=150°C, and aging time t=20min. Fig. 1 (a) is the XRD patterns of $Ca(OH)_2$ nanoparticles using different NaOH and $CaCl_2{\cdot}2H_2O$ molar ratios, respectively. Sharp XRD peaks imply that single phase $Ca(OH)_2$ is of high crystallinity. Morphology and size of particles are shown in Fig. 2. It is clear that the size and shape of $Ca(OH)_2$ nanoparticles are strongly dependent on NaOH and $CaCl_2{\cdot}2H_2O$ molar ratio N. Nanoparticles synthesized at N=2.0 show a relatively wide but inhomogeneous size distribution. When N decreases to 1.8, $Ca(OH)_2$ nanoparticles with relatively homogeneous particle size were synthesized accompanying those with larger average particle size, ~50nm. When N further decreases to 1.6, particle size becomes finer (~40nm) and size distribution is more uniform. However, it is observed that particle size begins to increase and a high degree of particle agglomeration occurs when N is 1.4. Therefore, the optimal molar ratio is 1.6.

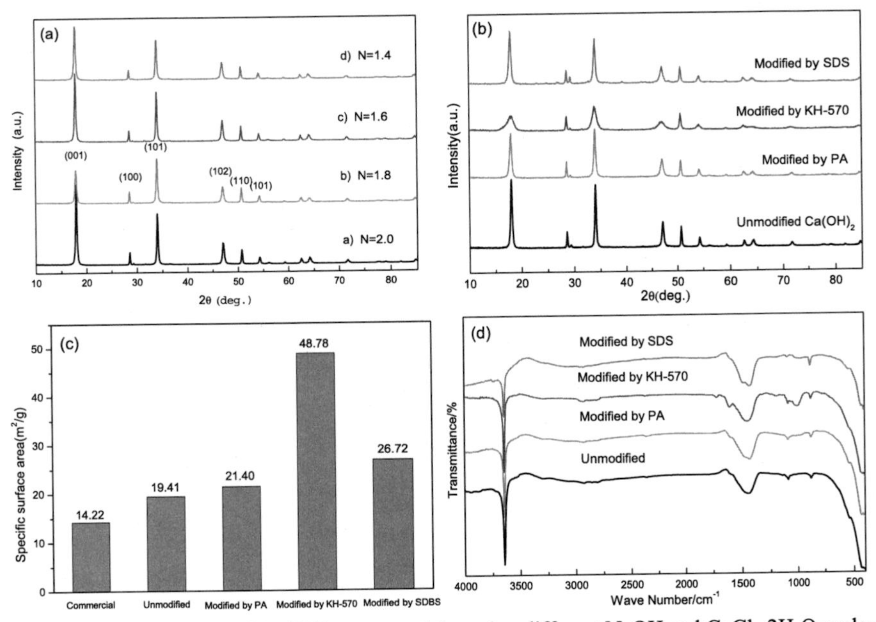

Fig. 1. (a) XRD patterns of $Ca(OH)_2$ nanoparticles using different NaOH and $CaCl_2{\cdot}2H_2O$ molar ratios; (b) XRD patterns of unmodified and modified $Ca(OH)_2$ (N=1.6); (c) Specific surface area of unmodified and modified $Ca(OH)_2$ (N=1.6); (d) FTIR spectra of $Ca(OH)_2$ nanoparticles before and after modified by various surfactants. Note that KH-570 has peak broadening (b) and greater specific surface area.

We further explored the influence of the NaOH concentration (C) on the morphology and size of particles where C is 1.8mol/L, C=1.2mol/L, C=0.9mol/L, and C=0.6mol/L, respectively. The other parameters are set as follows: NaOH and $CaCl_2 \cdot 2H_2O$ molar ratio N=1.6, oil bath temperature T=150 °C, and aging time t=20min. In addition to the formation of single phase $Ca(OH)_2$, the morphology and size of particles can also be significantly affected by NaOH concentration. In particular, $Ca(OH)_2$ particles synthesized using 1.2mol/L NaOH have the smallest size and a more uniform size distribution (not shown here).

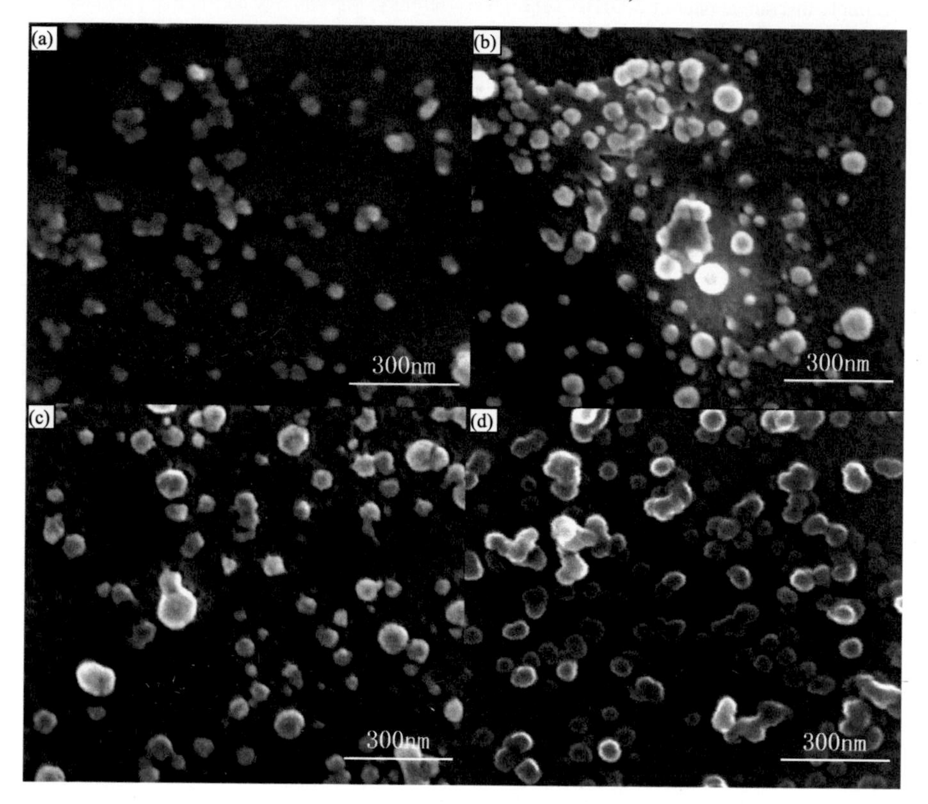

Fig. 2. SEM images of $Ca(OH)_2$ nanoparticles setting NaOH and $CaCl_2 \cdot 2H_2O$ molar ratio (N) as: (a) 1.4, (b) 1.6, (c) 1.8, and (d) 2.0.

Previous reports [21] show that it is difficult to control microstructure of $Ca(OH)_2$ nanoparticles (i.e., particle size and shape, particle size distribution, and degree of agglomeration), which plays an important role in optimizing physical properties of $Ca(OH)_2$ dispersions (i.e. dispersions stability, viscosity, and carbonation processing of $Ca(OH)_2$). Moreover, agglomeration of nanoparticles is usually unavoidable due to high Van der Waals forces between nano-particles. Addition of surfactant is considered as an effective and cheap way to tailor and modify the morphology and particle size.

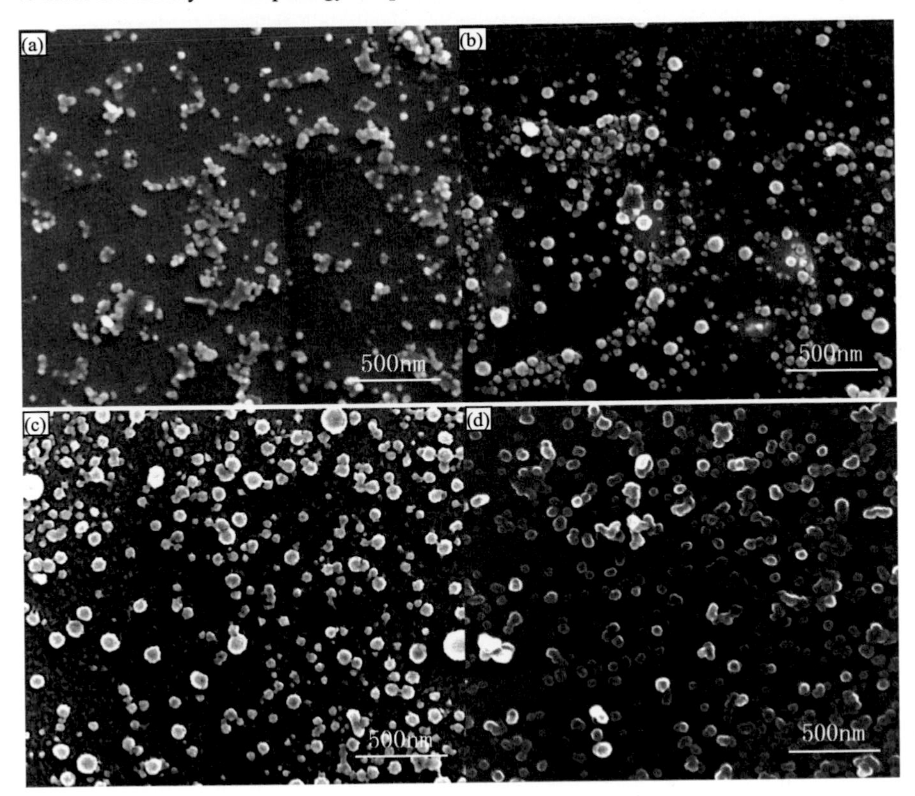

Fig. 3. SEM images of $Ca(OH)_2$ nanoparticles: (a) unmodified; (b) modified by PA; (c) modified by KH-570; (d) modified by SDBS

Fig. 1(b) shows XRD patterns of unmodified and modified $Ca(OH)_2$ nanoparticles by SDS, KH-570, and polyvinyl alcohol (PA), respectively. It is noticed that the hexagonal $Ca(OH)_2$ is still attained by the addition of surfactants. However, obvious broadening of XRD patterns is observed, implying that grain size of modified $Ca(OH)_2$ was significantly reduced. It is especially obvious for KH-570 modified $Ca(OH)_2$ nanoparticles. Since the particle size of nanoparticles modified by surfactants was not significantly changed as shown in Fig. 3 below, broadening of XRD patterns is mainly due to much less degree of aggregation of particles.

Fig. 3 shows SEM images of $Ca(OH)_2$ nanoparticles modified by different surfactants. Average particle size of modified $Ca(OH)_2$ nanoparticles (~50nm) increases slightly, but its dispersibility increases as well. However, particle size distribution of nanoparticles modified by SDS was wide, and obvious aggregates can be observed. In contrast, particle size distribution of nanoparticles modified by KH-570 is more homogenious. Specific surface area of nanoparticles was shown in Fig. 1(c). Specific surface area of nanoparticles modified by KH-570 is as high as $48.78m^2/g$, about 2.5 and 3.4 times higher than that of unmodified and commercial ones, respectively, implying that degree of agglomeration was significantly reduced.

Kinetic stability of $Ca(OH)_2$ nanoparticles dispersions

Table 2. Kinetic stability of dispersions by gravity sedimentation experiment

Samples	0h	12h	24h	36h	48h
Unmodifed Commercial $Ca(OH)_2$	10	8.4	7.1	6.0	4.8
Unmodified	10	8.8	8.2	7.2	6.4
Modified by PA	10	--	--	--	--
Modified by KH-570	10	9.4	9.0	8.5	7.8
Modified by SDBS	10	9.3	9.2	8.0	7.2

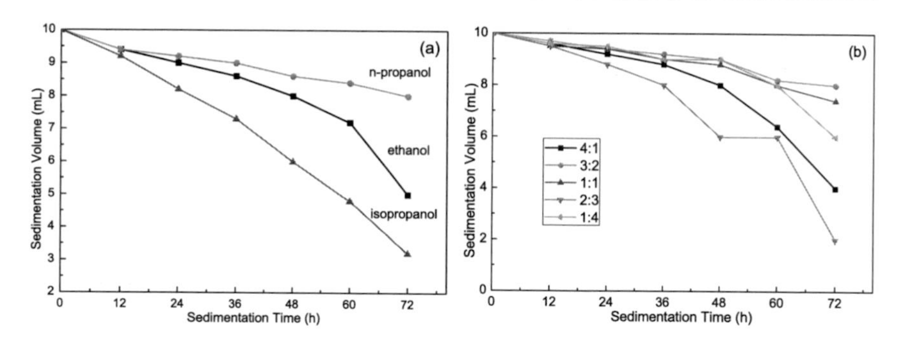

Fig. 4. Gravity sedimentation results of different dispersion media: (a) three nonaqueous dispersants; (b) mixing dispersants of ethanol and n-propanol with different volume ratios. Note that this nanoparticle dispersion was prepared from 1.6 molar ratio starting materials.

Fig.4 (a) shows kinetic stability of $Ca(OH)_2$ nanoparticles dispersions by gravity sedimentation measurements using n-propanol, ethanol, and isopropanol as a dispersion media. As shown, kinetic stability of n-propanol dispersion is higher than that of ethanol and isopropanol dispersion. It is reported that the stability of nanoparticle dispersions is related to hydrophobic chain length of dispersion medium [22]. Usually, the longer hydrophobic chain of alcohol results in higher dispersion stability. Gravity sedimentation technique was further used to measure kinetic stability of various $Ca(OH)_2$ dispersions at 25°C using n-propanol as dispersion medium, including commercial, unmodified, and surfactant modified $Ca(OH)_2$ nanoparticles. As shown in Table 2, $Ca(OH)_2$ dispersion modified by KH-570 shows the highest kinetic stability, as evidenced by higher sedimentation height measured at 12 hrs, 24 hrs, 36 hrs, and 48 hrs, respectively. Further investigation indicates kinetic stability of $Ca(OH)_2$ nanoparticles dispersions modified by KH-570 in n-propanol solution show still higher kinetic stability. In contrast, kinetic stability of dispersions in isopropanol solution is relatively low. As shown in FTIR results, the surface of $Ca(OH)_2$ nanoparticles in dispersions is wetted and covered by alcohol molecules. Especially, $-CH_3$ and $=CH_2$ functional groups can be absorbed on surface of KH-570 modified $Ca(OH)_2$ nanoparticles due to their strong affinity with alcohol molecules. This subsequently results in a decrease of electrostatic force and van der Waals forces on surface of particles and enhanced its kinetic stability in dispersant.

According to features of porous materials, different dispersants can be further selected and mixed to achieve ideal rheological properties and penetration depth inside the artifacts. Therefore, we prepared dispersant for $Ca(OH)_2$ nanoparticles dispersions by mixing different volume ratio of ethanol and n-propanol. Kinetic stability test results are shown in Fig. 4 (b). As expected, properties of kinetic stability can be further optimized by mixing different dispersants [12]. When the volume ratios of n-propanol and ethanol are set as 1:1 and 3:2, kinetic stability of $Ca(OH)_2$ nanoparticle dispersions is higher than that of dispersion using single n-propanol as dispersant. In contrast, kinetic stability of dispersion is relatively higher using 3:2 volume ratios of n-propanol and ethanol. On a basis of dispersion stability test, we preferably choose KH-570 modified $Ca(OH)_2$ nanoparticle dispersion using 3:2 volume ratio of ethanol and n-propanol as dispersant as testing materials in consolidating weathered sandstone from the Yungang grottoes.

Dispersions as consolidating materials

Table 3 summarizes the chromatic change, consolidation absorption rate, and capillary water absorption of samples treated by KH-570 modified $Ca(OH)_2$ nanoparticle dispersions using 3:2 n-propanol and ethanol solution. As shown, chromatic change of measured stone samples is less than 5 and satisfies the requirements of stone consolidation. However, it should be stressed that this value will further increase with increasing $Ca(OH)_2$ content in the dispersions. The consolidation absorption rate of treated samples increases with increasing $Ca(OH)_2$ content in the dispersions as well. We would like to point out that the consolidation absorption rate of KH-570 modified dispersion is slightly lower than that of unmodified ones. While water can penetrate into stone by capillary absorption, the capillary water absorption indirectly indicates the change of porosity and the resistance of the samples to water. As shown in Table 3, samples treated by dispersions decrease the capillary water absorption, implying enhanced resistance to water and decreased porosity of samples treated by dispersions.

Table 3. Testing results of consolidation performance

number	chromatic change (--)	consolidation absorption rate (%)	capillary water absorption ($g/(cm^2 \cdot h)$)
1#	--	--	8.25×10^{-3}
2#	2.55	0.031	3.74×10^{-3}
3#	2.62	0.044	6.13×10^{-3}
4#	3.78	0.069	6.30×10^{-3}
5#	3.31	0.081	1.85×10^{-3}

Comment: 1# is blank sample; 2#, 3# and 4# are KH-570 modified $Ca(OH)_2$ nanoparticle dispersions with a concentration of 1g/L, 3g/L and 5g/L, respectively; 5# is the non-modified $Ca(OH)_2$ nanoparticle dispersion with a concentration of 3g/L.

Fig. 5 shows the permeability of water vapor before and after dispersion treatment, which indirectly represents the breathability of porous stone. Usually, the permeability of water vapor of treated stone by dispersion is lower than that of untreated one, resulting from decreased porosity since $Ca(OH)_2$ nanoparticles fill up the pores and block the channels for water evaporation. As expected, samples treated by dispersions with higher $Ca(OH)_2$ concentration show lower permeability of water vapor. However, samples treated by KH-570 modified $Ca(OH)_2$ dispersions shows better water vapor permeability.

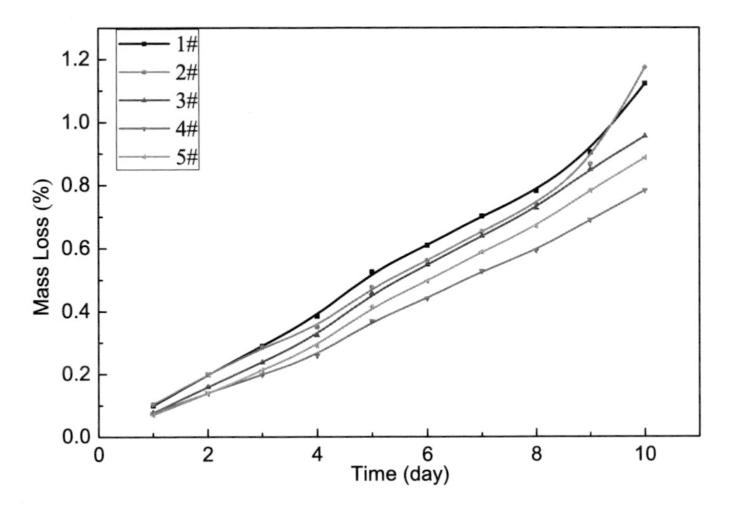

Fig. 5. Permeability of water vapor for different concentrations as described in Table3.

Application to weathered sandstone samples

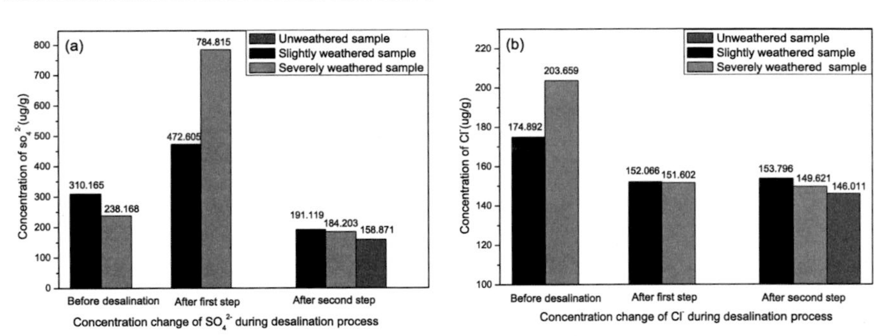

Fig. 6. Concentration change of (a) SO_4^{2-} and (b) Cl^- during desalination process. Note that unweathered sample shows no change from original to final desalination step.

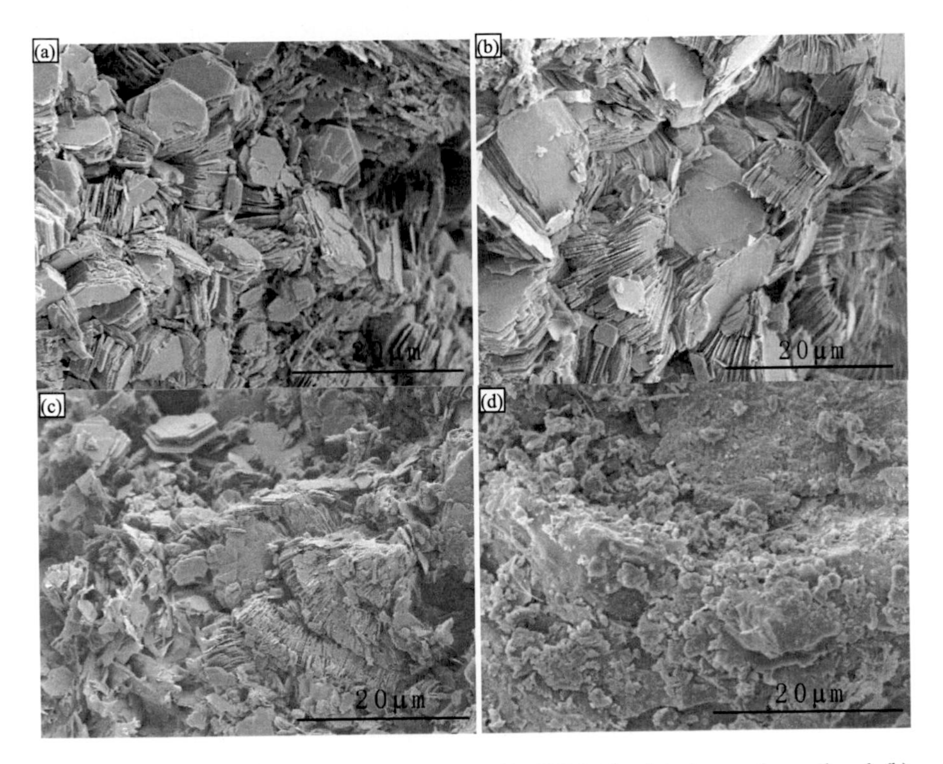

Fig.7. Microstructure of the stone characterized by SEM: (a) original severely weathered; (b) treated by saturated $(NH_4)_2CO_3$ solution; (c) treated by saturated $Ba(OH)_2$ solution; (d) consolidated by $Ca(OH)_2$ nanoparticle despersions. Note that fine bright points in (d) correspond to $Ca(OH)_2$ nanoparticles and/or partly transferred $CaCO_3$.

It should be pointed out that salt damage is considered one of the most crucial factors in the deterioration of the Yungang Grottoes. Although dispersions of $Ca(OH)_2$ nanoparticles can effectively consolidate weathered stone, the presence of soluble salts can damage sandstone continually. Therefore, it is necessary to carry out desalination before consolidation by $Ca(OH)_2$ nanoparticle dispersions. The Ferroni-Dini method has been successfully applied to the removal of detrimental salts and to the reinforcing of porous inorganic cultural heritage [11]. The method applied to stone from the Yungang Grottoes consists of two steps: the first is the application to the decayed stone of a saturated solution of ammonium carbonate, $(NH_4)_2SO_4$; the second is the treatment with $Ba(OH)_2$ solution. During the first step, the gypsum is converted into calcium carbonate with the formation of water soluble $(NH_4)_2SO_4$ according to the following chemical equation: $(NH_4)_2CO_3 + CaSO_4{\cdot}2H_2O \rightarrow (NH_4)_2SO_4 + CaCO_3 + 2H_2O$. The changes of SO_4^{2-} and Cl^- concentrations during desalination process by the Ferroni-Dini method is shown in Fig. 6. As shown, the concentration of Cl^- and SO_4^{2-} in the solution treated by $(NH_4)_2CO_3$ solution is much higher in the slightly weathered stone. The increase of SO_4^{2-} concentration is especially obvious. According to chemical equation above, $CaSO_4$ with relatively low solubility is transformed into easily soluble $(NH_4)_2SO_4$ in severely weathered stone from the Yungang Grottoes. However, the concentration of SO_4^{2-} in stone treated by supersaturated $Ba(OH)_2$ solution significantly decreases and is almost the same as that of fresh stone. This is due to the reaction $(NH_4)_2SO_4 + Ba(OH)_2 \rightarrow BaSO_4 + 2NH_3 + 2H_2O$. During this process, easily soluble $(NH_4)_2SO_4$ is transformed to insoluble $BaSO_4$ salts, which can fill up pores in stone and result in increase of strength. The byproducts (ammonia and water) evaporate into the air. Additionally, extra $Ba(OH)_2$ reacts with CO_2 to produce $BaCO_3$, which also fill pores in stone and further increases the strength of consolidated stone [11]. It should be mentioned that sandstone from the Yungang Grottoes are insensitive to base erosion [18].

This results in the reduction of concentration of SO_4^{2-} ions. In contrast, concentration of Cl^- in severely weathered stone is much higher than that of slightly weathered stone. It is noticeable that it significantly reduces for stone treated by supersaturated $(NH_4)_2CO_3$ solution. However, it become relatively lower until it reaches the values of fresh stone after treated by supersaturated $Ba(OH)_2$ solution.

The SEM images of severely weathered stone before and after treatment by combining the Ferroni-Dini method and $Ca(OH)_2$ nanoparticle dispersions are shown in Fig. 7. Stone particles in untreated samples are of lamellar structure. Stone desalinated by the Ferroni-Dini method show increased porosity, but internal structure is still loose. In contrast, significantly enhanced bonding between particles and enhanced porosity are observed after dispersions were applied. It is due to filling-up of insoluble $BaSO_4$, $BaCO_3$, and $CaCO_3$ salts inside stone.

CONCLUSIONS

Uniform hexagonal $Ca(OH)_2$ nanoparticles (~35nm) were obtained using homogeneous-phase reactions. KH-570 surfactant can significantly reduce agglomeration and improve specific surface area of $Ca(OH)_2$ nanoparticles. Specific surface area of modified $Ca(OH)_2$ nanoparticles by KH-570 reaches up to ~$48.78 m^2/g$, about 2.5 times higher than that of the unmodified dispersion. Kinetic stability of $Ca(OH)_2$ dispersion can be further enhanced and its viscosity decreased by optimizing the ratio of ethanol and n-propanol. The technique that combines the Ferroni-Dini method and $Ca(OH)_2$ nanoparticles dispersion can effectively desalinate and consolidate the decayed sandstone samples from the Yungang grottoes. With good compatibility with sandstone substrate and environmental friendly character, the technique described herein shows promising applications in sandstone or mortars heritage conservation.

ACKNOWLEDGMENTS

This work was financially supported by the China Science & Technology Polar Program (No. 2012BAK14B05 and 2009BAK53B06) and the Compass Special Program of the State Administration of Cultural Heritage of China (No. [2011]1806).

REFERENCE

1. N. Gómez-Ortíz, S. De la Rosa-García, W. González-Gómez, M. Soria-Castro, P. Quintana, G. Oskam, and B. Ortega-Morales, ACS Appl. Mater. Inter. **5**, 1556 (2013).
2. P. Tiano, E. Cantisani, I. Sutherland, J. Paget, J. Cult. Herit. **7**, 49 (2006).
3. V. Daniele, G. Taglieri, R. Quaresima, J. Cult. Herit. **11**, 294 (2008).
4. M. Ambrosi, L. Dei, R. Giorgi, C. Neto, P. Baglioni, Langmuir. **17**, 4251 (2001).
5. P. López-Arce, L. Gómez, L. Pinho, M. Fernández-Valle, M. Álvarez de Buergo, R. Fort, Mater. Charact. **61**, 168 (2010).
6. T. Yasue, Y. Tsuchida and Y. Arai, Gypsum. Lime. **189**, 17 (1984).
7. A. Nanni, L. Dei, Langmuir. **19**, 933 (2003).
8. V. Daniele, G. Taglieri, J. Cult. Herit. **13**, 40 (2012).
9. B. Salvadori, L. Dei, Langmuir. **17**, 2371 (2001).
10. L. A. Perez-Maqueda, L. Wang and E. Matijevic, Langmuir. **14**, 4397 (1998).
11. M. Ambrosi, L. Dei, R Giorgi, C. Neto, P. Baglioni, Prog. Colloid Polym. Sci. **118**, 68 (2001).
12. P. Baglioni, R. Giorgi, Soft. Matter. **2**, 293 (2006).
13. J. Huang, J. Yanbei Normal Univ. **19** (5), 57 (2003).
14. J. Yuan, X. Feng, World Antiquity. **5**, 74 (2004).
15. Z. Li, World. Antiquity. **5**, 3 (2004).
16. R. Espinosa-Marzal, G. Scherer, Accounts. Chem. Res. **43**, 695 (2010).
17. G Scherer, R Flatt, G Wheeler, MRS Bull. **26**, 44 (2001).
18. S Tian, S Liu, F Gao, M Fan, J Ren, Mater. Res. Soc. Symp. Proc. **1319,** Boston, MA, 2010, DOI: 10.1557/opl.2011.
19. S. Liu, M. Sun, F. Gao, Y. Xiao, K. Liu, J. Central. South. Univ: Sci. & Tech. **44** (1), 46 (2011).
20. K. Yura, K. Fredrikson, E. Matijevic, Colloids Surf. A. **50**, 281 (1990).
21. G. Rees, R. Evans-Gowing, S. Hammond, B. Robinson, Langmuir. **15**, 1999 (1993).

Mater. Res. Soc. Symp. Proc. Vol. 1656 © 2015 Materials Research Society
DOI: 10.1557/opl.2015.3

Unraveling the Core of The *Gran Pirámide* From Cholula, Puebla. A Compositional and Microstructural Analysis of the Adobe

N. A. Pérez[1], L. Bucio[1], E. Lima[2], C. Cedillo[3], D. M. Grimaldi[4]

[1]Instituto de Física, Universidad Nacional Autónoma de México, México DF, México
[2]Instituto de Investigaciones en Materiales, UNAM, México
[3]Zona Arqueológica de Cholula, Centro INAH Puebla, Instituto Nacional de Antropología e Historia, Puebla, México
[4]Área de Conservación Arqueológica, Coordinación Nacional de Conservación del Patrimonio Cultural, Instituto Nacional de Antropología e Historia, México DF, México.

ABSTRACT

The *Gran Pirámide*, a Mexican cultural heritage site, is located at the archaeological site of Cholula, Puebla, Mexico. At the base of its platform this pyramid is the largest in the world. It was built in layers from 800 to 1100 AD by the Cholultecan pre-Hispanic culture. The archaeological site is famous by its great mural paintings that have been well-studied. The pyramid was built with earthen construction, a system of multiple bulding episodes with layers of adobe. The building material, adobe, has not been well studied. Due to its fragile condition, a more extensive study was conducted to understand the behavior of the building and the mural paintings substrate, in order to propose conservation strategies.

Geological context of the area was the starting point to propose the relevant materials used in its construction. That was a fundamental key for the interpretation of the experimental techniques used that include X-ray Diffraction (XRD), Particle-Induced X-ray Emission (PIXE), ^{29}Si and ^{27}Al Nuclear-Magnetic Resonance with Magic-Angle Spin (NMR-MAS), Thermal Analysis, Optical and Scanning Electron Microscopy (SEM) and colorimetric measurements.

The results obtained from the original adobes have been compared with fresh soils from horizons related with pre-Hispanic activity. The results indicate presence of amorphous materials and neo-mineral formation besides feldspars and opal. The amorphous phases have been identified by NMR-MAS and SEM.

Differences were found in the composition from the adobe used for the joints, mainly in the clay fraction, that can be distinguished by color and that guided to group the information acquired.

These results provide new information on the composition and microstructure of adobes from the *Gran Pirámide* of Cholula. Further studies will involve soil physics methods and erosion tests to complete the task of having a comprehensive knowledge of the earth architecture of the pyramid.

INTRODUCTION

Geological and cultural Setting

Cholula is located in the center of Puebla valley, in the Mexican Transvolcanic Belt to the west of the Sierra Nevada, 2000 m above sea level (Figure 1). The main physiographic characteristics are the mountains that surround and enclose the valley. To to the northeast it is bordered by *La Malinche* volcano, to the west by the Sierra Nevada. To the south is the Atoyac River Basin and on the north with the hill of San Lorenzo. In the center are located the hills of Tecajete and Zapotecas, that consists of a type of volcanic cone complex with lava spills at the base.

The annual rainfall at the lower part of the valley is 600-700 mm and has a temperate climate with dry winters and cool summers with the heaviest rains [1].

Figure 1. Location of the archaeological site of Cholula on the Puebla Valley. The marked green region is the Mexican Transvolcanic Belt. Image of the Great Pyramid of Cholula (right).

The most important geological events in the region have been the recurring Plinian eruptions of *Popocatépetl* volcano every 1000 to 3000 years. These eruptions have followed a similar pattern and start with the expulsion of small ash flows. The eruptions reached their maximum with the main Plinian pulse that caused the ash deposition, the emplacement of hot ash flows and finally extensive volcanic mud flows known as lahars. On each occasion, the devastated area was repopulated before the return of another Plinian event. Siebe has established the extent of areas flooded with lahars and regional stratigraphy according to the eruptive sequences and relatined them to pre-Hispanic activity [2].

Contemporary with Teotihuacan, Cholula is seen as the other major city in central Mexico during the Classic period. However, the reference to Cholula persists in several early sources showing that the *cholulteca*s were among the first Mesoamerican people, as well as their centrality in the foundational stories of the highland cultures. The archaeological evidence of human occupation in the region since 300 BCE has provided further evidence and reinforced this conclusion. In addition, its unique geographic position allowed the city to be the connection point between different routes in ancient Mexico, for it was an obligatory crossing in the route to the Gulf Coast and to Oaxaca; therefore, a convergence of cultures occurred in this area [3, 4].

Tlachihualtépetl is The Great Pyramid's pre-Hispanic name which means the "handmade hill". It is the largest pyramid that was built in Ancient Mexico by the people of Cholula using adobe bricks. This magnificent and imposing ritual building has eight major construction phases that were carried out in such a way that previous stages were hidden each time. The Great Pyramid was mutilated starting in colonial times, when the last construction stage was dismantled to use the stone and building elements for the new catholic buildings.

Currently, the *Tlachihualtépetl* is seen as an artificial hill with some original adobe visible, and on top of which is based and was built the Sanctuary of the Virgen de los Remedios [3].

The main tunnel that tourists now use to visit in the interior layers covers 280 m of more than ten kilometers of tunnels that were required for archaeological exploration (Figure 2). The Great Pyramid in its final construction phase reached 65 m high and 400 m per side at the base as a result of the various building phases, constituting the largest volume of a pyramid in the American continent [3, 5].

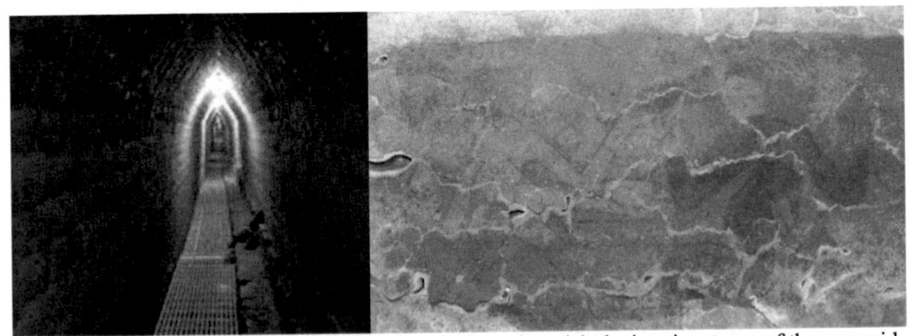

Figure 2. The main tunnel of The Great Pyramid used to visit the interior stages of the pyramid (left). Detail of the "Bebedores" mural painting (right).

The Great Pyramid as well as some other minor constructions around it, had facades extensively decorated with mural paintings executed over earthen support. Among the remnants the more important ones are "Chapulines" (Grasshopers), "Estrellas" (Stars) and "Los Bebedores" (The Drinking Ones), all of them painted in a *secco* technique, using inorganic pigments [6]. "Los Bebedores" is located in Building 3, to the south of the Great Pyramid. Its archaeological exploration that took place at the end of the 1960's uncovered 120 square meters of murals, which nowadays are not open to public visit due to conservation requirements. A tunnel was constructed in order to reach "Los Bebedores" mural without dismantling two other construction stages of the same building. The pyramid was constructed with different types of adobes that can be observed from area to area of the building, with different color, texture and dimensions which suggests that the building material was brought from different locations to the site.

Causes of on site deterioration

The permanence of the adobe construction sites is evident since there are structures that have lasted hundreds of years. However, they are also fragile structures that depend on context and environment. Proper maintenance, repairs and restoration that are compatible with the original buildings are critical for the survival of earthen archaeological sites [7].

In the archaeological site of Cholula, several factors promote the deterioration and decay of the adobes. From the natural context, the site is subjected to frequent earthquakes since it is a seismic zone. Additionaly, the adobes are made of soil, and are vulnerable to changes in the humidity not only directly but also from the various changes on the underground water level primarily due to the urban expansion of the city of Cholula.

Different changes in quantity and diffusion velocities of the water going in and out of the adobes generates alterations in the construction system that shows effects, such as fungi on the walls and adobe collapses (Figure 3). Recent studies have shown that the main collapses take place when the rainy season starts, followed by over three months to absorb all the water. When it starts again, the adobe is porous, friable and often collapses although this is not a predictable periodic cycle. Variations in the water content of the adobes underlies the need to understand the role that water plays in the microstructure of the adobes, and therefore, affects its properties and stability. At present, we do not know if the adobes had or not, an additive to enhance the cohesion and strength.

Plants covering the pyramid which have deep roots penetrating into the adobe bricks and walls, and they are a major factor in deterioration. Roots tend to expand or contract with changes in humidity, causing fractures that end in the collapse of the adobes and that leave micro-fractures and disaggregated material.

Figure 3. Detail of a collapsed earthen left upper wall of one of "Los Bebedores" mural that affects the stability and durability of the mural paintings.

Soluble salts have also affected the adobe. The cement extensively used to consolidate exterior walls as well as more recent construction stages at some buildings has added deleterious sulfates, as happened at Building 1. Broken pipes from the catholic church on the top of the

Great Pyramid has added nitrates [8]. Furthermore, some adobe was grouted or surfaces were treated with synthetic polymers beginning 1930 but mostly during the 1960's and 1970's, changing its original behavior [9]. Many of these polymers act as impermeable surface layers that do not breathe. The lack of stability of the adobe not only threatens wall paintings due to collapse from the upper part of the tunnels, but also because of cracking and delamination of the paintings and deterioration of the fragile support. Diagnosis of the condition of the murals at "Los Bebedores" recorded plenty of voids in the earthen support that have required of immediate intervention to prevent the paint layer collapse [8, 10, 11].

Results of monitoring during the conservation showed that the main deterioration factors exist from the top of the pyramid to the bottom. Therefore, an integral, long-term and comprehensive proposal needs to be made in order to assess correctly the conservation and preservation problems.

Purpose of the research

Our focus is placed on materials analysis, behavior and properties of the adobe, as the constituent material in the archaeological site of Cholula, is a brittle material which affects the stability of the buildings. Knowledge about the composition, their properties and how they interact with their environment is necessary to establish comprehensive strategies for its conservation.

EXPERIMENTAL DETAILS

Sampling

Twelve adobe brick samples were taken from the main tunnel of the interior of the Pyramid in a collapse area, based on a sampling selection that emphasized color and textural differences. Another five samples were taken from "Los Bebedores" tunnels four and six from areas nearby the mural paintings (Figure 3). Three fresh soils samples just offsite from horizons related to pre-Hispanic activity were selected due to their current use in brick manufacture at the Cholula region.

Composition and microstructural characterization

Optical microscopy was first employed to observe the main characteristics of the adobe surfaces and different phases in the soils. The identification of the mineral phases was made through petrographic analysis and X-ray Diffraction (XRD) of powder samples in a Bruker D8 Advance with Cu Kα radiation. Measurements were carried out in the 2θ angular range 6-90 degrees, the results obtained were complemented with chemical compositions analyzed by Particle Induced X-ray Emission (PIXE) spectrometry, using a 3 MeV proton external beam with 1.5 mm diameter beam spot and measurements of 600 s per region on each sample.

To study the local chemical environments and to identify the semi-crystalline and amorphous phases, nuclear magnetic resonance using solid state ^{29}Si and ^{27}Al (MAS-NMR) was used. The MAS-NMR spectra were acquired under magic angle conditions ($\theta = 54.74$ degrees) in a Bruker Advance II 300 spectrometer with a 7.05 T magnetic field. The chemical shift of ^{27}Al was referenced to an aqueous solution of $Al(NO_3)_3$ as an external standard and for the ^{29}Si nuclei tetramethylsilane (TMS) was used as reference. Simultaneous thermogravimetry (TGA) and differential thermal analysis (DTA) in a TA SDTQ600 instrument was performed with an air flux to complement the identification of the amorphous phases.

The textural properties determined by petrographic analysis were complimented with observation by Scanning Electron Microscopy (SEM), a JEOL JSM 5600-LV instrument, used in secondary electron mode in low vacuum conditions. Colorimetric measurements were done with an Ocean Optics USB4000 spectrometer with a halogen lamp (360-2500 nm) and were reported on the CIE Lab color space.

DISCUSSION

The thorough crystallographic study of this type of cultural heritage enables the expansion of our knowledge about the details of structure and properties. We identified the role that each mineral contributes on the properties of the adobe by integrating knowledge from mineralogy with that of materials science. Volcanic soils are important due to their unique properties. The presence of amorphous materials gives them distinctive engineering properties such as: low bulk density, high organic matter content, high porosity and high water retention capacity and high Atterberg limits, as the latter are very important for successful use and preservation of dynamic construction material such as adobe [12-15].

The XRD results of the research show that the adobe blocks from the Great Pyramid and the adobes from the mural painting have an amorphous phase and five main crystalline phases shows in Figure 4 as: calcium/sodium aluminosilicate (plagioclase, PLG), quartz (Qz), opal CT (Op CT), pyroxene (Pyx) and amphibole (Amp).

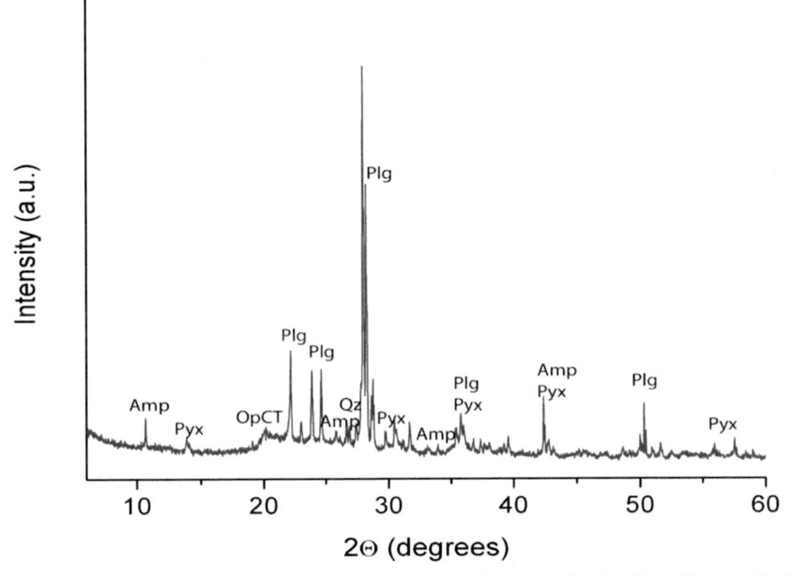

Figure 4. XRD pattern of an adobe block from the main tunnel in the Grand Pyramid, showing plagioclase (Plg), quartz (Qz), opal CT (Op CT), pyroxene (Pyx) and amphibole (Amp).

The results obtained from the original adobes were compared with fresh soils from horizons related to pre-Hispanic activity on which the same mineral phases have been identified. The main difference among the sample groups is the particle size distribution as observed by optical microscopy and petrographic analysis (Figure 5).

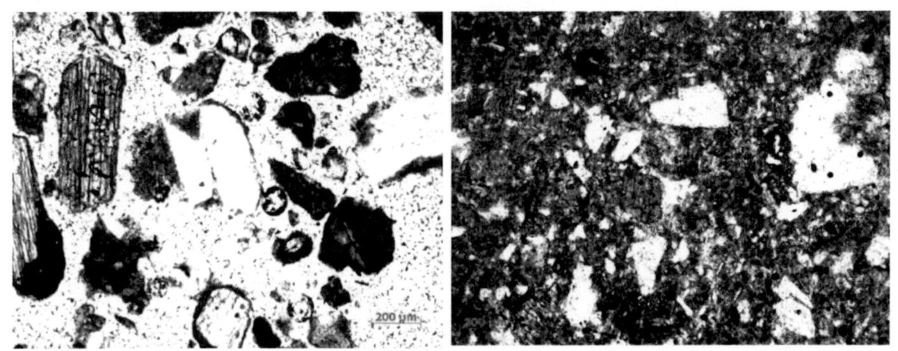

Figure 5. Petrographic images of representative local soil denominated tepetate (left) and adobe block from the main tunnel in the Grand Pyramid (right).

Based on quantitative elemental analysis accordingly to the semi-quantitative analysis of XRD results and observation by optical microscope, the plagioclase, pyroxene, amphibole and quartz phases constitute the larger aggregates forming the sand- and silt-sized phases of the mixture. A mixture of cristobalite and tridymite indicate the presence of opal CT, which together with the amorphous phases, corresponds to the clay-sized particles. The literature on volcanic soils suggests presence of allophane, glass and organic material. Based on these results it was of great interest to determine the amorphous materials since they have a strong influence on plasticity and on the shrink-swell behavior of the soils. Most of these clay-sized materials act to strengthen the contacts between crystalline particles, thus explaining the use of this material that increases plasticity and cohesion by ancient cultures [12, 16].

The MAS-NMR technique was useful to identify the allophane phase of the adobe samples and also in the soils (Table 1). This result was confirmed by the thermal analysis (TGA-DTA) that shows the exothermic peak at 900°C corresponding to allophane [17, 18].

Table 1. ^{29}Si MAS-NMR signals assignment for the adobes and soil samples.

	^{29}Si signal (ppm)	^{27}Al signal (ppm)
Allophane	-85	-4,39,54
Opal CT	-103, -96	54
Quartz	-115	---
Aluminosilicates	-92, -96,-103,-108	-4,54

The color measurements of dry and wet samples confirmed that the macroscopic textural difference in adobes is due to particle size variations and not to composition because the color is very similar in the adobes and soils. The color variation is more evident when the adobes are wet and have a larger dispersion of colors (more than five units in CIE Lab color space). In the dry samples, a difference of three units or less is not noticeable to the naked eye (Figure 6).

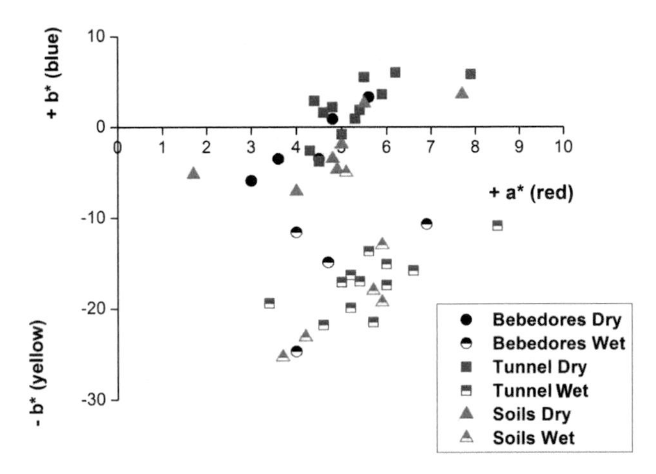

Figure 6. Colorimetric measurements of the dry and wet adobe samples of the main tunnel, mural paintings ("Los Bebedores") and local soils.

CONCLUSIONS

The relationship between structure and function is fundamental to understanding materials, and, from this knowledge, pathways of how to control and forecast the properties of earth building materials are developed. From this approach we determined that the materials used for the construction of the adobes are from local soils but the noticeable difference in color is due to different proportions of the various soils and selection of particle size for the different adobes according to the construction system. The results indicate presence of amorphous materials and neo-mineral formation besides a major presence of plagioclase and quartz. The amorphous phases have been identified by MAS-NMR and DTA. Therefore cohesive properties of the adobes are due to clay-sized amorphous materials such as glass, allophane and organic matter. The combination of cohesive, plastic materials, such as the regional soils fulfilled the requirements for building the cities with available material that still remain functional today.

ACKNOWLEDGMENTS

We thank Eréndira Martínez, Antonio Morales, Marco Vera, Angel Osornio, Jacqueline Cañetas and Mario Monroy for the technical assistance for this research. To the Laboratorio Central de Microscopia (IFUNAM) for the SEM facility. Thanks to CONACyT for the graduate scholarship and CONACyT project CB-2011/167624.

REFERENCES

1. M. Reyes in *Proyecto Cholula,* edited by I. Marquina (INAH, México, 1970), p. 9-15.
2. C. Siebe, M. Abrams, J.L. Macías, J. Obenholzner, Geology 24, 399 (1996).
3. F. Solís, G. Uruñuela, P. Plunket, M. Cruz, D. Rodríguez, *Cholula: La Gran Pirámide* (CONACULTA-INAH, México, 2006).
4. A. Ashwell, Revista Elementos 54, 39 (2004).
5. P. Plunket, G. Uruñuela, FAMSI Grantee Report, 2005.
6. A. Huerta, Report CNCPC-INAH, 1972.
7. F. Pacheco-Torgal, S. Jalali, Constr. Build. Mater. 29, 512 (2012).
8. D.M. Grimaldi, M. Aguirre, C. Ramirez, Report CNCPC-INAH Informe del Proyecto de Conservación e Investigación de la Pintura Mural de la Zona Arqueológica de Cholula, Puebla, temporada de campo, 2012.
9. D.M. Grimaldi and T. López, Report CNCPC-INAH Resumen de los tratamientos realizados en las pinturas murales de la Zona Arqueológica de Cholula, Puebla (1967-1970), 2006.
10. D. M. Grimaldi, M. Aguirre, C. Ramirez, Report CNCPC-INAH Informe del Proyecto de Conservación e Investigación de la Pintura Mural de la Zona Arqueológica de Cholula, Puebla, temporada de campo, 2010.
11. D. M. Grimaldi, J. Porter, C. Ramirez, Report CNCPC-INAH Informe del Proyecto de Conservación e Investigación de la Pintura Mural de la Zona Arqueológica de Cholula, Puebla, temporadas de campo, 2011.
12. R. Horn, H. Taubner, M. Wuttke, T. Baumgartl, Soil & Tillage Research 30, 187 (1994).

13. Y. Wan, J. Kwong, H. G. Brandes, R. C. Jones, J. Geotech. Geoenviron. 128, 1026 (2002).
14. B. Prado, C. Duwiga, C. Hidalgo, D. Gómez, H. Yee, C. Prat, M. Esteves, J.D. Etchevers, Geoderma 139, 300 (2007).
15. A. J. Mehta, E. J. Hayter, W. R. Parker, R. B. Krone, A. M. Teeter, Journal of Hydraulic Engineering 115, 1076 (1989).
16. Y. Wan, J. Kwong, Eng. Geol. 65, 293 (2002).
17. B.A. Goodman, J.D. Russell, B. Montez, E. Oldfield and R.J. Kirkpatrick, Phys Chem Minerals 12, 342 (1985).
18. A. F. Plante, J. M. Fernández, and J. Leifeld, Geoderma 153, 1 (2009).

Mater. Res. Soc. Symp. Proc. Vol. 1656 © 2014 Materials Research Society
DOI: 10.1557/opl.2014.812

Environmental Monitoring of Volatile Organic Compounds Using Silica Gel, Zeolite and Activated Charcoal

Molly McGath[1], Blythe McCarthy[2], Jenifer Bosworth[2]

[1] Heritage Science for Conservation, Department of Conservation and Preservation, Johns Hopkins University, Baltimore, MD 21218 USA
[2] Freer Gallery of Art and Arthur M. Sackler Gallery, Smithsonian Institution, Washington, DC 20013 USA

ABSTRACT

Volatile organic compounds (VOCs) can be hazardous to human health and can negatively impact the long-term stability of art objects. This research evaluated the VOC adsorbent properties of three materials commonly used in museums as humidity regulating or air filtering agents. Silica gel, activated charcoal, and zeolite powder, materials often placed in proximity to art objects, were analyzed using Thermal Desorption GC-MS to qualitatively identify adsorbed VOC's from model environments. This research compared the adsorbing capabilities of these materials with a solid-phase micro-extraction (SPME) carboxen/polydimethyl siloxane fiber to frame their adsorbing powers. It was found that different adsorbents have very different ranges of adsorption for the chemicals tested. Silica gel powder and zeolite powder have the greatest sensitivity for acetic acid over a 24 hour exposure period. Zeolite powder and activated charcoal were more sensitive for identification of naphthalene. Silica gel powder proved to be the most sensitive adsorbent overall. This research discovered that the methods used to condition silica gel pellets for reuse need to be re-examined in light of fact they trap VOC's, especially as it was observed that VOC's desorb from the silica gel pellets under ambient conditions.

INTRODUCTION

Volatile organic compounds (VOCs) are hydrocarbon rich chemicals with varying functionalities that typically have low boiling points and high vapor pressures. Some materials release VOCs over time, and in some cases this is a planned release of VOCs in a drying/curing step as seen with paint or caulking. In other cases the VOCs are given off over the life-span of the material, either as compounds that migrate out of the material: as is seen with plasticizers in some plastics [1]; or as the material ages or breaks down: as is seen in the aging and degradation of wood [2], wool and other proteinaceous materials [3], or in food as it spoils [4] [5]. Monitoring VOCs provides information on the presence of chemicals which can be hazardous to both humans [6][7][8] and art [9] [2] [1].

As VOCs enter the air they are diluted, often to concentrations below the limits of direct measurement. Some compounds, many sulfur containing, present observable odors at low ppb concentrations [10]. Some compounds are health hazards even at low concentrations [8]. VOC damage to art has been seen in the darkening of lead white by exposure to sulfur-containing compounds that migrate out of materials[11], and the tarnishing of silver objects to highlight just a few instances [12]. Acetic and formic acids, which are the breakdown products of wood [13], are both hazards to health and can cause damage to objects [14]. VOCs can be adsorbed onto a solid substrate as done in solid-phase micro-extraction (SPME) which concentrates them sufficiently to be measured [1].

Commonly encountered VOCs in the museum setting include: acetic acid [14], formic acid [13], naphthalene [15], camphor [16], toluene [14], many plasticizers [1], and other chemicals. Methods for testing for VOCs in the museum include: passive air sampling techniques [17], Oddy testing [2], and pollutant strips [17]. Direct air sampling, headspace analysis or passive adsorption are useful for monitoring a broad-spectrum of VOCs, while Oddy testing and pollutant strip testing monitor for the presence of specific types of VOCs.

In Oddy tests one exposes metal coupons to a material suspected of releasing hazardous VOC's and after a set time evaluates the deterioration observed on the metal coupons [2]. Though a generally robust test, the results are limited to identifying the presence or absence of chemicals that would degrade the metals tested. This does not usually allow for the identification of the chemicals present or provide a ready test for VOCs that may be hazardous to other material types. Pollutant strips are excellent in identifying the chemical tested for, but are limited to testing for certain chemicals.

The goal of this research was to investigate the adsorption of VOCs by materials that are commonly used within an art object's environment and to compare the results with what is seen using SPME fibers exposed to the same environment. SPME fibers are commonly used to measure VOCs in the museum environment. By examining materials which often are already housed with objects, scientists may ultimately gain a better understanding of which VOCs are present in an enclosure. This work focuses on determining the detection limits of common VOCs like acetic acid and naphthalene after an exposure time of 24 hours when measured using 5-8mgs of silica gel, zeolite, and activated charcoal.

EXPERIMENTS AND PROCEDURES

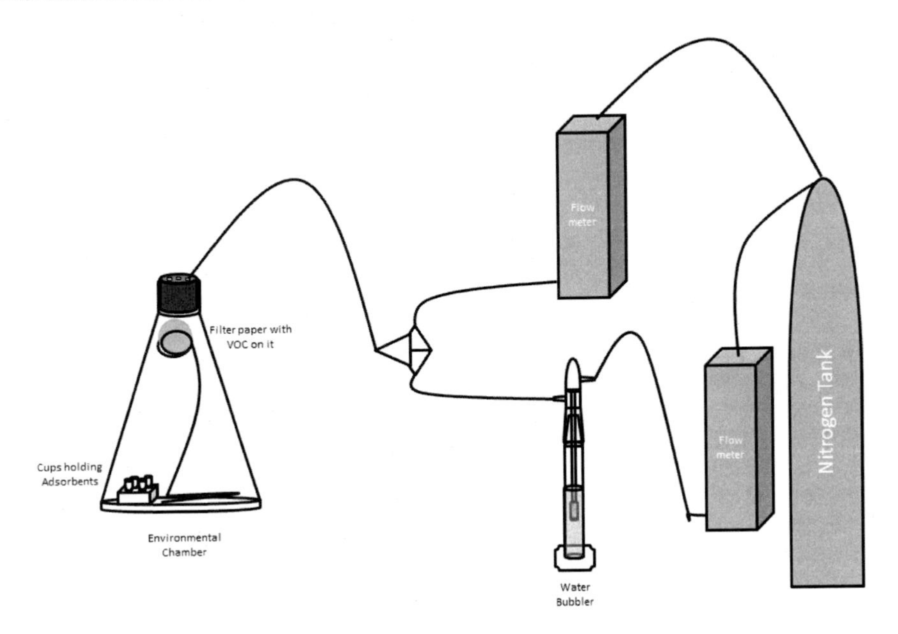

Figure 1. Schematic of an environmental exposure chamber: Each sample of adsorbent (5mg) was placed in a stainless steel "Eco-cup" produced by Frontier Laboratories. The cups in a plastic holder were put into the glass container that then was closed. The chamber then was flushed with humidified nitrogen (50% relative humidity) for 2 hours to ensure complete exchange of nitrogen in the chamber, prior to introduction of a target VOC.

Table 1. Adsorbent materials and the concentrations of chemicals to which they were exposed. Each sample was made in triplicate excepting SPME.

	Silica Gel Powder with 5% $CaSO_4$, Chemie-Erzeungnisse und Adsorptionstechnik AG	**Activated Charcoal, Cocoanut, 50-200mesh,** Fisher Scientific	**Zeolite Powder,** Sigma-Aldrich 96096	**Silica Gel Pellets**	**SPME – 75um Carboxen/PDMS *(done in duplicate)**
Acetic acid (Thomas Scientific)	1ppb, 10ppb, 100ppb, 100ppm	1ppb, 10ppb, 100ppb, 100ppm	1ppb, 10ppb, 100ppb, 100ppm	1ppb, 10ppb, 100ppb, 100ppm	100ppb, 100ppm
Naphthalene (Enoz old-fashioned mothball 98+%)	1ppb, 10ppb, 100ppb, 100ppm	1ppb, 10ppb, 100ppb, 100ppm	1ppb, 10ppb, 100ppb, 100ppm	1ppb, 10ppb, 100ppb, 100ppm	100ppb, 100ppm

Samples of activated carbon, silica gel powder, silica gel pellets and zeolite powder (about 5-8mg) were placed in stainless steel Eco-cups in the chamber shown in Figure 1 for 24 hours and then analyzed using GC/MS as described below to measure the presence of any background chemicals. This procedure was repeated with the addition of a volatile organic compound (VOC) to create a known concentration in the chamber. The VOC's examined were naphthalene and acetic acid. These adsorbents were exposed for 24 hours in the closed container shown in Figure 1. All chemicals were used as purchased. Temperature was that of the room, nominally 70°C, the relative humidity inside the chamber was controlled to be 50%.

After 24 hours the adsorbents were analyzed using a Shimadzu GC/MS QP2010Ultra with a Frontier Laboratory Pyrolysis Pyr-2020iD unit on one of the GC inlets. On the other inlet there was a low volume inlet liner that was used for solid phase micro-extraction (SPME) fiber analysis. The column used was Agilent's DB5MS-UI methyl 5% phenyl polysiloxane column I.D. 0.25mm with film thickness of 0.25μm and length of 30m and was attached through Frontier Laboratories vent-free attachment to the mass spectrometer. The same column was used for both the inlets. The sample was desorbed at 250°C for 10min using the pyrolysis unit. The Pyr-GC interface was held at 320°C, the column oven started at 40°C held for 5 minutes and then increased at a rate of 20°C/minute to 300°C and held for another 10 minutes. The GC/MS interface was held at 250°C and the ion source at 200°C. The voltage for the ion source was 0.7kV, the m/z range was 35 to 600 m/z at a scan rate of 10,000 and with single ion monitoring (SIM) of the molecular ion peaks of the chemical being analyzed. A SPME fiber was conditioned and exposed to each of the VOC's used in the adsorbent experiments. The SPME fiber underwent desorption in the inlet of the GCMS at 300°C for five minutes. The GC/MS parameters were otherwise the same.

As a case study a cabinet filled with cellulose acetate film was monitored over time with silica gel powder, activated charcoal and zeolite powder.

RESULTS AND DISCUSSION

The materials tested adsorbed VOCs when exposed to them in the test chamber (Figure 1). They thermally-desorbed the VOCs when exposed to 250°C in the pyrolysis oven of the Frontier Pyr2020iD, and the VOCs were identified by mass spectrometry. No quantitative relationship between concentration of the VOC present in the test chamber and the adsorption of VOC compounds by the materials tested, was seen with the concentrations of VOC's tested but futher tests may show a quantitative relationship. These adsorbents are useable for the qualitative identification of the presence of acetic acid at 1ppb after 24hours if monitoring for the molecular ion peak using silica gel powder or zeolite powder and the presence of naphthalene at concentrations of 100ppb or greater using activated charcoal or zeolites after an exposure of 24 hours as summarized in Table 2.

Table 2. Limits of detection observed for the adsorbents tested. See Figure 5 for more information about silica gel pellet and acetic acid detection.

	Silica Gel Powder with 5% $CaS0_4$	Activated Charcoal	Zeolite Powder	Silica Gel Pellet	SPME – 75um Carboxen/PDMS
Acetic acid	1ppb (SIM) 100ppb (TIC)	10ppb (SIM), 100ppb (TIC)	1ppb (SIM) 100ppb(TIC)	N/A	100ppb<x<100ppm (TIC) or (SIM)
Naphthalene	100ppb<x<100ppm (TIC)	100ppb (SIM) ~100ppm (TIC)	100ppb (SIM) ~100ppm (TIC)	100ppb (SIM)	>100ppm (TIC) or (SIM)

Each of the adsorbents was analyzed to provide a baseline for the chemicals they had adsorbed from their containers. The chromatograms of zeolite powder before exposure, shown in Figure 2, and of activated charcoal before exposure, shown in Figure 3, reveal that these adsorbents did not have very many chemicals already adsorbed onto them, likely due to having been sold and stored in glass containers. The silica gel powder container was made of plastic and the result of this was that there were many chemicals already adsorbed onto the silica gel powder as seen in Figure 4, and the chemicals are summarized in Table 3.

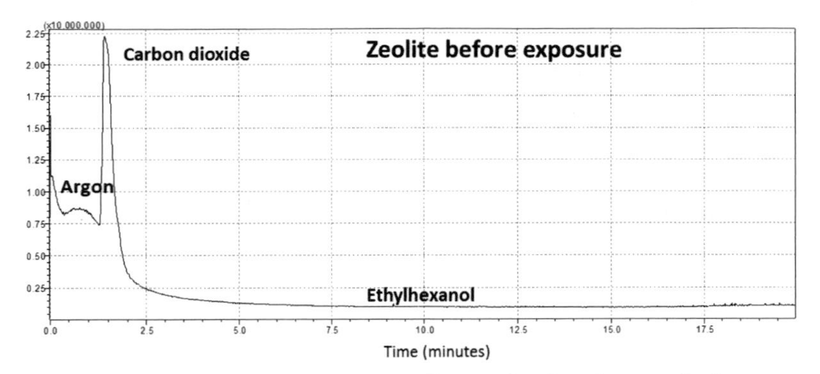

Figure 2. Chromatogram of unexposed zeolite powder that shows peaks for argon, carbon dioxide and a small peak for ethylhexanol.

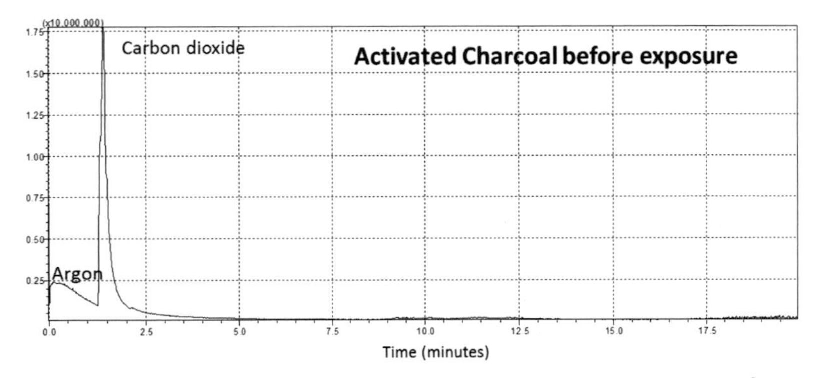

Figure 3. Chromatogram of unexposed activated charcoal that shows peaks for argon and carbon dioxide.

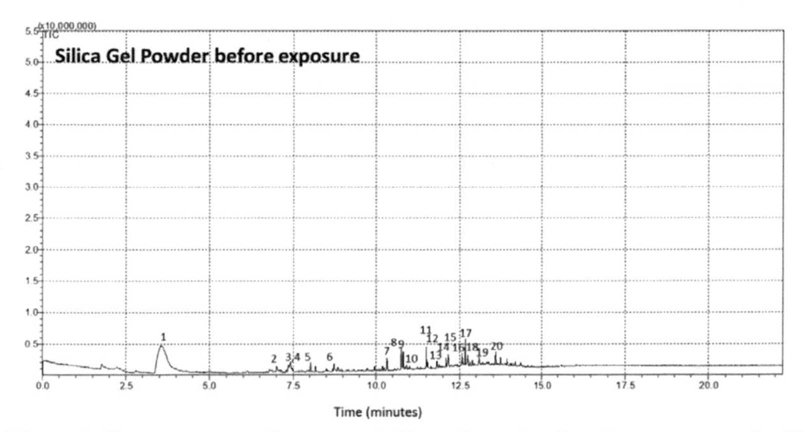

Figure 4. Chromatogram of unexposed silica gel powder that shows many peaks, likely from the plastic container that the silica gel is sold in.

Table 3. Identification of the chemicals found initially adsorbed onto silica gel powder.

Peak	Chemical
1	1-Butanol
2	8-methyl-1-Undecene
3	2,6,6-trimethyl-1-heptene
4	D-Limonene
5	Nonanal
6	1-Dodecene
7	Tetradecanal
8	Z-3-Hexadence
9	1s,4R,7R,11R-1,3,4,7-Tetramethyltricyclo[5.3.1.0(4,11)]undec-2-en-8-one
10	Pentadecane
11	n-Pentadecanol
12	Hexadecane
13	2,6,10-trimethyl-pentadecane
14	2-propenoic acid, tridecyl ester
15	2,6,10,14-tetramethyl-pentadecane
16	3,5-di-tert-Butyl-4-hydroxybenzaldehyde
17	n-Nonadecanol-1
18	Eicosane
19	Caffeine
20	Dibutyl phthalate

Tests of silica gel pellets that had been conditioned for re-use, revealed that the conditioning done to the pellets had not been sufficient to remove organic acids from the pellets (Figure 5). The literature surrounding the reuse of silica gel pellets focuses on conditioning the pellets from the perspective of removing all water and then introducing water to control humidity in a closed environment. In light of what was seen with these pellets and their ability to retain organic acids, the protocols that are used for conditioning silica gel pellets should be reconsidered.

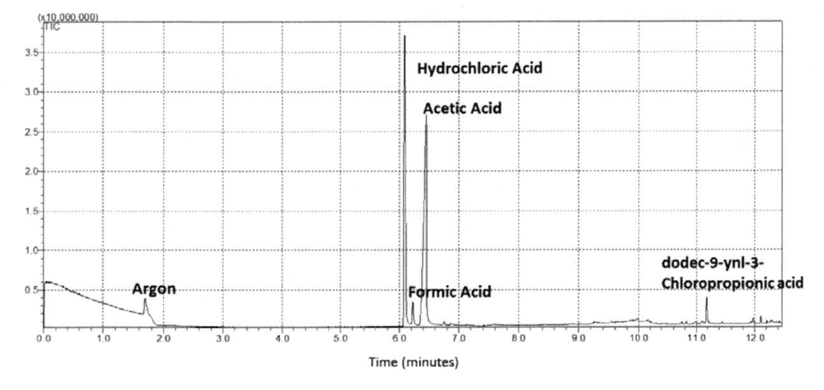

Figure 5. Chromatogram of silica gel pellet conditioned for reuse that shows peaks for argon, hydrochloric acid, formic acid, acetic acid and dodec-9-nyl-3-chloropropionic acid with other smaller peaks.

Other adsorbents picked up some of these acids, which were desorbed from the silica gel pellets during experiments conducted at room temperature (Figure 6). This result points to a potential hazard to art that placed in the vicinity of reused silica gel pellets.

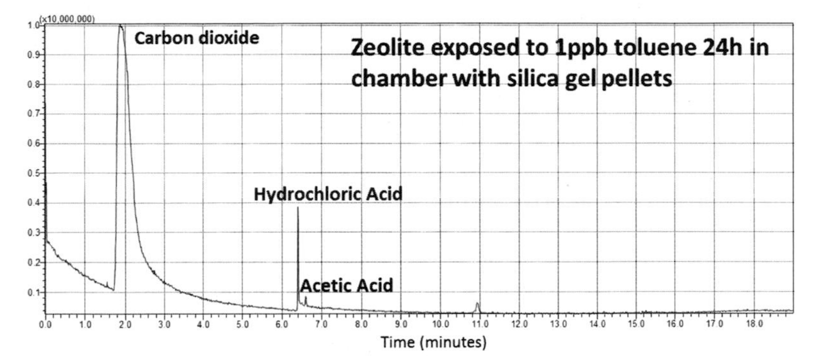

Figure 6. Chromatogram of zeolite powder exposed to 1ppb toluene for 24hours in same container with three silica gel pellets shows peaks for carbon dioxide, hydrochloric acid, acetic acid and dodec-9-nyl-3-chloropropionic acid.

While a protocol for reuse of silica gel pellets suggests temperatures up to 120°C for indicator dyed pellets and 200°C for undyed pellets and exposure times of four days, this does not consider the temperatures and time needed to drive off VOC's that may have adsorbed onto the silica gel pellets. Future research into the times and temperatures needed to drive off VOCs adsorbed by these materials should be conducted to address this issue. As was shown in Figures 7-9, low temperatures in a closed oven did not get rid of the acids adsorbed onto the materials even after 21 days. While there was an initial drop in the area/gram observed for these chemicals, there appeared to be a limit to how much actually desorbs at room temperature. This may be due to the fact that the oven used in this instance was a closed system with no active venting.

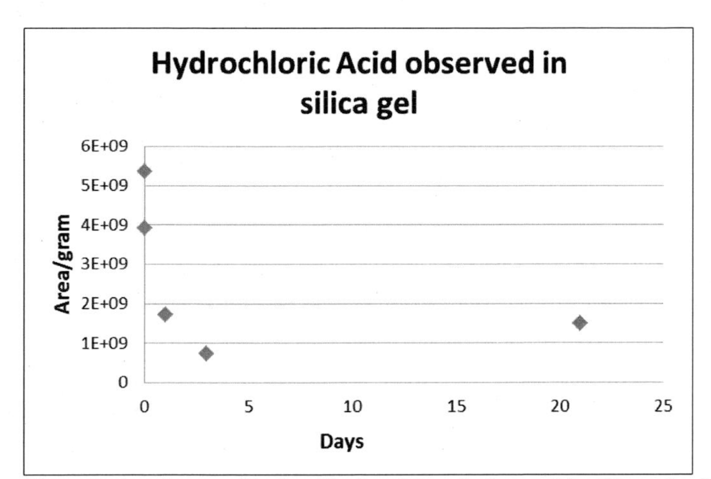

Figure 7. Area/gram of hydrochloric acid peak observed in silica gel pellets after exposure to an oven at 43°C over the course of 21 days.

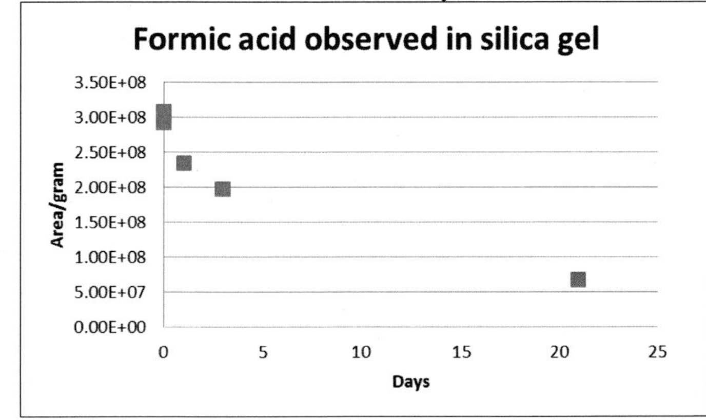

Figure 8. Area/gram of formic acid peak observed in silica gel pellets after exposure to an oven at 43°C over the course of 21 days.

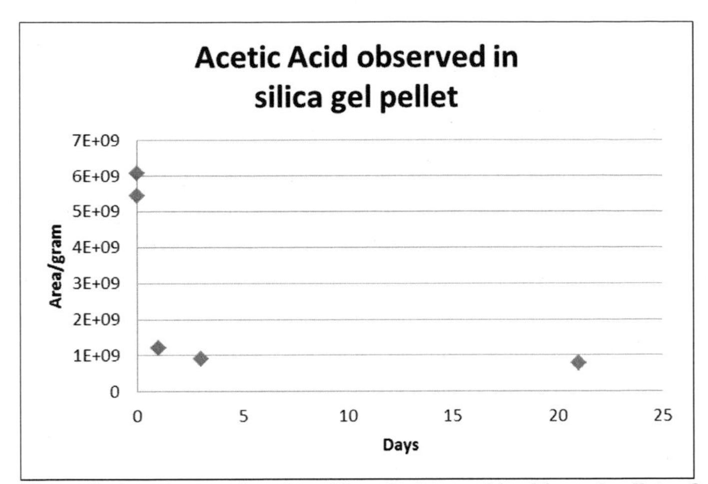

Figure 9. Area/gram of acetic acid peak observed in silica gel pellets after exposure to an oven at 43°C over the course of 21 days.

A case study was conducted to evaluate the ability of the adsorbents to adsorb chemicals in a relatively closed environment over time. These adsorbents were exposed to a closed cabinet filled with cellulose acetate negatives which smelled strongly of acetic acid for varying lengths of time. While there was an initial linear trend for the adsorption of acetic acid by silica gel over time, our results showed that between 24 and 48 hours this linearity began to fail (Figure 10A). After 120 hours there was an apparent decrease in the adsorption which may again be due to competitive adsorption, or due to saturation of the adsorbent. Other chemicals were adsorbed in this environment, the most prevalent other chemical being the plasticizer dibutyl phthalate. The linearity observed in the adsorption of dibutyl phthalate does not follow the same timeline, and the linear relationship seems to hold even at 160hours of exposure. The chromatogram of silica gel collected after 168 hours of exposure was compared to the chromatogram of silica gel powder collected prior to exposure (Figure 12). These chromatograms were subtracted to see which chemicals had been adsorbed, and those that show a positive peak in the subtracted spectrum(Figure 13) are labeled. Those that show a negative peak are the chemicals that desorbed from the silica gel powder.

Table 4. Weight of silica gel pellets, area of peaks and normalized area of peaks.

time in days	weight in grams	hydrochloric acid peak area	normalized to weight	formic acid peak area	normalized to weight	acetic acid peak area	normalized to weight
0	0.016134	63337432	3.93E+09	4971146	3.08E+08	98033837	6.08E+09
0	0.015838	84850961	5.36E+09	4619226	2.92E+08	86103748	5.44E+09
1	0.009432	16180692	1.72E+09	2211622	2.34E+08	11470139	1.22E+09
3	0.117136	84415523	7.21E+08	23173927	1.98E+08	1.06E+08	9.04E+08
21	0.011184	16700851	1.49E+09	751115	67159782	8901862	7.96E+08

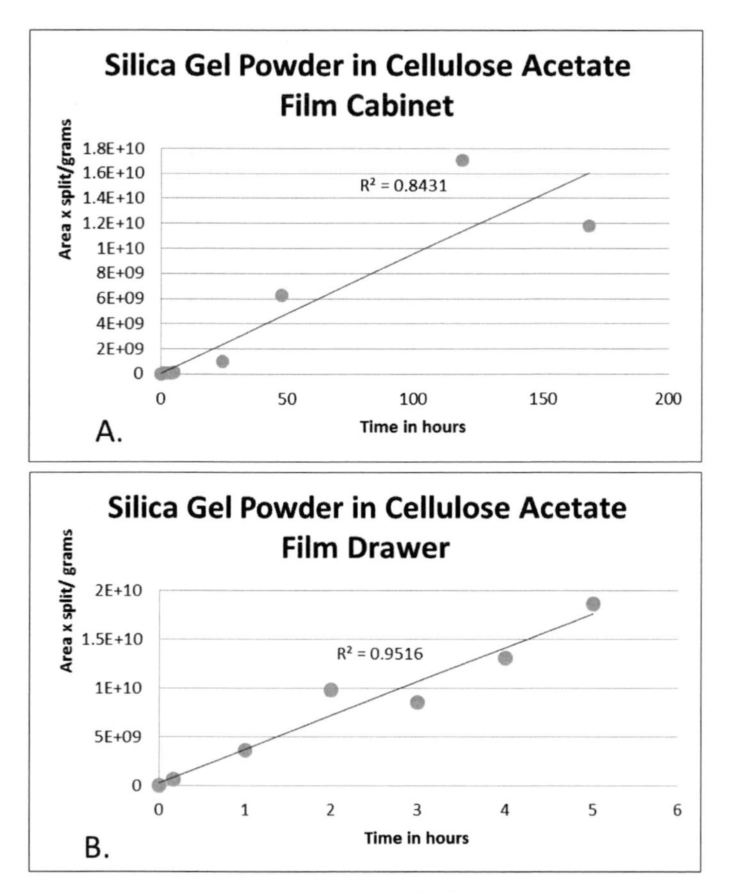

Figure 10. A: Normalized area (area multiplied by split/grams of adsorbent) for the acetic acid peak adsorbed onto silica gel powder adsorbent plotted against exposure time in hours in a cabinet filled with cellulose acetate film. B: Expansion of data for the first 5 hours of exposure shows linear behavior

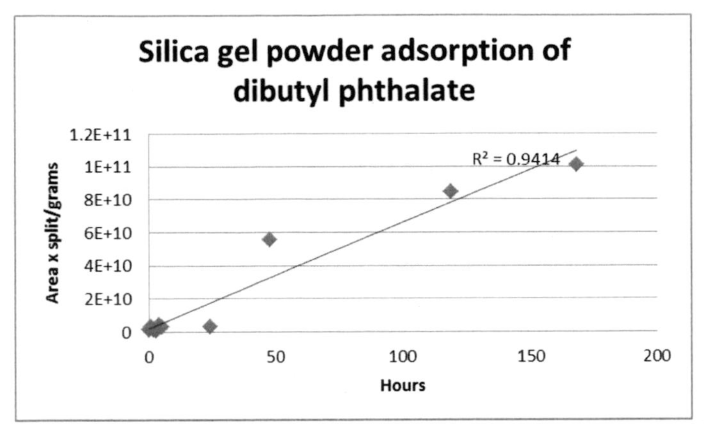

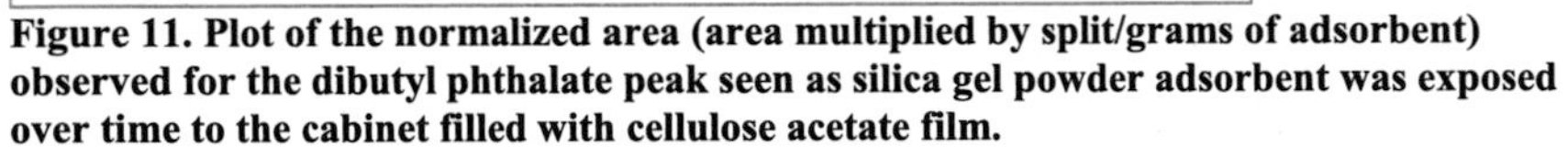

Figure 11. Plot of the normalized area (area multiplied by split/grams of adsorbent) observed for the dibutyl phthalate peak seen as silica gel powder adsorbent was exposed over time to the cabinet filled with cellulose acetate film.

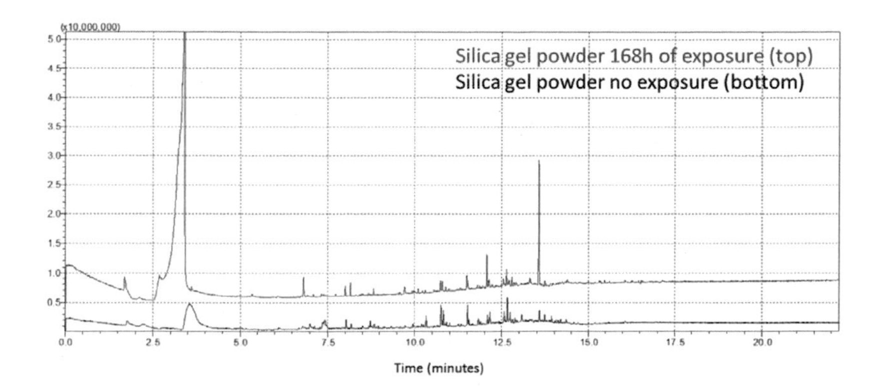

Figure 12. Comparison of the chromatogram of silica gel powder with no exposure to cabinet filled with cellulose acetate film (black) with the chromatogram of the silica gel powder exposed for 168 hours of exposure (red).

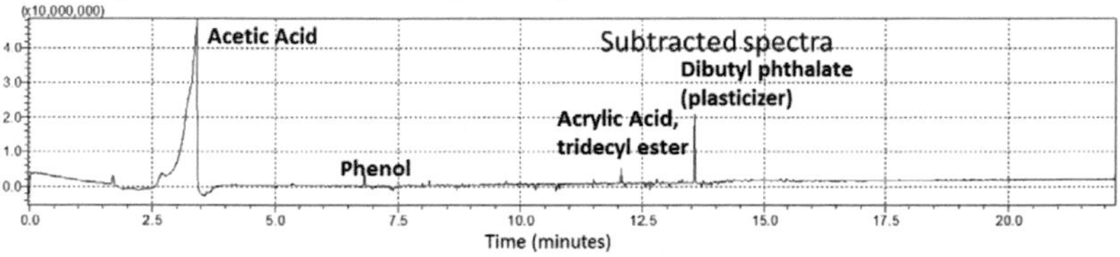

Figure 13. Subtraction of the chromatogram of silica gel powder with no exposure to cabinet filled with cellulose acetate film from the chromatogram of the silica gel powder exposed for 168hours of exposure shows that there is acetic acid, phenol, tridecyl ester acrylic acid and dibutyl phthalate observed after 168 hours.

While a linear relationship was observed between time and acetic acid adsorption with silica gel powder, there did not appear to be a linear relationship between time and acetic acid adsorption for activated charcoal (Figure 14). A linear relationship between time and dibutyl phthalate adsorption by activated charcoal held for at least the first 28.5 hours, as shown in Figure 15.

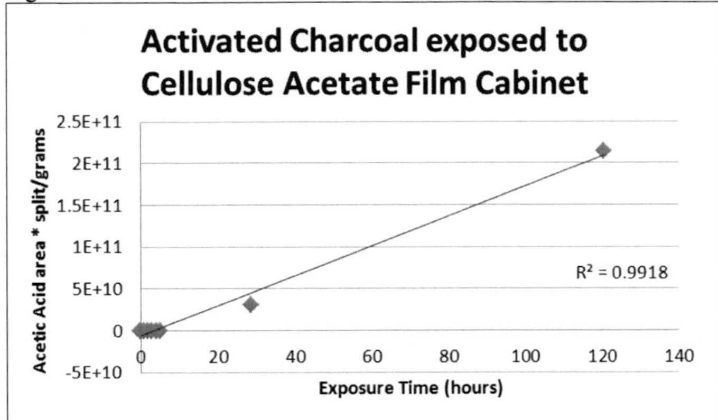

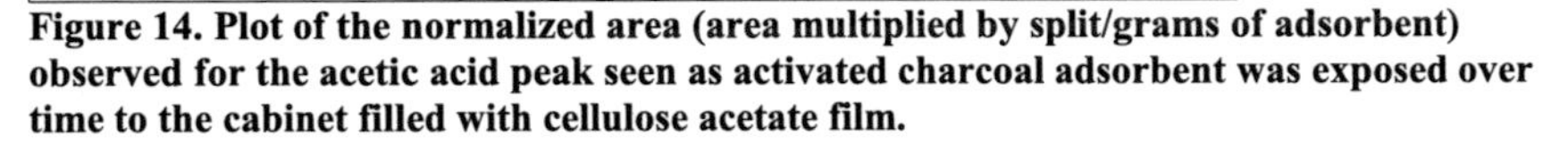
Figure 14. Plot of the normalized area (area multiplied by split/grams of adsorbent) observed for the acetic acid peak seen as activated charcoal adsorbent was exposed over time to the cabinet filled with cellulose acetate film.

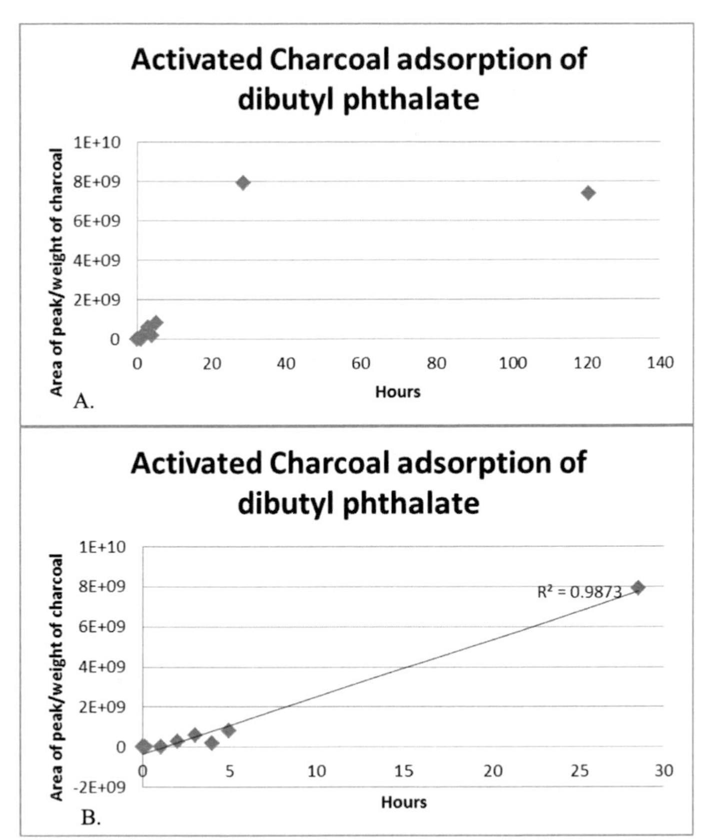

Figure 15. A. Plot of the normalized area (area multiplied by split/grams of adsorbent) observed for the dibutyl phthalate peak seen as activated charcoal adsorbent was exposed over time to the cabinet filled with cellulose acetate film, B. Close up of time points 0-28.5 hours of the plot of normalized area (area multiplied by split/grams of adsorbent) observed for the dibutyl phthalate peak seen as activated charcoal adsorbent was exposed over time to the cabinet filled with cellulose acetate film.

The chromatograms of activated charcoal after 120hours of exposure and after no exposure (Figure 16) were subtracted to see which chemicals had been adsorbed (Figure 17).

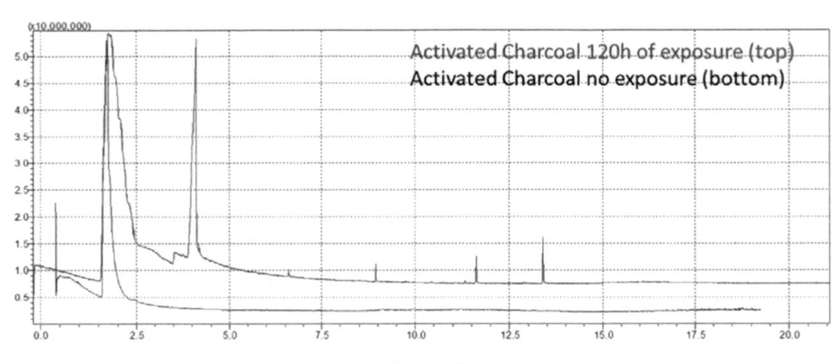

Figure 16. Comparison of the chromatogram of activated charcoal with no exposure to cabinet filled with cellulose acetate film (black) with the chromatogram of the activated charcoal exposed for 120 hours of exposure (red).

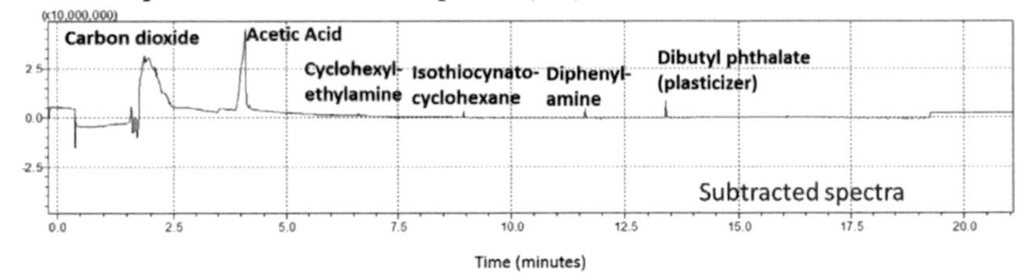

Figure 17. Subtraction of the chromatogram of activated charcoal with no exposure to cabinet filled with cellulose acetate film from the chromatogram of the activated charcaol exposed for 120 hours of exposure shows that there is acetic acid, cyclohexylethylamine, isothiocynatocycohexane, diphenyl amine and dibutyl phthalate observed after 120 hours.

A linear relationship between time and acetic acid adsorption was not observed with zeolite powder (Figure 18). However, a linear relationship was observed for the adsorption of dibutyl phthalate (Figure 19) by zeolite powder.

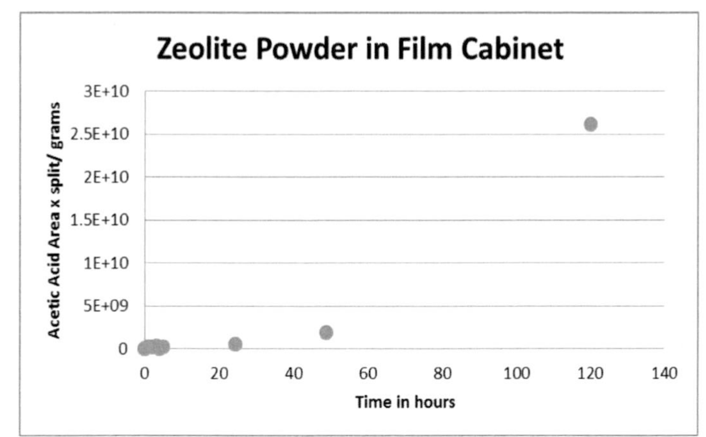

Figure 18. Plot of the normalized area (area multiplied by split/grams of adsorbent) observed for the acetic acid peak seen as zeolite powder adsorbent was exposed over time to the cabinet filled with cellulose acetate film.

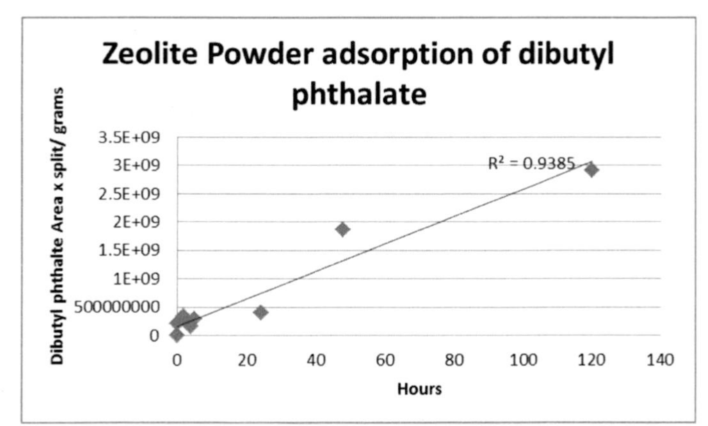

Figure 19. Plot of the normalized area (area multiplied by split/grams of adsorbent) observed for the dibutyl phthalate peak seen as zeolite powder adsorbent was exposed over time to the cabinet filled with cellulose acetate film.

Some similarities were observed between the types of chemicals adsorbed by the adsorbents. Acetic acid and dibutyl phthalate were both adsorbed by all adsorbents, while phenol was adsorbed by both silica gel and zeolite powder but not activated charcoal. The chromatograms of zeolite powder after 168hours of exposure and after no exposure were also compared in Figure 20 and subtracted in Figure 21.

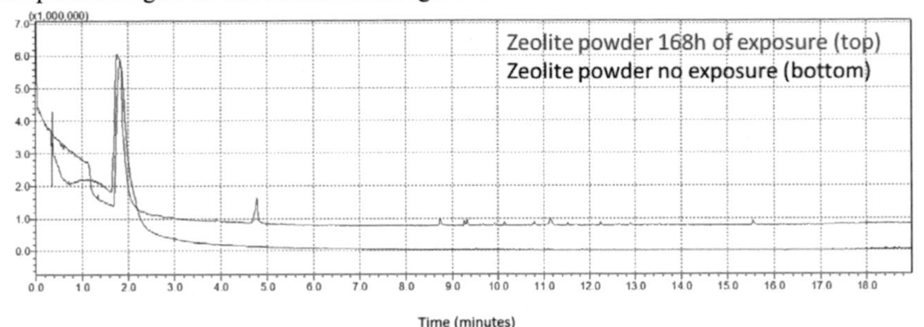

Figure 20. Comparison of the chromatogram of zeolite powder with no exposure to cabinet filled with cellulose acetate film (black) with the chromatogram of the zeolite powder exposed for 168 hours of exposure (red).

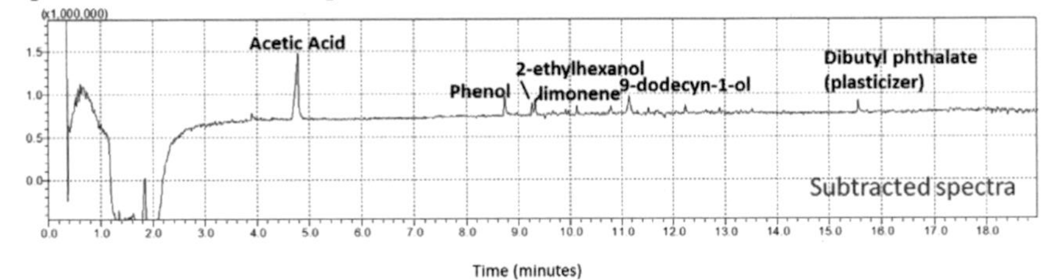

Figure 21. Subtraction of the chromatogram of zeolite powder with no exposure to cabinet filled with cellulose acetate film from the chromatogram of the zeolite powder exposed for 16 8hours of exposure shows that there is acetic acid, phenol, 2-ethylhexanol, limonene, 9-dodecyn-1-ol and dibutyl phthalate observed after 168 hours.

Differences were found between what was observed adsorbing to zeolite powder, activated charcoal and silica gel. Only silica gel adsorbed tridecyl ester acrylic acid. Only activated charcoal adsorbed cyclohexylethylamine, isocynatothiocyclohexane and diphenyl amine. Only zeolite adsorbed 2-ethylhexanol, limonene, and 9-dodecyn-1-ol. This data shows that preferential adsorption of certain chemicals by certain adsorbents does occur. The differences between the adsorbents are enhanced by their differing behaviors with regards to their adsorptions of acetic acid and dibutyl phthalate over time.

Future work should consider how differences in adsorbance would trend at longer time scales. One anticipates that either the adsorbent would become saturated or that the system would reach an equilibrium over a longer exposure time. This goes beyond the scope of this research but is of interest in continuing the study of these adsorbents as monitors of VOCs in a museum environment over longer periods of time than those studied here.

CONCLUSIONS

Materials commonly found in museums: silica gel, activated charcoal, and zeolite were analyzed using Thermal Desorption GC-MS to qualitatively identify adsorbed VOC's from museum environments at relatively low concentrations. It was found that different adsorbents have different ranges of adsorption for the chemicals tested: silica gel powder and zeolite powder have the greatest sensitivity for acetic acid, while zeolite powder and activated charcoal were more sensitive for identification of naphthalene. Silica gel powder proved to be the most sensitive overall. The non-linearity seen at longer exposure times (120 hours) may be due to competitive adsorption issues or saturation of the adsorbent. Activated charcoal desorbs most readily at room temperature and pyrolysis needs to be done within an hour after removal from the environment.

Additionally it was found that the procedures used to condition silica gel pellets for reuse need to be re-examined as the silica gel pellets adsorb VOC's and have been seen to release these VOCs at ambient temperature over time. The methods currently in use for conditioning silica gel pellets for reuse focus on the water content of these pellets and the percent humidity they are conditioned for, rather than having addressed the issue of these pellets as potential VOC sinks and sources.

ACKNOWLEDGMENTS

Chika Mori, of the Freer Gallery of Art and Arthur M. Sackler Gallery, Smithsonian Institution, for her feedback and assistance. Chris Maines, National Gallery of Art, for his ideas regarding the adsorbents used. Suzanne Lomax and Robyn Hodgkins, National Gallery of Art; Jennifer Giaccai, Museum Conservation Institute, Smithsonian; Mark Ormsby, National Archives and Eric Breitung, Library of Congress for their technical advice about GCMS.

REFERENCES

[1] M. Ormsby, "Analysis of Laminated Documents Using Solid-Phase Microextraction," *J. Am. Inst. Conserv.*, vol. 44, no. 1, pp. 13–26, 2005.

[2] A. L. R. Green, D. Thickett, and L. R. Green, "Testing Materials for Use in the Storage and Display of Antiquities : A Revised Methodology," *Stud. Conserv.*, vol. 40, no. 3, pp. 145–152, 1995.

[3] A. M. Lisovac and D. Shooter, "Volatiles from sheep wool and the modification of wool odour," *Small Rumin. Res.*, vol. 49, no. 2, pp. 115–124, Aug. 2003.

[4] D. Ellis and R. Goodacre, "Rapid and quantitative detection of the microbial spoilage of muscle foods: current status and future trends," *Trends Food Sci. Technol.*, vol. 12, no. 2001, pp. 414–424, 2001.

[5] Y. Xu, W. Cheung, C. L. Winder, and R. Goodacre, "VOC-based metabolic profiling for food spoilage detection with the application to detecting Salmonella typhimurium-contaminated pork.," *Anal. Bioanal. Chem.*, vol. 397, no. 6, pp. 2439–49, Jul. 2010.

[6] D. Norbäck, E. Björnsson, C. Janson, J. Widström, and G. Boman, "Asthmatic symptoms and volatile organic compounds, formaldehyde, and carbon dioxide in dwellings.," *Occup. Environ. Med.*, vol. 52, no. 6, pp. 388–95, Jun. 1995.

[7] G. Wieslander, D. Norbäck, E. Björnsson, C. Janson, and G. Boman, "Asthma and the indoor environment: the significance of emission of formaldehyde and volatile organic compounds from newly painted indoor surfaces.," *Int. Arch. Occup. Environ. Health*, vol. 69, no. 2, pp. 115–24, Jan. 1997.

[8] F. Lestremau, F. A. T. Andersson, V. Desauziers, and J.-L. Fanlo, "Evaluation of solid-phase microextraction for time-weighted average sampling of volatile sulfur compounds at ppb concentrations.," *Anal. Chem.*, vol. 75, no. 11, pp. 2626–32, Jun. 2003.

[9] J. Zhang, D. Thickett, and L. Green, "Two Tests for the Detection of Volatile Organic Acids and Formaldehyde," *J. Am. Inst. Conserv.*, vol. 33, no. 1, pp. 47–53, 1994.

[10] R. Schneider, Y. Kotseridis, J.-L. Ray, C. Augier, and R. Baumes, "Quantitative Determination of Sulfur-Containing Wine Odorants at Sub Parts per Billion Levels . 2 . Development and Application," *J. Agric. Food Chem.*, vol. 51, pp. 3243–3248, 2003.

[11] D. Goltz, J. McClelland, A. Schellenberg, M. Attas, E. Cloutis, and C. Collins, "Spectroscopic studies on the darkening of lead white.," *Appl. Spectrosc.*, vol. 57, no. 11, pp. 1393–8, Nov. 2003.

[12] V. Daniels and S. Ward, "A rapid test for the detection of substances which will tarnish silver," *Stud. Conserv.*, vol. 27, no. 2, pp. 58–60, 1982.

[13] M. Ryhl-Svendsen and J. Glastrup, "Acetic acid and formic acid concentrations in the museum environment measured by SPME-GC/MS," *Atmos. Environ.*, vol. 36, no. 24, pp. 3909–3916, Aug. 2002.

[14] I. Bonaduce, M. Odlyha, F. Di Girolamo, S. Lopez-Aparicio, T. Grøntoft, and M. P. Colombini, "The role of organic and inorganic indoor pollutants in museum environments in the degradation of dammar varnish," *Analyst*, vol. 138, no. 2, p. 487, 2013.

[15] M. J. Linnie and M. J. Keatinge, "Pest control in museums: toxicity of para-dichlorobenzene, 'Vapona' (TM), and naphthalene against all stages in the life-cycle of museum pests, Dermestes maculatus Degeer, and Anthrenus verbasci (l.) (Coleoptera: Dermestidae)," *Int. Biodeterioartion Biodegrad.*, vol. 45, pp. 1–13, 2000.

[16] A. Schieweck, W. Delius, N. Siwinski, W. Vogtenrath, C. Genning, and T. Salthammer, "Occurrence of organic and inorganic biocides in the museum environment," *Atmos. Environ.*, vol. 41, no. 15, pp. 3266–3275, May 2007.

[17] C. Grzywacz, *Monitoring for gaseous pollutants in museum environments*. Los Angeles: The Getty Conservation Institute, 2006.

The Narrative Interpretation of Process Reconstruction and Materials Characterization in Technical Art History and Archaeological Science

Mater. Res. Soc. Symp. Proc. Vol. 1656 © 2015 Materials Research Society
DOI: 10.1557/opl.2015.1

Study of Mexican Colonial Mural Paintings: An In-situ Non-Invasive Approach

José Luis Ruvalcaba-Sil, Malinalli Wong-Rueda, Maria Angelica Garcia-Bucio, Edgar Casanova-Gonzalez, Mayra Dafne Manrique-Ortega, Valentina Aguilar-Melo, Pieterjan Claes, and Dulce Maria Aguilar-Tellez.

Instituto de Física, Universidad Nacional Autónoma de México, Circuito de la Investigación Científica s/n, Ciudad Universitaria, Mexico D.F. 04510, Mexico. e-mail: sil@fisica.unam.mx

ABSTRACT

Colonial mural painting developed in Mexico in XVI century after the conquest of the pre-Hispanic cultures following the evangelization process little information exists about the chronology of the paintings and workshops, the painters, the pictorial techniques or the materiality of this art work.

In this work, we present the non-invasive methodology of study of the pigments and other components of nine mural paintings in three colonial Augustinian ex-convents located in Epazoyucan, Actopan and Ixmiquilpan, in the state of Hidalgo, central Mexico. These places were selected not only because of the inherent value and iconographic characteristics of the paintings, which date to the XVI and XVII century, but also because they are in the same region and are well preserved and in good condition. Then it is possible to compare their materiality and get new information to answer to some of questions related to these paintings.

X-ray Fluorescence (XRF) and Raman spectroscopy were conducted using portable equipment on scaffolds after a global examination under ultraviolet light. We were able to distinguish between different pigments used for different colors such as vermillion, orpiment, and a copper pigment, for the bright red, gold yellow, and green, respectively. These pigments are characteristic of the known Mexican Colonial color palette. Apart from this, we also found the presence of indigo in the blues, minium, and cochineal. A first comparison among the mural paintings of the three sites indicates different palettes and painting periods.

INTRODUCTION

The Spanish colonization of what is currently known as North and Central America was initiated by the discovery of the continent at the end of XV century. From this moment on, Spanish conquerors arrived in the 'New World' in search of wealth, both from precious metals and other valuable objects (such as, new species and new pigments) and from the discovery of new trade routes. A catholic evangelization rapidly followed the military conquest when missionaries of different orders came to bring the word of God to the indigenous inhabitants of the country [1]. The collapse of the Aztecs capital, Tenochtitlan, in 1521 marks the beginning of 300 year hegemony of Spanish rulers in Mexico, which from that time is known as New Spain.

From the beginning of the conquest, the evangelization of the peoples was a priority for the Spanish, and the first Catholic orders arrived in Mexico shortly after the fall of the Aztec rule. Their role was to obtain a closer relationship with the indigenous community than was possible for the conquerors. It was assumed that by their vows of poverty, obedience, and chastity, these mendicant orders would not resemble the occupying power, and that religious teaching would facilitate the expansion into the mainland of Mexico.

While the Franciscans and Dominicans arrived in Mexico shortly after the conquest (1524 and 1526, respectively), it took a decade before the Augustinians started their evangelization of the indigenous people (1533). They were sent to spread the Word into the wild where the others did not travel. According to the beliefs of the doctrine of the order, it seems that the Augustinians were able to reach the native people through an interaction with and respect for their culture. The blend of both worlds was captured in the painted murals of Augustinians church-temples where local techniques and themes were used to evangelize the native population [2, 3]. Paintings were made through different religious iconography in order to overcome the differences of language and spread the word of God in the community. Nevertheless, very little information exists about the chronology of the paintings, the painters, the pictorial techniques, and their materiality. These are some of the most important questions related to this art work. These Colonial cloisters and paintings are important cultural heritage in Mexico.

We show an overview of the study of different mural paintings present in three Augustinian churches in the state of Hidalgo, central Mexico. By means of non-destructive techniques, such as imaging under ultraviolet light, X-Ray Fluorescence (XRF) and Raman measurements, the pigments used in these paintings were identified. A comparison among the main results for the mural paintings of the three convents is also presented.

This interdisciplinary research has been performed in close collaboration between the Aesthetics Research Institute (IIE) and the Physics Institute (IF) of the Universidad Nacional Autonoma de Mexico (UNAM). It is part of larger project considering also other aspects of the convents.

THE MURAL PAINTINGS

The three selected Augustinian convents are located in the state of Hidalgo [3], central Mexico, within a region at 130 km from Mexico City. The distance between the convents is 35 to 70 km. These places were selected not only because of the inherent value and iconographic characteristics of the paintings, which date to the XVI and XVII century, but also because they are in the same region and are well preserved and in good condition.

Prior to these analyses, we did not have information about techniques and technological variability, for instance, about the fresco and dry paintings techniques as they are related to pigment use in these paintings. We did know that several were executed in grisaille, but we did not know the materials involved. From our study pigments, mixtures of pigments, and binders were identified and comparisons of materiality made. We obtained new information to answer questions related to Mexican Colonial mural paintings and its technological style that is similar to studies done in other regions [4-7].

Epazoyucan

The first study involved the paintings of the church and ex-convent of San Andres Apostol in Epazoyucan founded around 1540 [3, 8-9]. Five paintings are located in the lower cloister. For several years the paintings were covered by multiple layers of calcite and after being rediscovered in the XX century, they were restored in 1960 by the National Institute for Anthropology and History (INAH). Four of these paintings are located in frames and the corners, while the last painting is located above the entrance to the *refectorium*, the common use room for the monks. Two representative paintings were chosen: *The Descent from the Cross* and *The*

Death of the Virgin Mary (Figure 1 and 2, respectively). The first painting, *The Descent from the Cross* is located in the left corner of the lower cloister and belongs to what is supposed to be different and earlier stage of painting that *The Death of the Virgin Mary,* that is located above the entrance of the refectory.

Figure 1. *The Descent from the Cross* at Epazoyucan, Hidalgo.

Figure 2. *The Death of the Virgin Mary* at Epazoyucan, Hidalgo.

Actopan

The second church and ex-convent is San Nicolas Tolentino, located in Actopan and founded around 1550 [3, 9-10]. One of the most famous colonial mural paintings in Mexico is located here, in the stairway hall. It consists of 12 paintings which tell the story of the Augustinian order chronologically from placement below to the highest part of the composition. The paintings were covered by several layers of $CaCO_3$ but in 1927 an integral restoration of the scenes was performed by R. Montenegro [3]. At this ex-convent, three paintings were analyzed. Two of them belong to the same time period, namely, *Saint Augustin Kneeling before Sompliciano* and *Unidentified Bishop with Miter* (Figure 3), while the third painting, called *Juan or Ivan of Inica and Pedro of Izcuincuitlapilco Caciques of Actopan with the Friar Martín of Asebeido*, is believed to belong to a later time in the convent's history (Figure 4). These paintings were made using a different technique compared to the one from Epazoyucan. This technique is called grisaille, and consists of a main picture made entirely in black, grey and white colors with few color additions at special positions.

Figure 3. *Unidentified Bishop with Miter* at Actopan, Hidalgo, executed in grisaille technique.

Figure 4. *Juan or Ivan of Inica and Pedro of Izcuincuitlapilco Caciques of Actopan with the Friar Martín of Asebeido* at Actopan, Hidalgo.

Ixmilquilpan

The third study was made at the ex-convent of San Miguel Arcangel in Ixmiquilpan, founded around 1560 [3, 9, 11]. The main nave of the church was decorated by a wide variety of mural paintings. However, not all of them were discovered underneath several layers of $CaCO_3$ until 1992. These paintings belong to a bigger composition representing an allegory of the battle between good and evil forces. The style and composition of these murals are completely different from the ones found in the other convents at Epazoyucan and Actopan. In Ixmiquilpan, the images and iconography show several warriors dressed as traditional pre-Hispanic warriors fighting different anthropomorphic and fantastic figures [6]. Three paintings were selected for this study and are named according to the main character in the scene: *The Dragon* (Figure 5), *The Coyote Warrior* and *The Jaguar Warrior* (Figure 6). A fourth and last painting was also investigated, *The Last Judgment,* located in the lower cloister and belongs to a completely different stage and style. This last painting is more closely related to the ones from Actopan because of its manufacturing technique of grisaille.

Figure 5. *The Dragon* at Ixmiquilpan, Hidalgo.

Figure 6. *The Jaguar Warrior* at Ixmiquilpan, Hidalgo.

ANALYTICAL METHODS

Portable and non-destructive techniques were carefully chosen in order to obtain as much information as possible without the need to acquire even micro-samples from the mural paintings. This approach has proven its effectiveness for similar cultural heritage studies [12-15]. Previous to the analysis, an examination at night under UV light was performed. Photographs of the paintings were taken at night under UV light illumination (both at short 254 nm and long 365 nm wavelength) using a Nikon 3000 camera without any additional filter in order to register the image and locate the regions of interest. Three different types of areas were observed: original pigments were distinguished from deteriorated areas as well as from repainted areas, as has been shown by studies of murals in other countries [16-18]. Modern interventions (XX century) have a different response to UV light and can be clearly differentiated from original pigments.

This imaging study was followed by an elemental and molecular analysis *in situ* of the pigments in the mural paintings via X-ray Fluorescence (XRF) measurements and Raman spectroscopy, respectively. These measurements were a challenge because the experimental devices cannot move during the analysis. For this purpose, special modifications were carried out on the scaffolds in order to minimize the movement of the experimental setup during the measurements.

For this analysis, our portable XRF equipment SANDRA [19] was used in order to identify the different inorganic elements [20] of the material composition in all the paintings (Figure 7). Because of the dimensions of the murals and also the architecture of the convents the equipment was mounted over different supports, including the scaffolding to reach the highest points. The X-ray tube of Mo was used under 35 kV and 0.2-0.4 mA, each spectrum required 120 to 180 s/region.

Figure 7. *In-situ* measurements using the XRF system in the hall with stairway at Actopan.

Raman spectroscopy, on the other hand, identifies the molecular structure of materials used in the objects under study [21]. Each material has a unique molecular structure and characteristic bands in the spectra. For this analysis, the device used was the Inspector Raman from Delta-Nu with a red laser of 785 nm (Figure 8). With this analysis technique, we corroborated some of the XRF identifications as well as identified organics compounds that cannot be determined by XRF analysis. These results helped to identify specific pigments [15]. This equipment was set also on scaffolding. Measurements require several minutes; the minimum laser intensity was 20 mW.

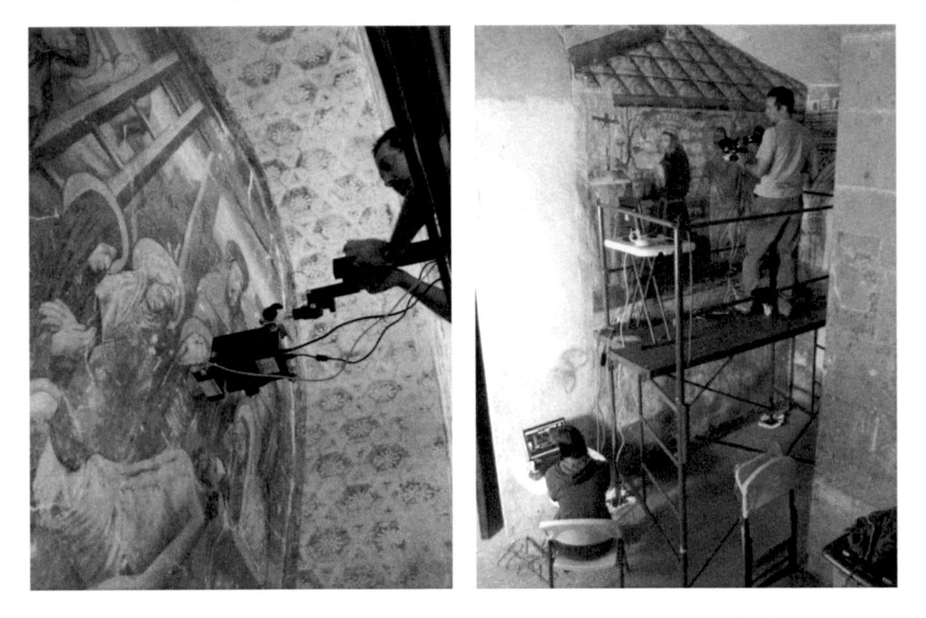

Figure 8. *In-situ* measurements using the Raman system in the Epazoyucan cloister (left) and in the stairsway hall in the Actopan convent (right).

RESULTS AND DISCUSSION

San Andres Apostol, Epazoyucan

Figure 9 shows the different elements found within the color palette analyzed by the two paintings of the ex-convent of San Andres Apostol. In this case Epa1 corresponds to *The Descent from the Cross* and Epa2 corresponds to *The Death of the Virgin Mary*, respectively.

In the first painting, *The Descent from the Cross*, two different red pigments were identified. The darker red consists of iron oxides which lead to the identification of an earth pigment, while the principal compound of the brighter red color is lead (Pb). This led to the identification of the pigment known as minium. On the other hand, the darker red color found in the painting *The Death of the Virgin Mary,* contains mercury (Hg) which leads to the identification of vermillion. These results were confirmed by Raman spectroscopy (Figure 10). Skin tones and brown colors were made using earth pigments, rich in iron oxides.

The XRF analyses of the blue pigments revealed that the darker blue color contained a higher proportion of iron (Fe) indicating that this color was made of Prussian blue. However, the lighter blue color contained a higher proportion of lead and very low amount of Fe, too low to consider a mixture of Prussian blue with lead white. By contrast, in the painting of *The Death of the Virgin Mary* other pigments such as orpiment (As) and Smalt (Co, Ni, As) [22] were identified in the yellow-golden and blue-gray colors, respectively. Both pigments were mixed with lead white to produce lighter colors.

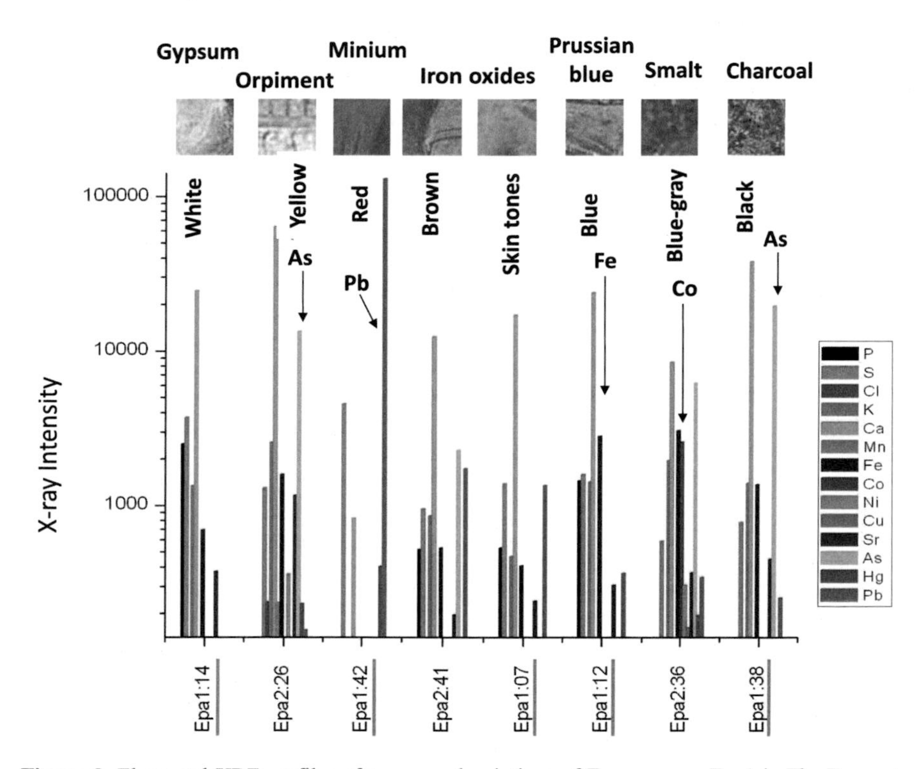

Figure 9. Elemental XRF profiles of two mural paintings of Epazoyucan. Epa1 is *The Descent from the Cross* and Epa2 is *The Death of the Virgin Mary.*

The problem of the light blue colors explained above was resolved through the use of Raman spectroscopy performed on the same regions. The light-blue color is not a mixture of Prussian blue with a white pigment as it may be expected, but is it most likely a mixture between indigo, an organic pigment, with white lead pigment and white gypsum. The pigment indigo [23] is an artificial pigment made of an organic strong blue colorant extracted from a plant called *Indigofera suffructosa,* that cannot be identified by XRF analysis. It was often possible to observe sharp and intense bands of indigo in the Raman spectra (Figure 10, Table 1), mainly the ones around 545 cm^{-1} and 1570 cm^{-1}.

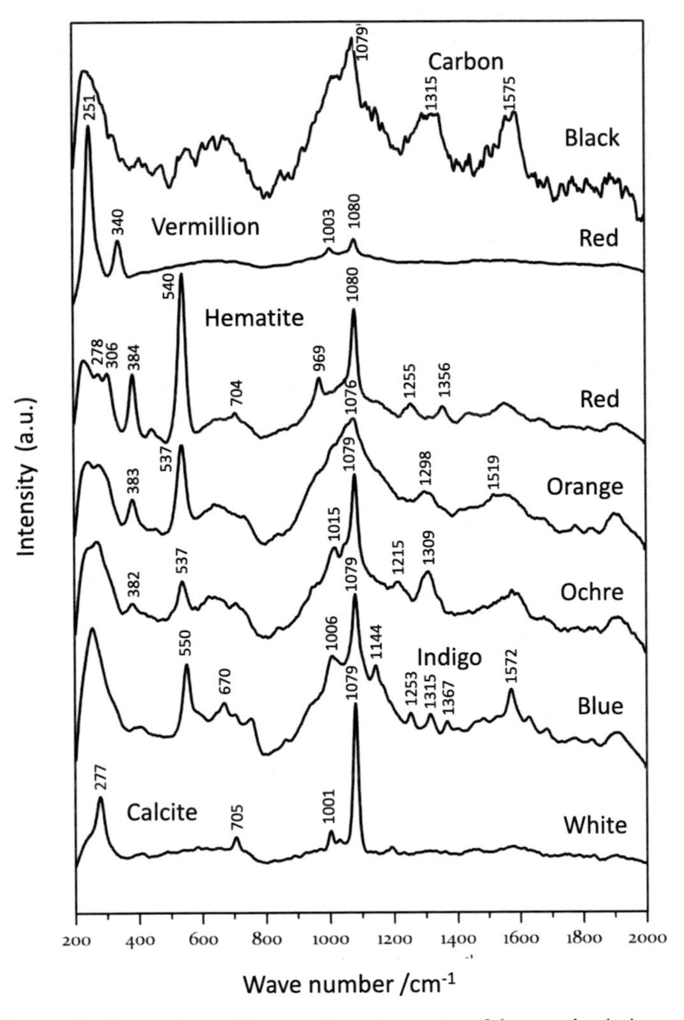

Figure 10. Comparison of Raman pigments spectra of the mural paintings.

Using Raman (Table 1), calcite bands were also identified on a high percentage of the spectra, due both to its use as a white pigment and to the fact that the murals were covered with lime paint for several years. Gypsum was detected by Ca and S X-rays and confirmed by Raman. The gypsum was used to achieved whites and clear shades, and probably as part of the preparation layer.

Table 1. Main Raman bands identified in the spectra of the pigments of the mural paintings.

Identified pigments	Raman bands (cm^{-1})
Charcoal black	1305, 1544
Gypsum white	1007, 1138
Calcite white	704, 1078
Minium	390, 537
Vermillion	251, 341
Cochineal	448, 1221, 1307
Iron oxides	283, 398, 601
Indigo	547, 660, 1312, 1570

The carbon bands observed in the black color were linked to the use of this substance to yield dark shades, and in some cases, to small burns (around 50 μm in diameter) caused by the laser beam on the analyzed surface. Another interesting aspect was the presence of bands that could be attributed to an oil or resin [24].

Historical records for the cloister paintings indicate that at the beginning of XX century the paintings were repainted [8]. The results for the *The Descent from the Cross* indicate that the Prussian blue is related to the repaint while indigo may correspond to the previous polychromy, probably earlier than XVIII century. On the other hand, the differences in the use of pigments between both paintings suggest different stages or painters for the original polychromy. These results also suggest that the early XX century repaint may follow the original patterns of polychromy and color remains.

San Nicolas Tolentino, Actopan

Three paintings were analyzed in the ex-convent of San Nicolas Tolentino in Actopan. The main difference between these paintings and the ones found in the previous convent of Epazoyucan is related to the technique used known as grisaille [16], and to a different color palette. These mural paintings were the first ones to be painted by the grisaille technique in Actopan. The quantity of analysis points was less for than in Epazoyucan because the range of colors was less than in the other case. The Figure 11 shows an overview of the color palette and the pigments used in these three paintings from XRF results. Act1 belongs to the regions of the first painting (*Saint Augustin Kneeling before Sompliciano)* while Act2 and Act3 to the *Unidentified Bishop with Miter* and *Juan or Ivan of Inica and Pedro of Izcuincuitlapilco caciques of Actopan with the Frair Martín of Asebeido*, respectively. The identification of the pigments was completed by Raman spectra.

The white colors were mainly calcium carbonates but also some gypsum has been found. In retouches areas was the white color identified as lead white. The black color of the grisaille is charcoal.

In the painting of the *Unidentified Bishop with Miter* a high concentration of copper was found in the green color of the miter of the bishop (Figure 3). XRF data indicate a pigment rich in copper and ultraviolet fluorescence suggests the use of verdigris or copper resinate. Raman analysis made over the blue color let us to identify remains of indigo, while the analysis made with XRF did not show the presence of the Prussian blue identified in Epazoyucan.

The skin color have at least two tones, a darker brownish color used for indigenous people is composed by earth pigments rich in iron oxides, while the light pink for the religious people consists of lead (minium) and calcium with a presence of mercury (vermillion) in the darker areas. Raman spectroscopy confirms the identification of the vermillion in these skin tones, more precisely in the third painting contained a variety of skin tones in larger areas that made them easier to identify.

The yellow pigment in these paintings is an ochre, rich in iron (Figure 11), perhaps goethite or limonite in contrast to the orpiment detected at Epazoyucan.

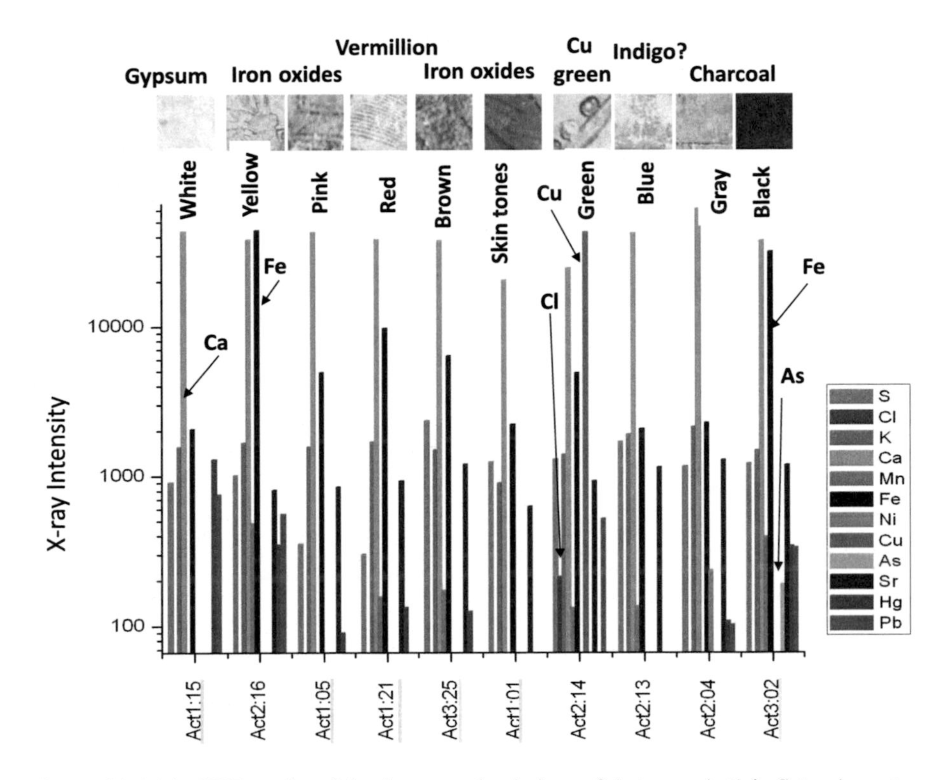

Figure 11. Main XRF results of the three mural paintings of Actopan. Act1 is *Saint Augustin Kneeling before Sompliciano,* Act2 is *Unidentified Bishop with Miter* and Act3 is *Juan or Ivan of Inica and Pedro of Izcuincuitlapilco caciques of Actopan with the Frair Martín of Asebeido.*

Based on the analytical results we suggest that two different stages of painting were executed. In the *Unidentified Bishop with Miter* painting, the blue of indigo may has been repainted by the copper pigment in a later stage.

San Miguel Arcangel, Ixmiquilpan

The church and ex-convent of San Miguel Arcangel, Ixmiquilpan was the third place we worked. Results from Ixmiquilpan are remarkable as they differ from the ones obtained in both Epazoyucan and Actopan. At first glance, the paintings located in the main nave of the church with their colorful images suggested a great variety of pigments were used. On the other hand, the painting of the *Last Judgment*, located in the lower cloister, clearly belongs to a completely different painting stage and more similarities can be found between this painting and the grisailles from Actopan, although this last painting was only analyzed with Raman spectroscopy.

The XRF analysis of the three selected mural paintings in the interior of the church shows a color palette that included ten colors (Figure 12). *The Dragon* is labeled Ixmi0, *The Jaguar* Ixmi1 and Ixmi2 relates to points located in *The Coyote Warrior.*

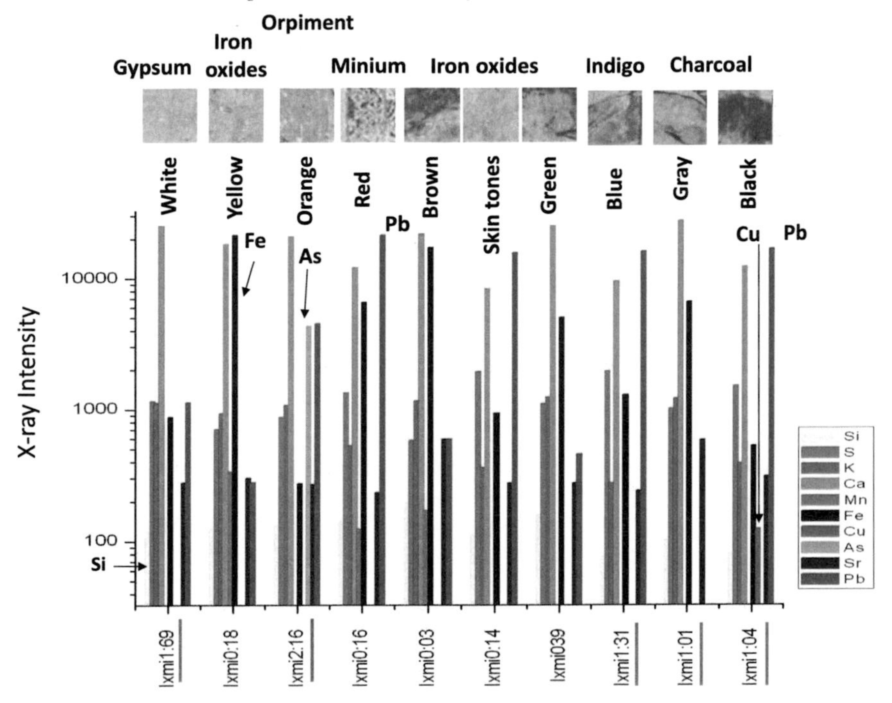

Figure 12. XRF results of the three mural paintings of Ixmilquilpan. *The Dragon* is Ixmi0, *The Jaguar* is Ixmi1 and *The Coyote Warrior* is Ixmi2.

The presence of silicon indicates that sand is present in the plaster. The yellow colors have a higher concentration of iron, while the analyzed red and orange colors have a higher lead concentration. Almost no change was observed between the intensities of the different elements found in the blue colors compared with the white ones which lead us to believe that blue colors most likely originated from an organic material. The blue colors are dissimilar to Prussian blue. A high concentration of copper indicates the use of some kind of earth pigment in the black color.

Raman spectroscopy was able not only to confirm previous statements but also additional information regarding different compounds of the pigments was obtained. For instance, cochineal was found both in the yellow and orange colors, corroborating to the theory that some local painting elements may be used during the elaboration of these murals. The orange colors show a clear spectrum related to minium while the blue and green colors show low amounts of indigo. The black colors are made of charcoal, as confirmed by their Raman spectra. In the case of these paintings it is more difficult to determine different painting stages.

Comparison among the Convents

The data recollected at the three ex convents proved to be valuable for research on Colonial Mexican mural paintings. Apart from the great difference of the painting styles, the material compositions were different in each case.

The plaster in each convent is a remarkable difference among the convents. In Epazoyucan the main elements found through XRF were calcium (Ca), potassium (K), sulfur (S), strontium (Sr) and iron (Fe); in Actopan there was calcium (Ca), potassium (K), sulfur (S), strontium (Sr) and iron (Fe) but this last element appeared with a lower intensity than the one in Epazoyucan; lastly, in Ixmiquilpan the same elements were found in almost the same intensity as the ones found in Epazoyucan, however, there was a presence of silicon (Si) which was not found in the other two convents leading to the specific answer that the main compound, sand, in the plaster was different from the one used in the other two places.

The pigment analyses lead to several conclusions about the process of painting. First the identification using both techniques allowed us to classify the painting pigments used for each color as those characteristics of the first part of the Novo-Hispanic period (XVI-XVII centuries), as shown in Table 2. In Epazoyucan and Actopan, specific pigments, such as orpiment, smalt and copper resinate, are known to have been used before the XVIII century.

Most of the paintings were made using a dry technique. Raman confirmed these results in which the presence of oils and other organic materials that could suggest different bindings agents were used, this kind of material can only be found in dry techniques [6-7, 14, 16]. An original fresco painting is a technique that requires the previous hydrated pigments to get mixed with the fresh calcium carbonate. We deduce that although the main drawing was made as a fresco and some base pigments were as well applied with this technique, all the details were likable added later with a dry technique.

Overall, even if the techniques in the drawings and the use of the pigments are very different in these three convents, some of the pigments coincide in the paintings. That is the case of the red of minium, the iron oxides or earths used in the yellow-earth tones and in the reds, the white of calcium carbonate, the blue of indigo and the red of vermillion. The use of European painting materials is implicit in the three cases; however, at least in the case of Ixmiquilpan, the probable use of local materials such as the cochineal was identified. There is no evidence of the use of Maya blue and clays associated with indigo [25-26].

The differences in the palette and the use of pigments between the convents indicate that the hypothesis that a same group of painters were involved in the paintings is no longer supported. Moreover, several stages of paintings and several painters are in evidence.

Table 2. Pigments were identified for each convent using XRF and Raman data in comparison with reference spectra and previous information [20, 27-28].

Convent	Color					
	White	**Red**	**Pink**	**Skin tones**	**Brown**	**Orange**
Epazoyucan	White lead, Gypsum, Calcite	Minium, Vermillion, Iron oxides	-	Ochres, White lead	Ochres, Vermillion, Charcoal black	-
Actopan	White lead, Gypsum, Calcite	Vermillion, Iron oxides	-	Ochres, White lead, Vermillion	Ochres, Vermillion, Charcoal black, White lead	-
Ixmiquilpan	White lead, Gypsum, Calcite	Minium, Vermillion	Minium, Cochineal	Ochres, White lead	Ochres, Minium	Minium, Cochineal

Convent	Color					
	Yellow	**Green**	**Blue**	**Black**	**Gray**	**Retouches**
Epazoyucan	Orpiment, Ochres	-	Indigo, Prussian blue, Smalt	Charcoal black	-	-
Actopan	Ochres	Copper resinate	Indigo	Charcoal black	Charcoal black, White lead	Red, Pink, Skin tones
Ixmiquilpan	Ochres, Cochineal	Ochres, Indigo, Calcite	Indigo	Charcoal black	Charcoal black, White lead, Calcite	Orange

CONCLUSIONS

The study made with portable equipment of non-destructive spectroscopic techniques lead to a very complete analysis of the material composition of the nine mural paintings in the three Augustinian convents in Hidalgo without the need of sampling them. The great size and importance of the murals at these Mexico's heritage site churches and ex-convents and the diversity of materials required a portable laboratory in order to conduct the research in the most efficient way. The balance between both atomic and molecular techniques provided a wide range of data which complimented each other in order to obtain not only quantity, but good quality of the information. The different pigments and materials used in the three convents can be cataloged between inorganic and organic pigments. In the case of binding agents and oils, these were classified as organic materials that contributed to the identification of the painting technique as a dry mural painting method used in all the three monasteries.

After the first investigation made with UV light the elements of the inorganic materials found with XRF were confirmed with Raman spectroscopy. At the same time organic materials which the XRF cannot identify were found through their bands in Raman spectra. There were similarities found in the pigments used in the paintings, specifically the earth based tones, browns, yellows and reds pigments, as well as the pigments like red minium, white lead, blue indigo and also vermillion. In the specific case of the blue indigo, it consistently co-occurred with calcium carbonates. This still needs to be studied. However, the finding of specific pigments like copper resinate and pigments like orpiment and smalt, helped to establish the Novo-Hispanic period of time, before XVIII century, in which the murals were most probably made.

Concerning the paintings, the remarkable differences in the palette and the use of pigments between the convents indicate several stages of painting and several painters involved in the making of these art works.

Further analyses of a set of samples for the mural paintings, as well as experimental work and non-invasive measurements on replicas of mural paintings are in progress in order to understand the role of various materials in the colors.

ACKNOWLEDGMENTS

The study the mural painting of Colonial convents of Hidalgo was supported by the project PAPIIT UNAM IN401710 of the Instituto de Investigaciones Esteticas (IIE) of the UNAM while the *in situ* analysis was funded by grants from CONACyT Mexico 131944 MOVIL, PAPIIT UNAM IN402813 ANDREAH II and ICyTDF PICCO10-57. Figures 1 to 6 were provided by E. Hernandez Vazquez from IIE-UNAM. Figures 7 to 8 were provided by J.L. Ruvalcaba.

REFERENCES

1. I. Fernández, *Historia de México*. (Ed. Pearson, Mexico, 2004) pp. 75-78.
2. C. Reyes-Valerio, *Arqueología Mexicana* **3**, 62-67 (1995).
3. M. Toussaint, *Pinturas Murales en los Conventos Mexicanos del S. XVI*. (Ediciones de Arte S. A., Mexico, 1948) pp.1-4.
4. A. Middleton, K. Uprichard, *The Nebamun Wall Paintings: Conservation, Scientific Analysis and Display at the British Museum* (Archetype, London, 2008) pp. 144.
5. E. Aquilia, G. Barone, V. Crupi, F. Longo, D. Majolino, P. Mazzoleni, V. Venuti, *Journal of Cultural Heritage* **13**, 229-233 (2012).
6. I. Kakoulli, *Greek Painting Techniques and Materials: From the Fourth to the First Century BC* (Archetype, London, 2009) pp.167.
7. H. Howard, *Pigments of English Medieval Wall Painting*, (Archetype, London, 2003), pp. 345.
8. T. Falcón, *La Pintura mural de Epazoyucan: Consideraciones sobre sus Etapas Pictóricas, Técnica de Manufactura y Estado de Conservación* (Laboratorio de Diagnóstico de Obras de

Arte, Instituto de Investigaciones Estéticas, Universidad Nacional Autónoma de México, Mexico, 2012).

9. M. Toussaint, *Pintura Colonial en México*. (Universidad Nacional Autónoma de México, Mexico, 1965) pp. 309.
10. E. Arroyo, *La Pintura Mural de Actopan: Consideraciones sobre sus Etapas Pictóricas, Técnica de Manufactura y Estado de Conservación* (Laboratorio de Diagnóstico de Obras de Arte, Instituto de Investigaciones Estéticas, Universidad Nacional Autónoma de México, Mexico, 2012).
11. A. Vergara, *Las pinturas del templo de Ixmiquilpan ¿Evangelización reivindicación indígena o propaganda de guerra?*, (Universidad Autónoma del Estado de Hidalgo, Pachuca, 2010) pp. 11-31.
12. C. Miliani, F. Rosi, B.G. Brunetti, A. Sgamelotti, *Accounts of Chemical Research* **43**, 728-738 (2010).
13. N. Vornicu, C. Bibire, E. Murariu, D. Ivanov, *X-ray Spectrometry* **42**, 380-387 (2013).
14. K. F. Gebremariam, L. Kvittingen, F.G. Banica, *X-ray Spectrometry* **42**, 462-469 (2013).
15. M. Irazola, M. Olivares, K. Castro, M. Maguregui, I. Martínez-Arkarazo, J.M. Madariaga, *Journal of Raman Spectroscopy* **43**, 1676-1684 (2012).
16. M. Gómez, *La Restauración*: *Examen Científico Splicado a la Conservación de Obras de Arte*, (Cuadernos de Arte Cátedra, Madrid, 2004) pp. 55-63.
17. F. Castro, C. Pelosi, "Study of Wall Paintings and Mosaics by means of Ultraviolet Fluorescence and False Colour Infrared Photography" in *Proceedings of the International Meeting YOCOCU* or Youth in Conservation of Cultural Heritage, Rome, Italy, 2008.
18. A. Aldrovandi, M. Picollo, *Metodi di Documentazione e Indagini non Invasive sui Dipinti* (Collana i Talenti, Il Prato, Padova, 1999) p. 67-110.
19. J.L. Ruvalcaba Sil, D. Ramírez Miranda, V. Aguilar Melo, *X-Ray Spectrometry* **39**, 228-345 (2010).
20. D. Fontana, M:.F. Alberghina, R. Barraco, S. Basile, L. Tranchina, M. Brai, A. Gueli, S. Olindo Troja, *Journal of Cultural Heritage* **15**, 266-274 (2014).
21. H.G.M. Edwards, J.M. Chalmers, *Raman Spectroscopy in Archaeology and Art History*, Analytical Spectroscopy Series (Royal Society of Chemistry, London, 2005), pp 530.
22. B. Mühlethaler, J. Thissen, "Smalt" in *Artists' Pigments*, vol. 2, ed. Ashok Roy, (Oxford University Press, 1993) p.113-130.
23. E. Casanova-González, A. García Bucio, J.L. Ruvalcaba-Sil, V. Santos-Vasquez, B. Esquivel, T. Falcón, E. Arroyo, S. Zetina, L.M. Roldán, C. Domingo, *Journal of Raman Spectroscopy* **43**, 1551-1559 (2012).
24. P. Vandenabeele, M. Ortega-Avilés, D. Tenorio Castilleros, L. Moens, *Spectrochimica Acta* A **68**, 1085–1088 (2007).
25. P. Vandenabeele, S. Bodé, A. Alonso, L. Moens, *Spectrochimica Acta* A **61**, 2349-2356 (2005).
26. A. Doménech, M.T. Doménech-Carbó, H.G.M. Edwards, *Journal of Raman Spectroscopy* **42**, 86-96 (2011).
27. N. Eastaugh, V. Walsh, *The Pigment Compendium. Microscopy of Historic Pigments*. (Routledge, London, 2004), pp. 960.
28. R. Mayer, *The Artist's Handbook of Materials and Techniques* (Viking Penguin, New York, 1973) pp. 165, 253.

Mater. Res. Soc. Symp. Proc. Vol. 1656 © 2014 Materials Research Society
DOI: 10.1557/opl.2014.822

Investigating a Moche Cast Copper Artifact for Its Manufacturing Technology

Aaron Shugar[1], Michael Notis[2], Dale Newbury[3], Nicholas Ritchie[4].
[1] Art Conservation Department, SUNY Buffalo State, Buffalo, New York, USA.
[2] Department of Materials Science and Engineering, Lehigh University, Bethlehem, Pennsylvania, USA.
[3] Materials Measurement Science Division, National Institute of Standards and Technology, Gaithersburg, Maryland, USA
[4] Microanalysis Research Group, National Institute of Standards and Technology, Gaithersburg, Maryland, USA

ABSTRACT

A Moche cast copper alloy object was investigated with focus on three main areas: the alloy composition, the casting technology, and the corrosion process. This complex artifact has thin connective arms between the body and the head, a situation that would be very difficult to cast. The entire artifact was mounted and polished allowing for complete microstructural and microchemical analysis, providing insight into the forming technology. In addition, gigapixel x-ray spectrum imaging was undertaken to explore the alloy composition and the solidification process of the entire sample. This process used four 30 mm^2 SDD-EDS detectors to collect the 150 gigabyte file mapping an area of 46 080 × 39 934 pixels. Raman analysis was performed to confirm the corrosion compounds.

INTRODUCTION

The Moche civilization flourished along the Northern coast of modern day Peru from approximately 100 C.E. to 800 C.E. The culture is named after the type site of Moche located in the Moche valley. The culture evolved out of local polities who developed their own political entity and recognizable materials culture. Their lands extended along the coast from the Huarmey river valley in the south to the Lambayeque river valley in the north, before extending inland (NNW) to the Piura river valley. The Moche are best known for their elaborate ceramic vessels and their ability to work metals, in particular gold.

The range of metals used in the region is extensive and have a long history. As early as 1800 B.C.E. in the northern/central coast region copper and gold were worked [1]. By the time the Moche culture was established, a long history of working metals including gold, copper, copper-gold alloys, copper-sliver alloys, and tumbaga (depletion gilding copper-gold-silver alloy). The newer alloy worked by the Moche appears to be a copper-arsenic alloy [1]. The introduction of arsenic into the metalwork's repertoire may be directly linked to better trade networks extending along the coast and over the Andes into Bolivia, Brazil and to the North with Ecuador.

The range of fabrication techniques mastered by the Moche includes shaping metals by hammered sheet, casting, and extensive joining techniques including folding, riveting, granulation, and soldering.

The association of gold with the sun, and silver with the moon, in Incan mythology is well known; but the association with, and status of copper is less understood. Pease [2] records, and Urton [3] translates the following: "Vichama asked his father, the Sun, to create a new race of humanity. The sun sent three eggs, of gold, silver and copper. The golden egg was the origin of the curacas (high officials) and nobles; the silver egg gave rise to women; and from the copper egg came the commoners and their families". It may thus be understood that copper was the metal of common use in Incan and earlier society. Falchetti [4] summarized previous archaeological finds and pointed out that graves of high-status individuals contained ornaments of solid gold and silver, tombs of nobles had gilded or silvered ornaments of copper or copper-based alloys, and graves of commoners contained objects made only of copper.

Many of these copper objects are cast, but the casting methods are in question whether being piece-mold or lost wax (*cire perdue*) casting, and whether the object was made from a single piece or if it was joined from multiple sections [5-9]. Answering such questions requires investigation of the possible joining technology and comparison of the microstructure and chemistry of the joining wires to the bulk materials. The analytical problem that arises in conjunction with these questions is how to perform both micro-chemical and macro-structural investigation over a large enough area to follow solidification development during the casting process. The full investigation of these problems has been limited by the analytical instrumentation available to date. Yet, significant advances in the development of high intensity sources and high efficiency detectors now enable SEM EDS mapping with gigapixel resolution which can open new avenues of interpretation.
Modern advancement in detectors (SDD) including quad systems, or the ability to combine detectors can help solve this problem. For example, the high throughput and stability of the thermal field emission gun scanning electron microscope (tFEG-SEM/Silicon Drift Detector (SDD)-EDS) and the ready availability of large scale data storage enables a new approach to this "macro-to-micro" problem: the collection of gigapixel x-ray spectrum images in which the entire macroscopic structure is mapped at microscopic resolution, thus capturing all possible compositional information about the macroscopic object within the limitations imposed by the electron dose per pixel [10].

We have applied this new technology in conjunction with several other techniques of analysis to investigate a small (5 cm high by 3 cm across and 2 cm thick) cast copper alloy Moche figurine (Figure 1). This figurine is part of an unprovenanced study collection at Lehigh University. It was donated with the intent of possible destructive analysis, for the very reason that most objects such as these are not available for such extensive testing. There is no way that we can verify the provenance of the object but to our knowledge it is part of a collection that existed in the United States prior to 1970, and was attributed to the Moche.

The surface of the artifact has green and black corrosion products. It has fine detail with thin wire-like arms and thin wire-like connections between the body and the head at the back; these wire-like features are approximately only 2-3 mm in diameter. Although this artifact is rather small, it has a complex set of features that make manufacturing difficult. Because of this, several questions arose as to its manufacture; how was it made? Is it a single casting (i.e. lost wax) or were the arms soldered on? Did the metalsmiths have good control of the procedure? Is there evidence of casting flaws (i.e. variable cooling rates and clear alteration to the microstructure)?

These specific question related to manufacturing are addressed here along with an investigation of the corrosion products which provide a clearer picture of the lifespan of the artifact.

Figure 1: The small (5x3cm) figurine in question, showing the exterior green/black corrosion products and the thin wire connectors in question.

Traditional examination of metal artifacts entails removing a small 'representative sample' from the artifact for metallographic and elemental analysis. The questions surrounding this artifact required a more aggressive approach. To fully investigate this artifact one would need to study details of the compositional microstructure with micrometer resolution at various locations throughout the macroscopic structure. We considered a traditional approach of taking multiple samples from the figurine but decided that the entire object should be mounted, allowing for a comprehensive macro investigation of its microstructure and elemental composition. This is unusual when investigating archaeological artifacts, but deemed valuable enough to sacrifice this one sample, and resulted in invaluable information being recovered

EXPERIMENT

The entire object was mounted in cold setting, two-part epoxy, and ground from the back side using SiC abrasive grinding papers (320 – 800 grit) until the full body and arm connection joints were exposed. The mounted specimen was then polished using 6 μm diamond suspension, 0.3 μm Al_2O_3 slurry and finishing with 0.05 μm Buehler MasterPrep – Colloidal Al_2O_3. The specimen was then examined in the unetched condition, first using light optical microscopy and then a variety of electron and x-ray imaging methods.

Optical light microscopy was performed as a first approach to investigate the microstructure of the artifact. In particular we were interested in confirming the casting technology used, and assessing the manufacturing technology employed by these metalsmiths. Using PAXcam™ software, we were able to stitch images together to gain a more global view of the entire artifact's microstructure.

Elemental analysis was performed in a number of stages. Macro elemental mapping was performed using an EDAX Eagle III Milliprobe X-ray fluorescence spectrometer with X-ray Spectrum Imaging (mXRF-XSI) mapping capability [11]. The Eagle III milliprobe X-ray fluorescence instrument was initially used because it has the capability of spectrum imaging over a large area (10 cm × 10 cm). The process is fairly rapid, but the incident beam size is approximately 50 micrometer from a capillary optic, resulting in 100 micrometer lateral resolution (that is, a 100 micrometer step length between sampling locations). Mapping with mXRF-XSI allows access to spatial range of millimeters to centimeters. This is useful in relating microscopic measurements to macroscopic phenomena.

X-ray excitation by primary photons (XRF) is an order of magnitude more efficient than electrons (SEM) in creating the photoelectric effect which produces characteristic elemental x-rays. In addition, XRF has the advantage of much lower background compared to electron-excited x-rays. This results in detection limits in the parts per million range for XRF, as opposed to parts per thousand for SEM, allowing for the investigation of potential trace elements.

For elemental analysis of the various phases present and micro-elemental mapping a JEOL JXA-8500F thermal field emission gun scanning electron microscope (tFEG-SEM) equipped with a Bruker QUAD Silicon Drift Detector (SDD) with EDS. This system provides higher magnification to elucidate the microstructure. A maximum field width of ~1.75 mm is visible at a minimum magnification of 40x .

A TESCAN MIRA3 tFEG-SEM equipped with four separate 30 mm^2 PulseTor SDD-EDS detectors was used because it has the stage movement capacity and rapid x-ray throughput to create a gigapixel x-ray spectrum image of the whole artifact while recording fine scale lateral resolution. A map of the entire artifact was taken encompassing 46080 pixels by 39934 pixels with a 100 μs dwell per pixel. The entire scan took 51.2 hours to collect and the entire dataset totaled 150 gigabytes.

To investigate the corrosion in more detail a TESCAN VEGA3 XMU tungsten variable pressure scanning electron microscope was used. Line scans were collected and processed using an Oxford Instruments X-Max^n Silicon Drift Detector (SDD-EDS) with a 50 mm^2 detector and AZtecEnergy analysis software.

Finally, to fully characterize the corrosion, a Bruker Senterra Raman microscope with a 785 nm, 25 mW (at the source) was used. A 100 × ultra-long working distance objective was used to focus the excitation beam to an analysis spot of approximately 1 μm. The resulting Raman spectra are the average of 20 scans at 1 s integrations each. Spectral resolution was 3-5 cm^{-1} across the spectral range analyzed. Spectral spikes due to cosmic rays were removed and baselines adjusted as necessary using Opus 7.2 software. Sample identification was achieved by comparison of the unknown spectrum to spectra of the RRUF database[12].

RESULTS AND DISCUSSION

Optical Microscopy

Optical microscopy revealed an as-cast hypo-eutectic (primary Cu-phase) dendritic structure with shrinkage voids around the dendrite arms. Inter-dendritic eutectic is clearly seen with some minor rounded inclusions (Figure 2). Elongated dendrites as well as the absence of chill crystals at the edges of the sample indicate slow cooling. In addition, dendrites extended through the narrow (~500 μm) wire-like arms and connectors, further indicating that the artifact was cooled near equilibrium, as a single piece casting.

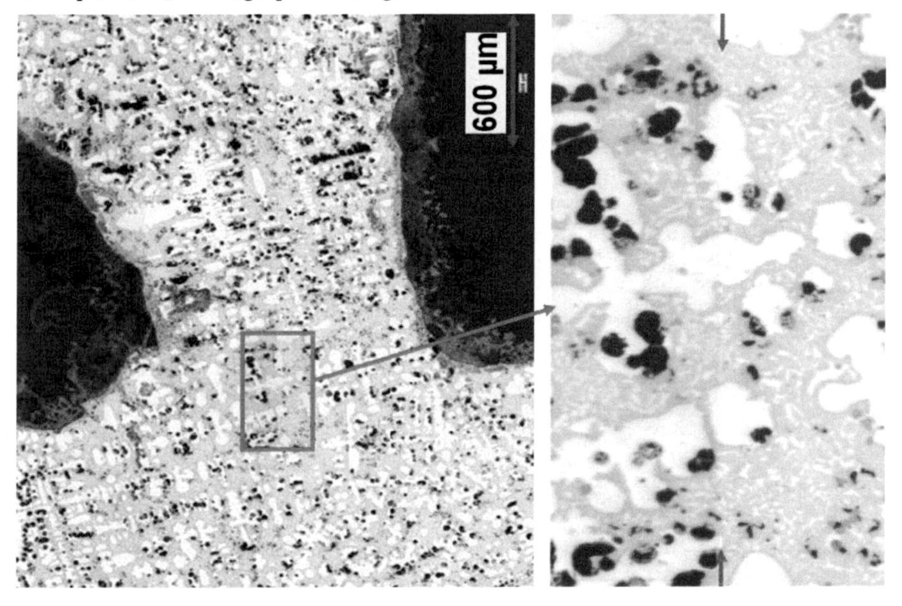

Figure 2: PAXcam image of one of the wire-like arms showing a hypo-eutectic (primary Cu-phase) microstructure with extended dendrites, shrinkage voids and additional inclusions.

The continuity of the microstructure throughout the intricate shape of the artifact, along with known traditional metalworking technology of the region provide strong indication that a single lost wax casting process was used.

X-Ray Fluorescence and Electron Microscopy

Determining the chemical composition of the artifact is imperative for a full examination. It provides significant detail on the fabrication technology (i.e. alloy used, melting temperature) which is important in characterizing the artifact as a whole. Although several techniques can be used when trying to understand the overall chemistry of the artifact (including INAA, XRF or SEM) elemental mapping can often be the most useful approach. However, until recently, there have been issues with this approach since mapping size has been restricted to micrometer resolution based on limited x-ray throughput of detectors; this results in small area maps that take long times to collect. Very often, situations occur where an object to be mapped has dimensions of several centimeters. In the past, the limited x-ray throughput in the mapping mode necessitated an analytical strategy that required careful selection of limited number of areas of interest for high magnification mapping which may not represent the entire macrostructure.

Milli X-ray Fluorescence X-ray Spectrum Imaging (mXRF-XSI)

mXRF-XSI was performed on the entire artifact. The resulting scans were analyzed using LISPIX , creating a data cube sum spectrum [13] to investigate the elemental distribution and identify any trace elements that may have been of interest. The artifact is found to be a complex alloy of copper, arsenic and nickel. Although fine microstructural details could not be determined, the resolution obtained revealed a macroscopically complex corrosion phenomenon centering on the depletion of nickel near the surrounding edges of the artifact. Even though the resolution was somewhat restricted, the detailed information about elemental distribution was confirmed by SEM elemental analysis using the JEOL JXA-8500F (tFEG-SEM) (see Table 1 for details).

Location	Ni wt. %	Cu wt. %	As wt. %
Head-1	2.6	77.5	19.9
Head-2	2.6	79.3	18.1
Head-3	2.6	77.8	19.6
Head-4	2.5	78.3	19.2
Head Avg	2.58	78.23	19.20
Torso-1	2.3	77.3	20.4
Torso-2	2.5	77.7	19.9
Torso-3	2.1	79	18.9
Torso-4	2.5	77.2	20.3
Torso Avg	2.35	77.80	19.88
RtArm-1	0.4	80.2	19.4
RtArm-2	0.6	79.6	19.8
RtArm-3	1	78.3	20.7
RtArm Avg	0.67	79.37	19.97

Table 1: Elemental composition of three locations on the figurine. The head and torso as well as the right arm, which is depleted in nickel.

This complex alloy with relatively high levels of nickel has not been documented before for Moche metalworking. There is evidence of complex copper arsenic nickel alloys occurring at Middle Horizon centers like Tiwanaku, but this dates after the Moche and is located to the south along the coast and into the highlands of Chile [14]. With these artifacts, the arsenic and nickel combine for a total range of 2 – 8 wt. % with one artifact having a combined concentration of 11.85 wt. %. These totals fall well below what we are seeing in this artifact with approximately a combined 22 wt. %.

It is clear that the overall composition of the artifact, approximately 78 wt. % Cu, 19 wt. % As, balance Ni, is very near the binary Cu-As eutectic composition of 20.8 wt. % (Figure 3).

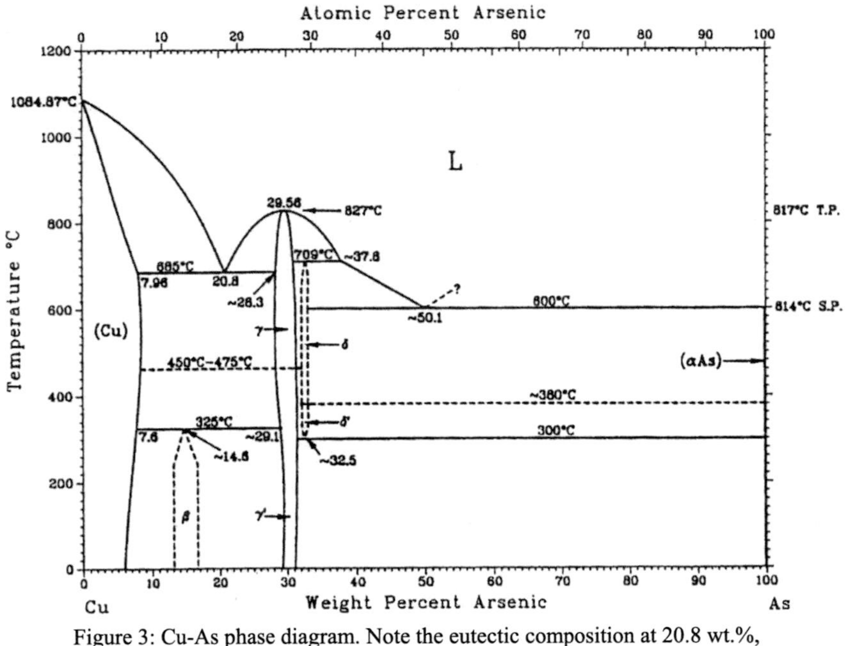

Figure 3: Cu-As phase diagram. Note the eutectic composition at 20.8 wt.%, the γ phase at ~29 wt.% and the δ phase at ~ 32 wt.% (after Brown [5]).

The composition of the individual phases gives an indication that an even more complex system is present (Figure 4). The dendrites have a nominal As composition of 6 wt. % (Table 2). This is a good match on the binary phase diagram for the solid state solubility limit of arsenic in copper in equilibrium at room temperature. The interdendritic phase has 27 wt. % As which matches well with the γ phase of the binary system.

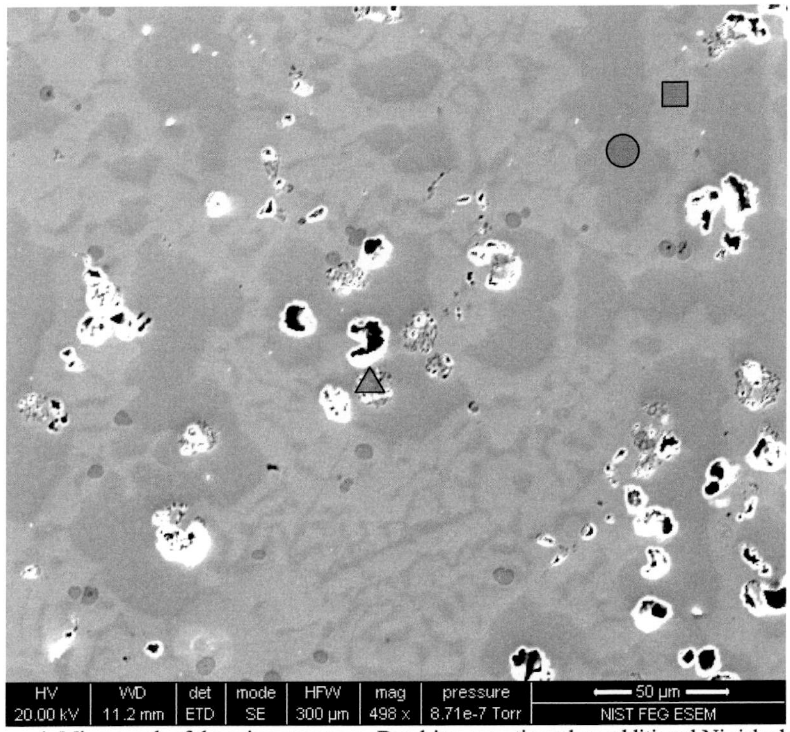

Figure 4: Micrograph of the microstructure. Dendrite, eutectic and an additional Ni rich phase are represented by a circle, square and triangle respectively.

	Ni wt. %	Cu wt. %	As wt. %
● Dendrite	1	93.1	6
■ Inter-Dendrite	0.5	72.5	27
▲ Ni-rich complex	8.8	58.5	32.7

Table 2: Compositional data for the microstructure of the artifact. Shapes relate to location of analysis in Figure 4.

A third, Ni-rich, complex region was identified that contained 8.8 wt. % Ni and 32.7 wt. % As. The Ni content in this region is high enough to consider the ternary system Cu-Ni-As [15] rather than the binary system, and this is shown in Figure 5. It may be seen that the composition of the measured complex region lies in a multiphase region of the ternary and requires more investigation before further elucidation becomes possible.

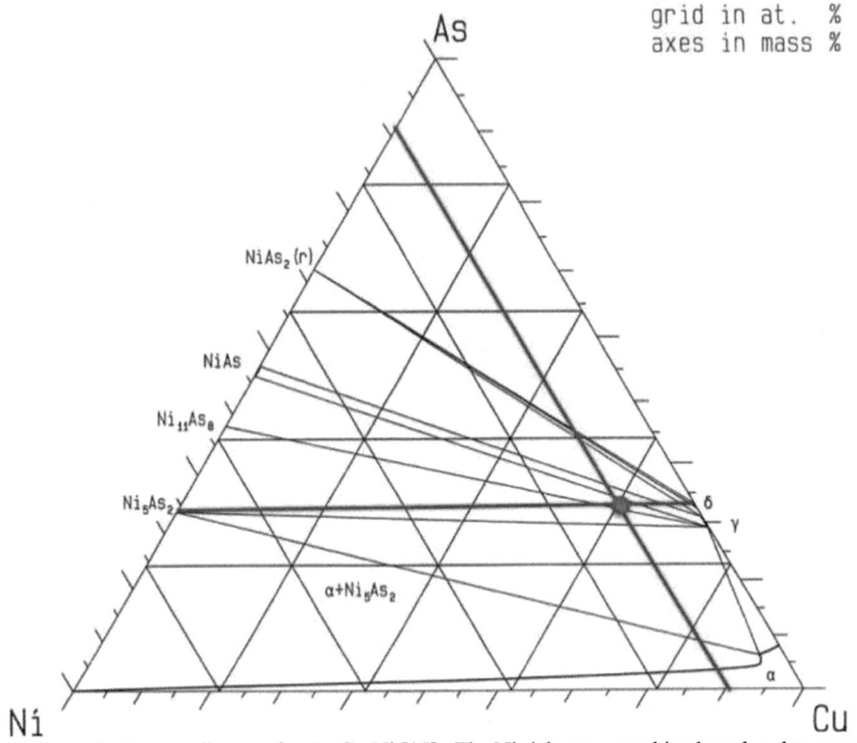

Figure 5: Ternary diagram for As-Cu-Ni [15]. The Ni rich compound is plotted at the cross section of the two red lines.

This rich nickel arsenic complex is not yet fully understood, but may have direct links to a local/regional ore source. There are extensive regions of arsenic-rich ores in the upper Andes, and evidence of nickel rich ores have been found in the same formations as far north as Cusco and Lima [14]. Additional sources have been identified along the Chilean coast. The trade routes needed to acquire these ores would have been well established in this time period, either from the south by way of marine trade, or through contacts over the Andes towards Bolivia and Brazil.

TESCAN MIRA3 tFEG-SEM

Based on the XRF-XSI mapping, it was determined that gigapixel mapping using a TESCAN MIRA3 tFEG-SEM equipped with four separate 30 mm^2 PulseTor SDD-EDS should be undertaken. This would give more detail concerning the depletion of nickel at the artifact surface and could help clarify the corrosion phenomenon in greater detail. The data was collected and background corrected elemental k-ratio maps were created using NIST DTSA-II [16].

The elemental maps show some characteristic corrosion phenomena, including increased surface concentrations of chlorine, oxygen and silica. In addition, there is clear evidence for some dissolution of copper and arsenic at the surface of the artifact. The extended depletion of nickel seen in the mXRF-XSI mapping was confirmed. The increased resolution provided by the gigapixel mapping allows for a more precise investigation of this particular corrosion phenomenon. When zooming into the arm or foot of the figurine, it became clear that the mobility of nickel causes an increased nickel concentration at the surface of the remaining metal. This tends to correspond with an increase in oxygen (Figure 6). The mobility of nickel is well documented [17, 18]; the preferential dissolution of nickel induces a decreased concentration of nickel within the sample, and has the potential to increase porosity as the nickel selectively dissolves out of the copper.

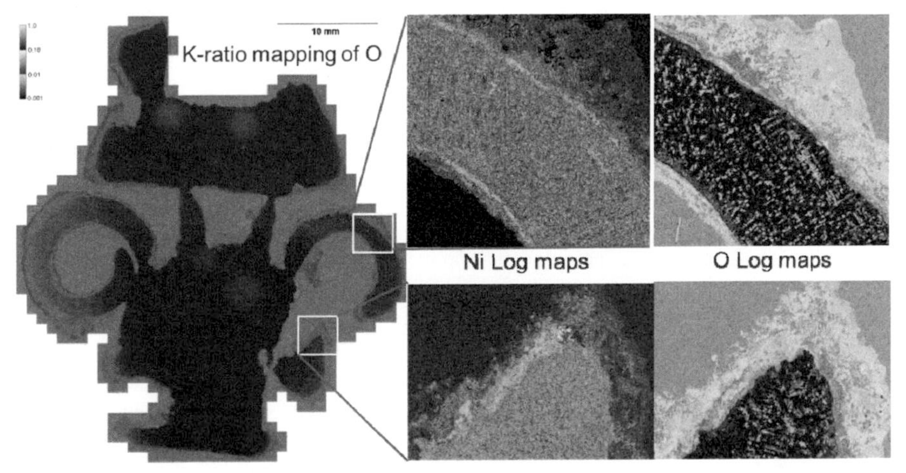

Figure 6: Gigapixel mapping image of the concentration of O by K-ratio. Close up of the arm and foot displaying the depletion of Ni and close association with O enrichment at the surface of the artifact.

This increased porosity generates a higher surface area which then interacts at an elevated rate with the local burial environment, resulting in a faster corrosion rate. As we have seen, there is a clear decrease in nickel content surrounding the entire artifact that extends through the arms, feet, and up to 4mm into the main body itself.

The higher concentration of nickel found at the surface of the remaining metal and bound in the corrosion layer was examined by SEM and Raman to determine the identity of the corrosion product. Line scans were taken using a TESCAN VEGA3 XMU, and which showed that the higher concentrations of nickel correlate with higher concentrations of oxygen and arsenic

(Figure 7). Raman analysis of this product identified it to be annabergite ($Ni_3(AsO_4)_2 \bullet 8H_2O$), a hydrous nickel arsenate.

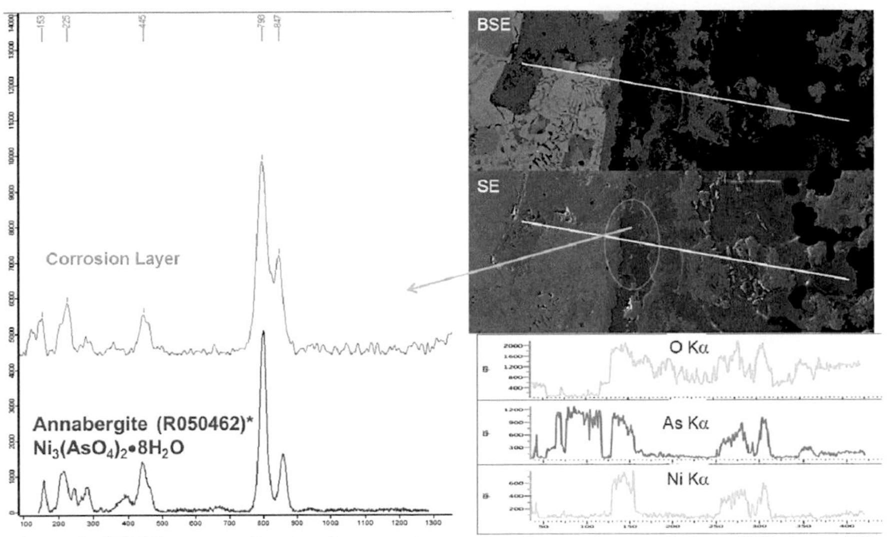

Figure 7: SEM line scans showing the corrosion product in question consisting of Ni, As, and O. Raman analysis of the area located in the oval indicated that annabergite is present.

CONCLUSIONS

Most analytical studies use a small micro-sample from the edge and hope to present evidence that is representative of the whole object but they do not test potential variability over the object. In addition, analysis that is considered non-destructive typically includes chemistry from the encasing corrosion products, the burial products, and occasionally excludes chemistry that is leached from the material itself. Due to increased throughput in modern detectors, combined with massive computer power we are now able to fully investigate artifacts taking into account the variables mentioned above. The combination of analytical techniques presented here answered unresolved questions about the casting technology of the Moche and raised new questions for archaeometallurgist to consider when deciding how, and where to sample artifacts for chemical analysis.

In this study, a series of analytical techniques were used to investigate a figurine attributed to the Moche culture. The entire specimen was examined by optical microscopy, mXRF-XSI mapping, SEM elemental analysis and gigapixel mapping, as well as Raman analysis to characterize corrosion products. The instrumental methods used in this study demonstrate the ability to examine both the microstructural and microchemical features of archaeological objects of macroscopic dimensions and highlights the benefits of doing so.

The figurine is a complex Cu-As-Ni ternary alloy. Although this is the first Moche artifact reported with this complex composition, it reflects the connection with Ni-rich ore sources adjacent to the region inhabited by the Moche people and may be directly linked to their source of arsenic used in the copper/arsenic alloys previously discovered

Analysis showed the presence of a continuous dendritic structure, at narrow connection points in the object, indicative of lost wax casting with an alloy of high fluidity and a well heated mold, and which allowed the object to cool slowly under near-equilibrium conditions. The determined composition places the melting temperature of the metal at approximately 685°C, an easily achievable temperature for these metalsmiths. In addition, the high As level, at the binary eutectic composition, indicates that the metalsmiths had the ability to smelt/remelt arsenical copper alloys without the excessive loss of highly volatile arsenic.

This suggests a lost wax casting method similar to those documented by Dhokra casters in India [19] or by Asahnti casters in Ghana, Africa [20]. In these methods, the lost wax casting is made in a traditional manner. Once the wax is melted out of the mold, an upper reservoir is made from clay and metallic metal is placed in the hold. It is then encased with clay and dried. The entire assembly is then turned upside down, with the reservoir at the bottom, and heated to about 1000°C. The mold is then turned over, allowing the molten metal to flow down into the voids of the lost wax shape, and placed in warm ashes and left to cool. This is an elegant solution for restricting the volatilization of arsenic, preheating the mold to a high enough temperature to avoid chill crystallization, and allow for slow equilibrium cooling and extended dendritic growth.

Elemental mapping revealed an extensive nickel depletion region in the base metal, demonstrating long term exposure to corrosion conditions. This corrosion phenomenon raises a concern for most traditional elemental analyses. INAA would require a small drilled sample that would likely not penetrate through the depletion layer. The resulting analysis would appear to be nickel depleted. Traditional metallurgical techniques of smaller size sample removal from easily accessible regions of the artifact would have similar results. More current non-destructive testing using XRF would encounter similar issues because the depth of depletion ensures that the higher nickel content would be totally attenuated but the surrounding layer. This raises concern as to the accuracy of previous analyses of Moche artifacts that indicated low levels of nickel. Is it possible that they also had extensive nickel depletion? Future analysis of pre-Columbian metal artifacts that show low levels of nickel should be considered for more invasive analysis, or to have deeper core drillings to ensure that a proper account can be made of the nickel composition.

REFERENCES

[1] T. Stöllner, "Gold in Southern Peru? Perspectives of Research into Mining Archaeology," in *New Technologies for Archaeology*, M. Reindel and G. Wagner, eds., ed: Springer Berlin, 2009, pp. 393-407.

[2] F. Pease G. Y, *El dios creador andino*. Lima: Mosca Azul Editores, 1973.

[3] G. Urton, *Inca Myths*: University of Texas Press, 1999.

[4] A. M. Falchetti, "The Seed of Life: Symbolic Power of Gold-Copper Alloys and Metallurgical Transformations," in *Gold and Power in Ancient Costa Rica, Panama, and Colombia*, J. Quilter and J. W.Hoopes, eds., Washington DC: Dumbarton Oaks, 2003, pp. 345-381.

[5] C. B. Donnan, *Moche Art of Peru: Pre-Columbian Symbolic Communication*: Museum of Cultural History, University of California, 1978.

[6] C. B. Donnan, D. A. Scott, and T. Bracken, "Moche Forms for Shaping Sheet Metal," in *The Art and Archaeology of the Moche: An Ancient Andean Society of the Peruvian North Coast*, S. Bourget and K. L. Jones, Eds., ed: University of Texas Press, 2009, pp. 113-128.

[7] D. T. Easby Jr., "Pre-Hispanic Metallurgy and Metalworking in the New World," *Proceedings of the American Philosophical Society,* vol. 109, pp. 89-98, 1965.

[8] H. Lechtmann, "Traditions and Styles in Central Andean Metalworking," in *The Beginning of the Use of Metals and Alloys*. R. Maddin, ed., Cambridge: MIT Press, 1988.pp. 344-378.

[9] H. B. Nicholson, A. Cordy-Collins, and L. K. Land, *Pre-Columbian Art from the Land Collection*: California Academy of Sciences, 1979.

[10] D. E. Newbury and N. W. M. Ritchie, "Elemental mapping of microstructures by scanning electron microscopy-energy dispersive X-ray spectrometry (SEM-EDS): extraordinary advances with the silicon drift detector (SDD)," *Journal of Analytical Atomic Spectrometry,* vol. 28, pp. 973-988, 2013.

[11] J. M. Davis, D. E. Newbury, N. W. Ritchie, E. P. Vicenzi, D. P. Bentz, and A. J. Fahey, "Bridging the Micro to Macro Gap: A New Application for Milli-probe X-ray Fluorescence," *Microscopy and Microanalysis,* vol. 13, pp. 410-417, 2011.

[12] R. T. Downs, *The RRUFF Project: an integrated study of the chemistry, crystallography, Raman and infrared spectroscopy of minerals.* in Program and Abstracts of the 19th General Meeting of the International Mineralogical Association in Kobe, Japan. 3-13, 2006.

[12]

[13] D. Bright and K. Milans, "Lispix: a public domain scientific image analysis program for the PC and Macintosh," Microscopy and Microanalysis New York -, vol. 6, pp. 1022-1023, 2000.

[14] H. Lechtman, "Middle Horizon bronze: centers and outliers," in *Patterns and Process: a Festschrift in honor of Dr. Edward V. Sayre*, L. vanZelst, Ed., ed Suitland MD, Smithsonian Center for Materials Research and Education, 2003, pp. 168-248.

[15] L. Guzei, "Arsenic-Copper-Nickel," in *Ternary Alloys: A Comprehensive Compendium of Evaluated Constitutional Data and Phase Diagrams. As-Cr-Fe to As-I-Zn.* vol. 10, G. Petzow, G. Effenberg, and F. Aldinger, Eds., ed: Wiley-VCH Verlag GmbH, 1994, pp. 92-97.

[16] N. W. M. Ritchie. (2013). *DTSA-II.* Available: http://www.cstl.nist.gov/div837/837.02/epq/dtsa2/

[17] L. Cassayre, P. Chamelot, L. Arurault, L. Massot, P. Palau, and P. Taxil, "Electrochemical oxidation of binary copper–nickel alloys in cryolite melts," *Corrosion Science,* vol. 49, pp. 3610-3625, 2007.

[18] F. Rioult, M. Pijolat, F. Valdivieso, and M.-A. Prin-Lamaze, "High temperature oxidation of a Cu-Ni based cermet: kinetic and microstructural study," *Journal of the American Cermaic Society*. vol 89, no. 3 pp. 996-1005 2006.
[19] D. J. Capers, "Dhokra: The Lost Wax Process in India," ed. Orissa, India, 1989, 26 mins. Film.
[20] F. R. Sias, *Lost-wax Casting: Old, New, and Inexpensive Methods*: Woodsmere Press, Pendleton, SC. 2005.

Mater. Res. Soc. Symp. Proc. Vol. 1656 © 2014 Materials Research Society
DOI: 10.1557/opl.2014.811

Hawaiian Barkcloth from the Bishop Museum Collections: A Characterization of Materials and Techniques in Collaboration with Modern Practitioners to Effect Preservation of a Traditional Cultural Practice

Christina Bisulca[1, 2], Lisa Schattenburg-Raymond[2], Kamalu du Preez[2]
[1]Department of Materials Science and Engineering, University of Arizona, Tucson Arizona 85721, U.S.A.
[2]Cultural Collections, Bernice Pauahi Bishop Museum, Honolulu HI 96817, U.S.A.

ABSTRACT

Hawaiian barkcloth ('kapa') is a traditional fabric made from beaten plant fibers. Because of its function in both utilitarian and chiefly ornaments, kapa is intimately tied to the history and traditions of Hawai'i. In the 19th century kapa was gradually replaced with imported textiles and the practice was lost. The traditional methods used to manufacture kapa are now only known from historic descriptions by early missionaries and explorers. Since the 1970s, cultural practitioners began an effort to revive this artform and are experimenting with materials and techniques to reproduce kapa with the quality of historic artifacts.

Research has been undertaken at the Bishop Museum using a multi-analytical approach to determine the colorants. The Bishop Museum holds the world's best collection of kapa, including some of the earliest pieces collected from Cook's voyage in the 18th century. The research has focused on a comprehensive survey of over 150 pieces of kapa with x-ray fluorescence spectroscopy (XRF). In some cases, samples were removed and analyzed with UV-Vis-NIR fiber optics reflectance spectroscopy, Fourier transform infrared spectroscopy (FTIR) and chromatographic techniques, including high performance liquid chromatography (HPLC). Scientific results document the use of traditional pigments and dyes as well as the incorporation of imported materials in the 19th century. Results are interpreted by period, design and use, as well as within the context of historic descriptions. An important aspect of this work is close collaboration with cultural practitioners experienced fabrication methods that have been successful in the recreation of kapa. With continued research, the goal is to ultimately gain a greater knowledge of historic materials and techniques for the continuation of this important tradition.

INTRODUCTION

Barkcloth ("tapa", or "kapa" in Hawaiian) is a fabric made from beaten plant fibers that was an integral aspect of cultures throughout the Asia-Pacific. The predominant fiber used is from the inner bark of paper mulberry (*Broussonetia papyrifera).* This tradition was once found throughout Polynesia, and was a central activity in everyday life. Kapa is unique in the Pacific due to its broad range of colorants as well as the application of a relief design ("watermark") pattern into the fabric. With Western contact and the arrival of missionaries in 1820s, the production of kapa and other traditional artifacts were gradually replaced with imported materials.

Much of the knowledge of historic kapa manufacture has been lost. In Hawaii, materials and methods once used are now known only through historic descriptions by early missionaries and explorers. However, these texts give only brief or vague accounts of fabrication techniques. Traditional plant materials commonly cited in these sources are listed in Table I. For mineral pigments, commonly cited are iron oxide earths (□alaea) and carbon blacks prepared from burning various plant materials [1, 2, 3].

Table I. Commonly cited plant sources for colorants used in kapa

Material	Common name	Scientific name	Plant part
Red dye	ʻākala or 'akalakala	*Rubus hawaiensis*	fruit/juice
	noni	*Morinda citrifolia*	root bark
	kōlea	*Mysrsine spp.*	bark
	kukui (candlenut)	*Aleurites moluccana*	bark
	kokia	*Kokia drynaroides*	bark
	koa	*Acacia koa*	bark, wood
	ʻamaʻu	*Sadleria cyatheoides*	fronds
Yellow dye	ma□o	*Gossypium tomentosum*	flowers
	noni	*Morinda citrifolia*	root, bark
	nanu, na□u	*Gardenia remyi, G. Brighamii, G. Manii*	fruit
	ʻōlena (turmeric)	*Curcuma longa*	root
Green dye	ma□o	*Gossypium tomentosum*	flowers
Blue dye	'uki'uki	*Dianella sandwicensis*	fruit
Oils	kukui (candlenut)	*Aleurites moluccana*	nut
	kamani	*Calophyllum inophyllum*	nut
	coconut	*Cocos nucifera*	nut
Resins	kukui	*Aleurites moluccana*	resin (gum)
	kōlea	*Mysrsine spp.*	resin (gum?)
	koa	*Acacia koa*	Resin (gum)
Black	kukui (candlenut)	*Aleurites moluccana*	nut (soot)
	pili grass	*Heteropogon contortus*	blades (soot)

Many of the techniques used today by cultural practitioners have "been acquired through the work, in recent years, of persons interested in experimenting with plants named as dye plants in the old literature" [2]. To date, there have only been limited scientific investigations into the materials used in Polynesian barkcloth [4], and no studies have yet been carried out on kapa specifically. The findings presented here are the results of an ongoing collaboration specifically for the analysis of colorants and binders used in decorated kapa.

EXPERIMENTAL DETAILS

Samples analyzed were from the Bishop Museum's reference kapa collection. The vast majority of these samples are from kapa pieces collected post-contact, primarily near the turn of the 19th century. Special attention was given to the Emerson Collection because the dates and location of collection were documented [5]. The reference collection also contains exchange samples of kapa from other institutions. Prior to analysis kapa samples were digitally photographed. Red, yellow, blue, and black colored areas were analyzed; brown colorants were not analyzed in this study.

X-ray fluorescence spectroscopy (XRF) was performed on a Bruker Tracer SD3 handheld XRF. Two settings were used: 15 keV, 25uA under vacuum with a 1 mil Ti filter and 40keV, 20uA in air with a 12 mil Al/1 mil Ti filter to get information on both low and high Z elements, respectively. In both settings spectra were recorded for 100 s live time.

UV-Vis-NIR fiber optics reflectance spectroscopy (UV-Vis-NIR FORS) and high performance liquid chromatography (HPLC) was conducted at the Department of Conservation and Scientific Research (DCSR), Freer Gallery of Art and A. M. Sackler Gallery of Art, Smithsonian Institution. FORS was performed using a Cary 50 ultraviolet-visible spectrophotometer (Varian Analytical Inc.) with a fiber optic reflectance probe, scanning rate of 30 nm/min at 1 nm intervals with a baseline correction. Labsphere white reference or areas of plain kapa were used for background correction. Identification was based on comparison of spectra to the reference spectra at the DCSR.

Areas where organic colorants were identified with XRF and FORS were sampled (approximately 0.5 mm fibers were removed) for analysis with HPLC. Samples were dissolved in 1:1 3N HCl: MeOH and centrifuged at 4k rpm for 10 minutes. All separations were performed on an Agilent 1100 HPLC with a binary pump system and diode array detector (DAD). Separations were preformed using a Wakosil RSC18 column (100mm x 2.1 mm I.D.), flow rate 0.300 mL/min at 38°C. The mobile phase consisted of A: 90%v/v 10mM oxalic acid (aq)/10% v/v acetonitrile, B: 90%v/v acetonitrile /10% v/v 10mM oxalic acid (aq). On the DAD, multiple wavelengths were recorded from 200-600 nm, using a slit with of 16nm to increase sensitivity. Wavelengths were chosen based on maximum absorbance of red and yellow dyes. Wavelenghts monitored: 1. 500 (b.w. 16), ref. 650 (b.w. 50), 2. 490 (b.w. 16), ref. 650 (b.w. 50) 3. 285 (b.w.16), ref. 650 (b.w. 100). All solvents were HPLC grade from Sigma Aldrich. The gradient used was previously developed at the DCSR for the separation of plant and animal anthraquinones found in dyes [6].

Attenuated total reflection Fourier transform infrared spectroscopy (ATR-FTIR) was performed at the Arizona State Museum with a Thermo-Nicolet FTIR spectrometer with an ATR attachment. Spectra were recorded from 3000-650 cm^{-1}, 64 scans at 8 cm^{-1} resolution using OMNIC ESP 6.1a software. For some analysis, a sample was removed and analyzed with Nicolet Nexus 670 Fourier transform infrared spectrometer with a SpectraTech IR-Plan Advantage infrared microscope attachment. Samples were removed and, compressed between diamond surfaces and analyzed in the microscope. Recorded spectra were then compared to various Infrared and Raman Users Group (IRUG) and commercial reference libraries. Fiber identification was done using polarized light microscopy (PLM) and comparison to reference fiber materials.

Modern kapa was used for comparison to historic kapa samples to assist in data interpretation. Kapa pieces were from mulberry fibers and decorated with traditional materials. Reference materials included paper mulberry (wauke), kukui oil, kukui resin, noni, ʻōlena, and carbon black prepared from burnt kukui nut meat. All materials were collected or purchased locally in Hawaii. Dye standards for HPLC analysis (carminic acid, alizarin and purpurin) were purchased from Aldrich.

RESULTS

Results for identified pigments, dyes and binders are shown in Table II, including the methods used for their identification.

Table II. Materials, their identification and methodology used for identification

Material	Identification	Methods
Red pigment	Iron oxide earth	XRF, FTIR
	Vermilion	XRF
Red dye	Noni	XRF, FORS, HPLC
	Natural madder	XRF, FORS, HPLC
	Synthetic alizarin	XRF, FORS, HPLC
Yellow dye	ʻŌlena (turmeric)	XRF, FORS, HPLC
	Noni	XRF, FORS, HPLC
Yellow pigment	Iron oxide earth	XRF
Black	Carbon black	XRF
Blue	Ultramarine	XRF, FORS, FTIR
	Unidentified synthetic	XRF, FTIR
Binder	Protein	FTIR, XRF

UV-Vis-NIR FORS

Twelve samples were analyzed with FORS prior to HPLC analysis. In areas of organic red, most were found to be anthraquinones. Anthraquinones have a distinctive UV-Vis pattern. Sub-absorption bands at ~510 and ~545 nm indicate an anthraquinone from a plant source [7, 8]. One sample of a blue kapa was also analyzed; based on reference spectra the pigment was identified as ultramarine. Yellow colorants were not identified using this method.

HPLC

All red samples showed the same elution pattern that contained primarily one component eluting at 9.4', and multiple components in minor quantities. All components were anthraquinones based on UV-Vis spectra. Chromatograms of select samples are shown in Figure 1a, and UV-Vis of the main component at 9.4' in Figure 1b.

Several red samples were identified as the imported materials natural madder and synthetic alizarin. Figure 3 shows chromatograms of the red dye from two kapa samples with a reference chromatogram of alizarin and purpurin eluting at t_R 8.4' and 8.8' respectively. Natural madder contains both alizarin and purpurin, along with other anthraquinones in minor amounts [9]. Synthetic alizarin contains only alizarin; in other studies where only alizarin is found are normally interpreted as synthetic alizarin [10]. The earlier eluting peaks in Figure 2 are also anthraquinones, and may be byproducts in the industrial process as they do not match other reference plant or animal anthraquinones in retention time or UV-Vis spectra.

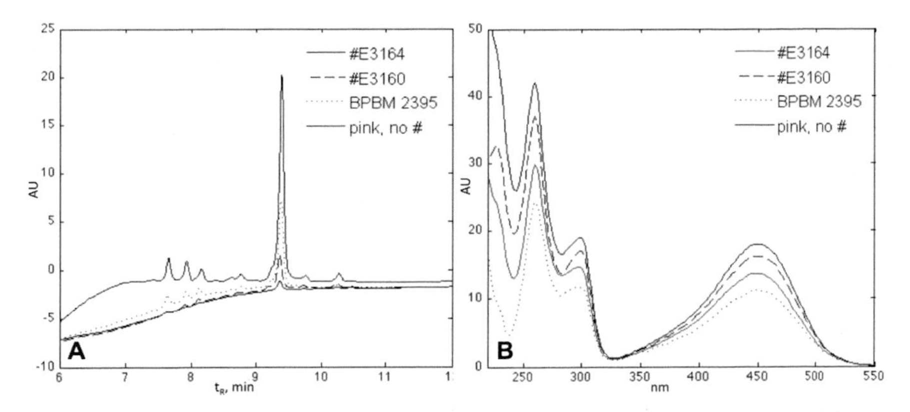

Figure 1. A. Chromatographs of red dyes from 4 samples at 450 nm (absorbance vs. t_R, min). #E3164, Peabody Essex, 1802; #E3164, Peabody Essex, prior to 1827; BPBM 2395, Emerson Collection, collected 1887. B. UV-Vis spectra (AU vs. nm) of the main peak at 9.4' of chromatograms in A.

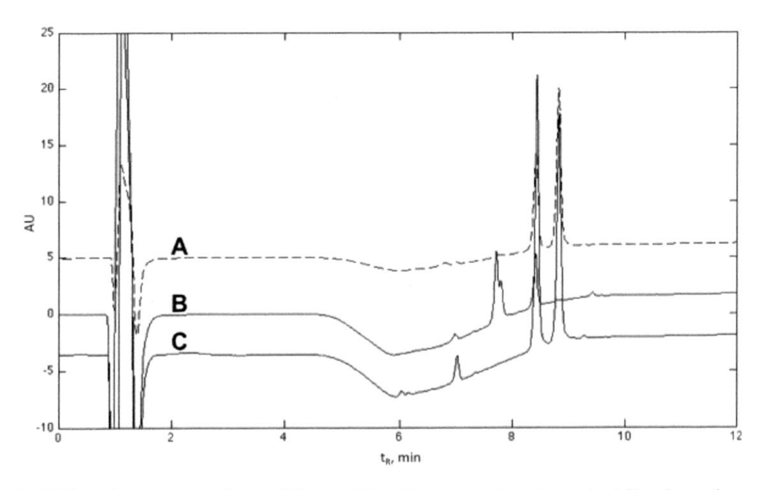

Figure 2. Offset chromatographs at 450 nm (absorbance vs. t_R, min). A. Alizarin and purpurin standards, the two main anthraquinones in natural madder. B. BPBM 2361, containing only alizarin. C. Sample from an unassociated kapa fragment containing natural madder.

Three of the yellow dyed samples were identified as ‘ōlena (turmeric) based on the elution of a three components and correlation of UV-Vis spectra. Turmeric contains three curcumenoid dyes curcumin, demethoxycurcumin and bisdemethoxycurcumin [11]. One of the yellow dyes was identified as noni yellow based on the main peak at 9.4’, identified as an athraquinone based on UV-Vis spectra, as is found in the noni red dyed samples. Unlike the noni red dyes, there were additional main components at 9.3’ and 10.3’, both of which are anthraquinones based on UV-Vis spectra. This may indicate that a different part of the plant was used in the production of the dye, as both the root and bark are known to have been used [12].

FTIR

Based on XRF and FTIR of the red pigments painted on the surface of five kapa samples, the material was identified as a high iron clay or iron oxide and clay. The identification of clay was based on the sharp water of hydration bands at ~3690 and 3620 $cm^{-1}$and Si-O and Al-O stretching at ~1000-1300 cm^{-1} [13]. Based on analysis using XRF alone, this red ochre was used extensively in kapa usually as a stamped or painted design. A similar clay component was sometimes found with black pigments, in particular where it was a thick layer painted on the surface. However, most black pigments appear to be carbon black only.

In samples with painted pigment on the surface, a protenaceous binder was identified in many samples based on FTIR and XRF. This proteinaceous material was only found on the pigmented side, indicating it was mixed with the pigment or else painted onto the pigment as a varnish.

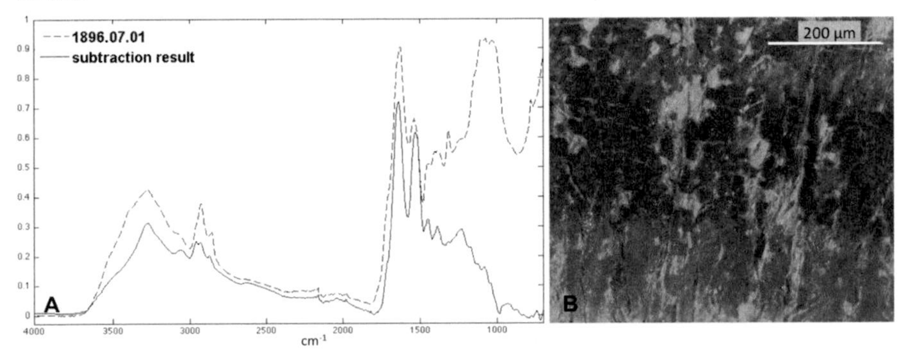

Figure 3. A. FTIR spectra (absorbance vs. cm^{-1}, normalized) from the surface black paint on BPBM 1896.07.01 (dashed line) and the subtraction result of paint layer and plain kapa on the reverse of the sample. B. Microphotograph of a resinous material on kapa also identified as a protein with FTIR analysis.

Figure 3a shows the FTIR spectra of sample 1896.07.01 and the subtraction result of the pigmented surface and the plain kapa reverse. A protein is clearly present based on the amide I C=O stretches at 1690-1600 cm^{-1}, amide II C-N stretching and N-H bending modes at 1600-1500 cm^{-1} and amide III bands at 1275-1240 cm^{-1} [14]. In some samples where a binder was visible on the surface of the kapa adjacent the pigmented area, these areas showed a higher sulfur signal in XRF which corroborates the presence of protein. This proteinaceous binder was found in 8 of 12 samples (67 %) of where a colorant was applied to the surface of kapa and analyzed with FTIR. The presence of a protein was also found in some black carbonaceous pigments as well as in one case of a painted organic red, presumed to be noni. Figure 3b shows a red, resinous material that was sampled for FTIR microscopy. This material was also identified as a protein.

No oil binders were found with FTIR analysis on historic samples. In analysis of modern kapa samples that used kukui nut oil mixed with or on pigmented areas, the oil was easily detectible even with bulk analysis based on the carbonyl stretch at ~1740-1750 cm^{-1}. In some samples there was indication that kukui resin may have been used based on FTIR of both bulk and microscopy samples. Based on analysis of locally collected reference materials, kukui resin was found to be a tree gum similar to acacia gum. However, because the contents of gum overlap with the cellulose substrate of kapa samples, the presence of a tree gum binder is not definitively confirmed in this study.

DISCUSSION

The use of noni

Although there are over 10 plant sources that are cited as historic red dyes used in kapa, noni appears to be the most common red dye based on the collections sampled. The red dye from noni was used both as a dye as well as a stamped decoration. Morindone is the primary anthraquinone colorant in the inner bark of noni root [15] and is yellow-orange in color. The Hawaiians added burnt coral (lime, $Ca(OH)_2$) to generate a red dye. The alkalinity of the lime solution was used to alter the color of noni root extracts to red: the color of morindone changes from yellow to red at pH 8.7 [16]. XRF analysis also detected significant amounts of Ca in many red organic areas, which is consistent with noni-lime red dye.

Noni is what is known as a "canoe plant", which is a non-endemic plant that was introduced into Hawaii approximately 1500 years ago by the Polynesians. The use of lime to generate a red dye from noni is described in early accounts for cultures throughout the Pacific [17] suggesting that this technology is very old.

The use of noni in all organic red samples analyzed is significant. Historic descriptions cite over 10 plant sources that were used as a source of red dye, including various hardwood extracts, fruits and flowers. However, noni root appears to be the most common red dye based on the collections sampled and other organic reds were not identified. In cases where the red design was not identified as organic, the red was found to be iron oxide pigment ('alaea) based on XRF, FTIR and UV-Vis-NIR FORS. Noni red dye was also found to be common throughout the post-contact period. Noni was found in samples #E3164 (Peabody Essex, exchange sample), collected in 1802 [18] and BPBM 2395, collected in 1887 [5].

The selective use of noni can in part be explained from based on its color intensity – noni is a deeper red than those derived from hardwoods (condensed tannins), which tend to be brownish. The anthraquinone colorants from noni are more light stable than the other reds and violets derived from flowers, fruits and leaves: these colorants are typically anthocyanins, which are prone to fading [19]. However, it is also possible that this has resulted in bias in the selected samples, as kapa previously decorated with other colorants may now be faded and consequently was not sampled for this study.

Binder

A protenaceous material was found as either a binder or varnish for painted pigments and also possibly as a colorant on kapa as shown in Figure 3. Proteinaceous binders were found in almost all samples where the pigment was a red or brown ochre, and were less frequently found in cases where the pigment was carbon black. The most probable source of protein binder is egg white, presumably from chicken, which were introduced into Hawaii by the Polynesians [20].

There are very few historic references to binders used in painted kapa. Malo (1903) cites the use of both hens and spider eggs in kapa production [3], and another manuscript refers to the use of beaten eggs brushed over the surface with a feather [21]. No oil binders were identified in the painted kapa samples analyzed to date. This was unexpected, as some early sources site kukui or kamani nut oil as binders for 'alaea [1].

It should be also noted that samples of "oiled" kapa were not analyzed in this study. Oiled kapa refers to kapa where the entire piece is coated with kukui or coconut oil to help protect it from water damage [1]. In an analysis of oils used in oiled kapa, Firnhaber and Erhart (1986) found that kukui nut oil was used [22].

Pa'i 'ula: Madder and synthetic alizarin

Natural madder and synthetic alizarin were only identified in a specific type of kapa known as pa'i 'ula ("beaten in red"). In this kapa, imported textiles are beaten into the top of uncolored kapa. "Turkey cloth" and "Turkey red" are cited as a favored material used in kapa making [3, 23, 22]. Turkey red cloth is a specific type of cotton cloth dyed with madder and mordanted with alum and oil; this process was used by European dyers as early as the late 18th century [24]. In 20 fiber samples of the textiles used in pa'i 'ula, all were found to be red dyed cotton; other fabrics like wool or silk were not identified based on examination with polarized light microscopy. Both were dyed with alizarin, confirming historic accounts that "Turkey red cloth" was the preferred material for pa'i 'ula.

Synthetic alizarin was identified in one sample of pa'i 'ula. Synthetic alizarin was not invented until 1868, and was in widespread use by the 1870's [25]. This is important because an outstanding question by historians is when in the 19th century the production of kapa was lost. The sample where synthetic alizarin was identified (BPBM #2361) was collected by J. S. Emerson in the 1887, showing that the kapa was made shortly before it was purchased. With further analysis of pa'i 'ula, it may be possible to gain greater insight into the production of kapa toward the end of the 19th century.

Imported blues

The most common blue pigment identified in blue kapa was synthetic ultramarine. This pigment ($Na_{8-10}Al_6Si_6O_{24}S_{2-4}$) was first synthesized in 1822, and by the mid-19th century it was mass produced and dominated the market as an inexpensive blue pigment [27]. Historic accounts refer to the use of laundry bluing agents to decorate kapa. Various blue pigments and dyes were used in laundry blue including synthetic ultramarine, Prussian blue, indigo and aniline blue [28]. Advertisements in the early newspapers the *Pacific Commercial Advertiser* (1856-1921) and *The Polynesian* (1840-1841; 1844-1864) contain many references to laundry blues, referred to as "wash blue", "blue mottled soap" [29] as well as the trade name "Colman's Blue" [30], showing that these materials were available throughout the mid to late 19th century.

Selective use of Western materials

In the kapa samples analyzed, the only western materials found were Turkey red cloth, ultramarine and blue dyes (probably laundry blue), and less commonly vermillion. In the case of vermillion, it was found in three cases only, and was used as a stamped decoration in the same manner as traditional ‘alaea. The use of these limited materials is notable in that many other imported pigments and dyes would have been available. Advertisements in the early newspapers *The Pacific Commercial Advertiser* (1856-1921) and *The Polynesian* (1840-1841; 1844-1864) show that other pigments were available in Honolulu including verdigris, emerald green, lead white, chrome yellow, yellow ochre, litharge, red lead, vermillion, Prussian blue, as well as various prepared artists' paints and inks. Many of these materials were advertised as early as the first publication of *The Polynesian* in1840 [**31, 32**]. With many traditional products, it is often assumed that they were replaced with imported materials shortly after contact, particularly with the arrival of the missionaries [26]. However, with the exception of these aforementioned materials, other imported Western pigments were not typically used to decorate kapa based on the samples analyzed. This selective use of imported materials could be due to a number of factors – availability outside Honolulu, cost, or a preference for specific colorants (blue and red).

CONCLUSIONS

Noni red and yellow, ‘ōlena yellow, carbon black and □alaea (iron earth red, yellow and brown) were identified in historic 19th century kapa samples in the Bishop Museum collections. Results to date suggest that although many plant colorants were available, noni and ‘ōlena were the preferred dye sources. Protein, possibly from egg white, was identified in some samples as a binder in painted decoration using both inorganic and organic colorants. These results document the use of traditional materials throughout the post contact period. Turkey red cloth and synthetic ultramarine (possibly as laundry blue) were common western materials found in kapa. Other pigments and dyes, while available in Honolulu during this period, were not found.

This is a unique cross-disciplinary project that to date has involved materials science, conservation, ethno-botany and cultural practitioners. This work is part of an ongoing research project with goals both to gain a greater understanding of historic kapa making as well as provide information to help in the revival of this important tradition. These results are the first study of Hawaiian dyes and pigments on kapa. As research progresses, we anticipate additional dyes, resins and oils used in kapa-making will be determined.

ACKNOWLEDGMENTS

The authors would like to thank Dr. Blyth McCarthy, Andrew W. Mellon Senior Scientist, Freer Gallery of Art and A. M. Sackler Gallery of Art, Smithsonian Institution, for use of equipment; Bruce Kaiser and Andrea Tullos at Bruker Elemental for loan of the Bruker XRF;

Clyde Imada, Research Specialist, Department of Botany, Bernice Pauahi Bishop Museum for useful discussions.

REFERENCES

1. W.T. Brigham, *Ka Hana Kapa: The Making of Bark-Cloth in Hawaii* (Bishop Museum Press, 1911).
2. B.H. Krauss, *Plants in Hawaiian Culture* (University of Hawaii Press, 1993).
3. D. Malo, *Hawaiian Antiquities:(Moolelo Hawaii)* (Hawaiian Gazette Co., Ltd., 1903).
4. G. Smith and R. Te Kanawa, Chem. N. Z. (2008).
5. B.P.B. Museum and C.C. Summers, *Material Culture: The JS Emerson Collection of Hawaiian Artifacts* (Bishop Museum Press, 1999).
6. C. Bisulca and J. Winter, in *Later Chin. Paint. Mater.* (Archetype Publications, in preparation).
7. C. Bisulca, M. Picollo, M. Bacci, and D. Kunzelman, in 9th International Conference on Non-Destructive Testing, Art Jerus. 25–30 May 2008 Jerus. (2008).
8. J. Giacci and J. Winter, in *Sci. Res. Pict. Arts Asia Proc. Second Forbes Symp. Freer Gallery Art* (Archetype Publications, London, 2005).
9. H. Schweppe, J. Winter, and E.W. Fitzhugh, Madder and Alizarin in *Artist's Pigments: A Handbook of their History and Characteristics.* Wash. Oxf. (1997).
10. E. West FitzHugh, M. Leona, and N. Shibayama, Stud. Conserv. **56**, 115 (2011).
11. G.K. Jayaprakasha, L. Jagan Mohan Rao, and K.K. Sakariah, J. Agric. Food Chem. **50**, 3668 (2002).
12. J.F.C. Rock, *The Indigenous Trees of the Hawaiian Islands* (J.F. Rock, 1913).
13. J. Madejova, Vib. Spectrosc. **31**, 1 (2003).
14. M.R. Derrick, D. Stulik, and J.M. Landry, *Infrared Spectroscopy in Conservation Science* (Getty Publications, 1999).
15. P. Aobchey, S. Sriyam, W. Praharnripoorab, S. Lhieochaiphant, and S. Phutrakul, Production of red pigment from the root of Morinda angustifolia Roxb. var. scabridula Craib. by root cell culture. Chiang Mai Univ. Journal. **1**, 66 (2002).
16. R.A. Jacobson and R. Adams, J. Am. Chem. Soc. **47**, 283 (1925).
17. A.W. Larsen, J. Anthropol. Archaeol. **30**, 116 (2011).
18. P.M. of Salem, *The Hawaiian Portion of the Polynesian Collections in the Peabody Museum of Salem: Special Exhibition, August-November, 1920* (Peabody Museum, 1920).
19. P. Markakis and L. Jurd, Crit. Rev. Food Sci. Nutr. **4**, 437 (1974).
20. P.M. Vitousek, L.L. Loope, and C.P. Stone, Trends Ecol. Evol. **2**, 224 (1987).
21.Notes on kapa. Transcripts and translations in HEN.I.76-93, 144-147, and 372. Bishop Museum Archives.
22. N. Firnhaber and D. Erhardt, in *Recent Advances in the Conservation and Analysis of Artifacts. Jubilee Conservation Conference, London 6-10 July 1987.* (National Museums of Canada, 1986), pp. 178–185.
23. L. Arthur, Cultural Authentication of Hawaiian Quilting in the Early 19th Century. Cloth. Text. Res. J. **29**, 103 (2011).
24. R.A. Peel, J. Soc. Dye. Colour. **68**, 496 (1952).
25. A.S. Travis, Hist. Technol. Int. J. **12**, 1 (1994).

26. J.M. Bayman, Ideology, political economy, and technological change in the Hawaiian Islands after AD 1778. Bull. Indo-Pac. Prehistory Assoc. **27**, 3 (2007).
27. J. Plesters, in *Artists Pigments Handb. Their Hist. Charact. Vol. 2* (National Gallery of Art, Washington, DC, United States, 1993), pp. 37–65.
28. N.N. Odegaard and M.F. Crawford, in *ICOM Comm. Conserv. 11th Trienn. Meet. Edinb. Scotl. 1-6 Sept. 1996 Prepr.* (1996), pp. 634–638.
29. H. Hackfeld & Co., Pac. Commer. Advert. 4 (1876). Bishop Museum Archives.
30. Savidge & May, Pac. Commer. Advert. 3 (1856). Bishop Museum Archives.
31. Pierce and Brewer, The Polynesian 180 (1840). Library of Congress.
32. Thomas Cummins, The Polynesian 4 (1840). Library of Congress.

Mater. Res. Soc. Symp. Proc. Vol. 1656 © 2014 Materials Research Society
DOI: 10.1557/opl.2014.710

TECHNOLOGY OF EGYPTIAN CORE GLASS VESSELS

Blythe McCarthy[1], Pamela Vandiver[2], Alexander Nagel[1], Laure Dussubieux[3]

[1]Freer|Sackler, Smithsonian Institution, Washington D.C.
[2]Materials Science and Engineering, University of Arizona, Tucson, AZ
[3]Anthropology, Field Museum of Natural History, Chicago, IL

ABSTRACT

Our knowledge of glass production in ancient Egypt has been well augmented not only by the publication of recently excavated materials and glass workshops, but also by more recent materials analysis, and experiments of modern glass-makers attempting to reconstruct the production process of thin-walled core-formed glass vessels. The small but well preserved glass collection of the Freer Gallery of Art in Washington, D.C. was used to examine and study the technology and production of ancient Egyptian core-formed glass vessels. Previous study suggests that most of these vessels were produced in the 18th Dynasty in the 15th and 14th centuries BCE, while others date from the Hellenistic period and later. In an ongoing project we conducted computed radiography, x-ray fluorescence analysis and scanning electron microscopy on a selected group of vessels to understand further aspects of the ancient production process. This paper will provide an overview of our recent research.

INTRODUCTION

Ancient Egyptian core-formed glass vessels have a layer of glass that was formed around a porous or porely sintered sandy clay core. Several methods have been proposed for forming these vessels including the application of threads of glass to a rotating heated core then marvering and heating, dipping the core into molten glass, and rolling a preheated core in powdered glass followed by marvering and heating steps.

Recent experiments of modern glass-makers have attempted to reconstruct the production process of thin-walled core-formed glass vessels [1,2,3]. Our knowledge of the production process has also been furthered by excavations of glass-workshops of the 2nd millennium BCE at Amarna, Lisht, Malkata and Gurob [4,5,6] and by the large number of excavated samples of core-formed vessels discovered since the nineteenth century [7]. Due to information from these studies, the use of powdered glass applied to a core is currently favored as the most likely production method [8,1].

While our knowledge about the process, beginning with the mounting of a prefabricated core and continuing through to the final glass product, has much improved through these excavations and experiments, further results may be achieved by analysis of the core-formed glass itself. In cases where subsequent working and annealing steps have not obliterated the evidence, the different forming methods should result in different patterns of inclusions, porosity and compositional variation. If the vessel is formed by wrapping threads around a core, porosity should be elongated and/or aligned in the direction of winding. Compositional variation would be expected to occur from thread to thread and also aligned with the threads. In the case where the core is dipped into molten glass, pores should be few in number and varied in size with no particular alignment. Compositional differences should cover broad areas and show evidence of flow following the path the molten glass would have followed due to gravity, similar to what is seen in blown glass. If the vessel were formed by rolling in a powdered glass, there should be large numbers of pores evenly distributed and composed of one or more size clusters.

In the summer of 1909, while on a trip in Egypt, the Detroit businessman Charles Lang Freer (1854-1919) acquired a collection of 1,388 ancient Egyptian glass objects from the dealer Giovanni Dattari (1858-1923) in Cairo, among them 21 core-formed vessels (F1909.412-436) [9], (Figure 1).

Figure 1. Ancient Egyptian glass vessels in the Freer|Sackler, Smithsonian. (F1909. 430, 413, 428, 423), dated to the 15th to 14th centuries, BCE.

Dattari was well connected to excavations in Egypt including those of Flinders Petrie (1853-1942) at the extensive archaeological site of Amarna on the East bank of the Nile River. Freer's Egyptian glass was gifted to the Smithsonian Institution officially in 1919. Among the outstanding pieces is a collection of glass vessels datable to the 15th to 14th centuries BCE [10]. Of these, many are made of bright light blue glass, sometimes translucent (e.g., F1909.430), with the decoration in lighter blue, yellow and white. Previous studies of these vessels by scholars who had access to the collections suggested that vessels and fragments like those in the Freer collection were excavated from royal glass workshops during the reigns of the pharaohs Amenhotep III (1391-1353 BCE) and Amenhotep IV, who changed his name to Akhenaten (1353-1335 BCE). While Egypt's climate favored the excellent preservation of ancient glass, important datable materials have also been excavated in 5th and 4th centuries BCE tombs in Greece and the Black Sea areas in more recent years [11,12], adding important new information for the chronology of vessels.

The shapes of the vessels in the Freer collection range from two-handled miniature krateriskoi (F1909. 430; F1909.421), four-handled vessels (F1909.413), and perfume flasks (alabastroi) to other miniature pitchers, fusiform jars with a flat top and base in cylinder forms

(F1909.420), bottles shaped as a pomegranate (F1909. 423), to vessels in lentoid forms (F1909.416; F1909.422; F1909.428) and kohl-tubes in the form of lotus columns (F1909.427; F1909.432; F1909.436). In comparison to excavated materials in other parts of the Mediterranean, some vessels from the Freer collection have been identified as of Eastern Mediterranean origin. In a previous paper on the Freer Gallery glass collections we discussed computed radiography (CR) in conjunction with analysis of colorants using qualitative x-ray fluorescence analysis (XRF) to understand further aspects of the ancient production process[13]. This paper will provide an overview of our recent research on selected glass vessels to document evidence of manufacturing technique using x-ray computed radiography, scanning electron microscopy with energy dispersive x-ray analysis (SEM/EDS) and scanning x-ray fluorescence analysis. Replicate glass vessels formed either by dipping in molten glass or the use of powdered glass were studied for comparison (Figure 2) with the glass vessels in the Freer Gallery of Art collection.

METHODS

The glass vessels were radiographed using an Isovolt Titan X-ray tube with a 1.9mm focal spot. Parameters used were 50-70 Kv, 1-2 mA and exposure times in the range of 30-40 seconds. Once scanned, the images were manipulated using the "Find Edges" filter in Adobe Photoshop CS5.

Surfaces of both ancient glass and replicate glass vessels were scanned with a Bruker Artax x-ray fluorescence (XRF) spectrometer with a rhodium tube and a polycapillary lens (0.60). The sample was flushed with helium and conditions of 15 KV, 250 μA and 40 seconds live time were used.

Carbon coated samples from selected vessels and replicate vessels were examined with a FEI XL30 scanning electron microscope using both backscattered and secondary electron imaging modes as well as x-ray mapping with an EDAX Phoenix EDS system. Spectra were collected at 15KV, 25.6 amp time for 100 seconds. Maps were acquired at 15KV, 25.6 amp time for 1000 to 1500 live seconds.

RESULTS

Macrostructure

The CR scans of a turquoise glass lentoid shaped vessel and replicate turquoise vessel made from powdered glass show similar distribution of porosity as well as variation in vessel wall thickness (Figures 3 and 4) Larger pores are visible in these images but there are also smaller pores visible at higher magnification. No striae or other differences in composition within the base glass were seen in the radiographs

CR scans of a replicate glass vessel formed by dipping the core in molten glass(Figure 5), lacks this porosity, has more even walls and a thicker area of glass at the lower edge as well as at the shoulder. A glass vessel fragment in the Freer collection, F1909.887, shows similar thickened glass at its lower edge and few pores in the computed radiograph.

Figure 2. Replications of Egyptian core vessels. Upper left vessel was formed by dipping in molten glass. The remainder were formed from powdered glass. Five produced by Pamela Vandiver; vessel at upper right produced by Dudley Giberson.

Variations in composition

Linescans across the base of the turquoise core-formed replicate vessel produced using powdered glass show irregular variations in composition on the order of a few tenths of a millimeter (Figure 6). Some of the variations are due to the presence of the opacifier, calcium antimonate, but observation of the antimony profile (Figure 7) shows that compositional variations are not solely due to calcium antimonate. (Similarly, recent analysis with laser ablation ICP-MS by the authors has found variations in alkali content with measurement depth for a turquoise glass, F1909.427).

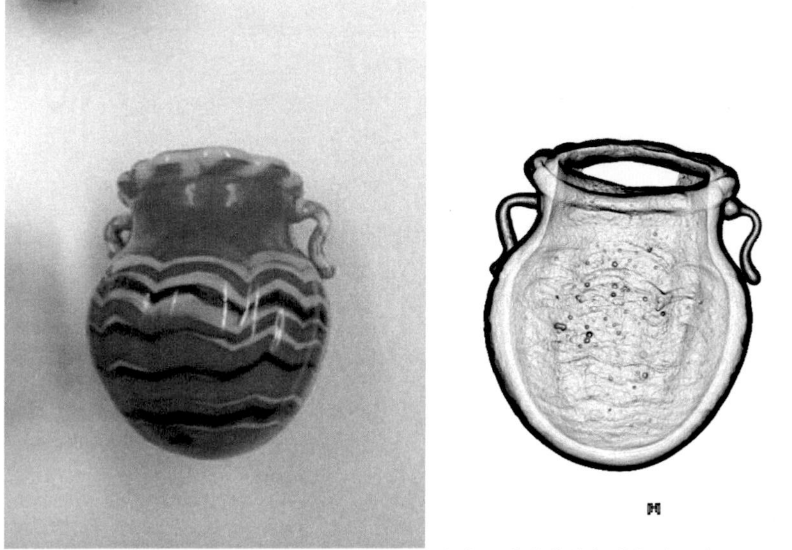

Figure 3. Replicate vessel formed using powdered glass, left (height 6.5cm) and computed radiograph processed with Adobe Photoshop "Find edges" filter, right.

Figure 4. Egyptian core-formed vessel, left (height is 8.4cm) and computed radiograph processed with Adobe Photoshop "Find edges" filter, right.

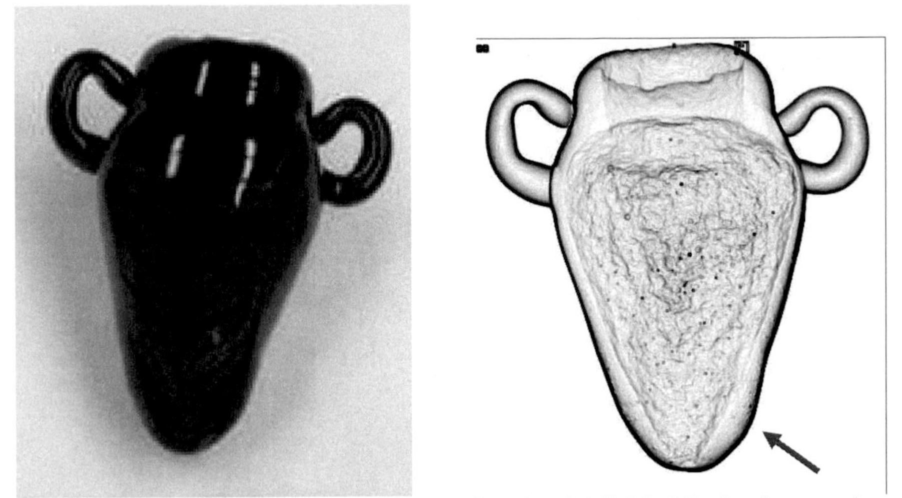

Figure 5. Replicate vessel formed by dipping in molten glass, left (height 9.7cm) and computed radiograph processed with Adobe Photoshop "Find edges" filter, right. Arrow indicates where molten glass has pooled.

Microstructure and microchemistry

Backscattered electron images of the turquoise replica show partially melted glass grains in a glass matrix. The centers of the grains in the image contain partially filled pores (Figure 8). EDS found the matrix to be higher in Cl, and P, and to have a higher Na:Si ratio than the grain glass. Material in the pores is higher in Al and P than that in the glass grains. A similar image of the turquoise glass in F1909.427 found angular grains with partially filled pores at their center and along edges (Figure 9). They are joined by a very thin amount of matrix glass; too thin to acquire a separate EDS spectrum. Relative to the glass composition of the grains, the material filling the pores is higher in Al and Na. EDS x-ray maps of this sample show increased sodium, calcium, chlorine and sulfur as well as decreased oxygen at particle interfaces.

A sample from the cobalt blue glass replica formed from powdered glass that is depicted at the far left in Figure 2 did not show evidence of glass grains. However, some of the inclusions (they appear dark in the gray glass matrix) follow a scalloped line at an angle to the vessel surfaces,, evidence of viscous flow that likely destroyed any residual evidence of the glass powder grains.

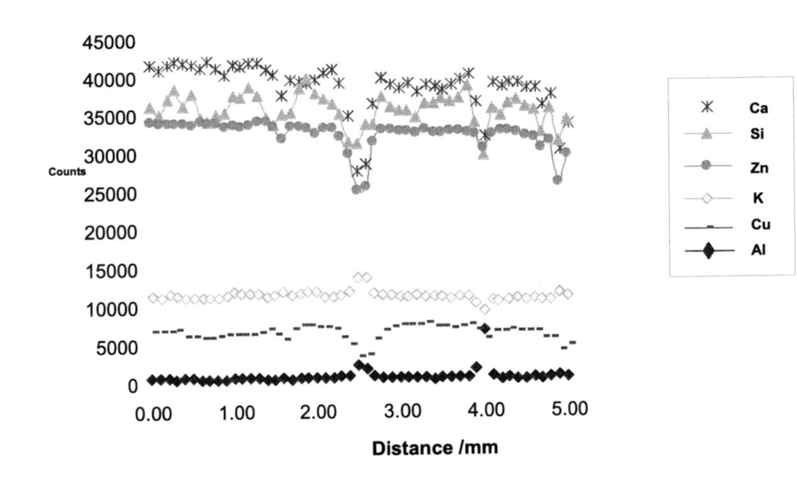

Figure 6. XRF line scan of turquoise glass replica made from glass powder.

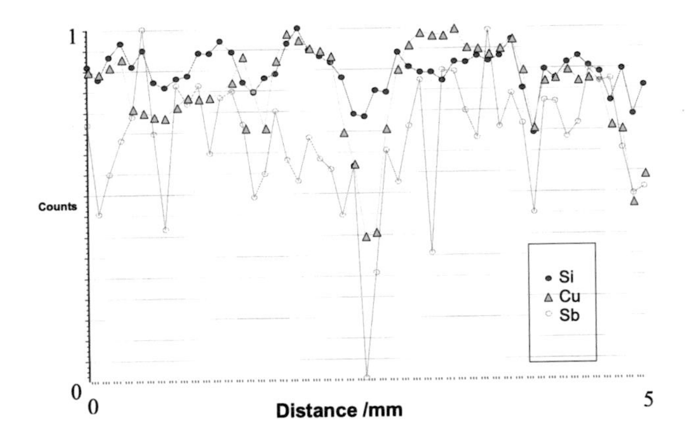

Figure 7. XRF linescan of antimony, normalized and plotted with silicon and copper acquired from same replica as that in Figure 6.

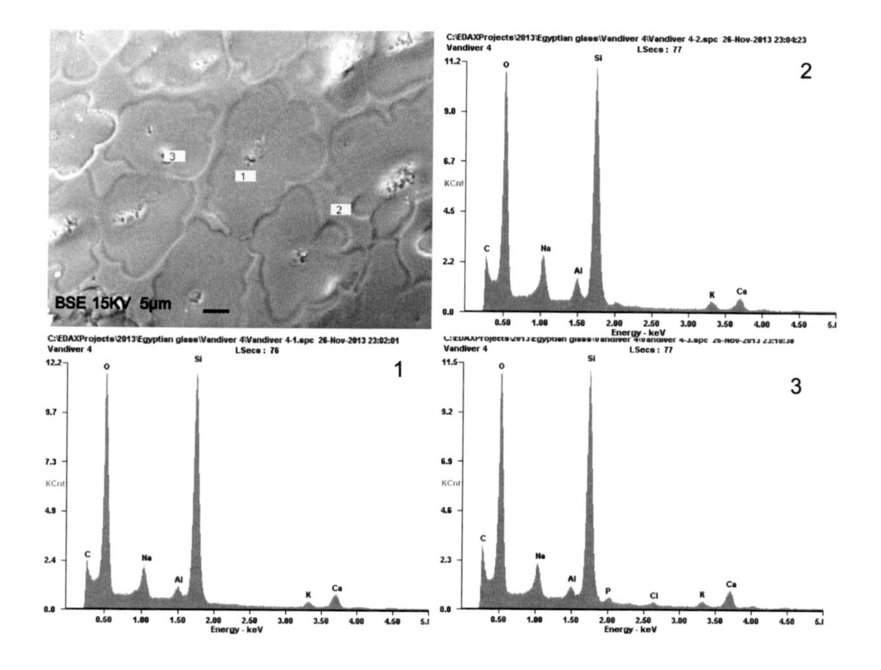

Figure 8. Backscattered electron image of surface of replicate turquoise core-formed vessel and EDS spectra corresponding to marked points in the image. The image shows partially dissolved glass particles, EDS 1; in a glass matrix, EDS 2; with pores and particles at their center, EDS 3. Spectra acquired at 15KV, amp time 25.6 for 100 seconds.

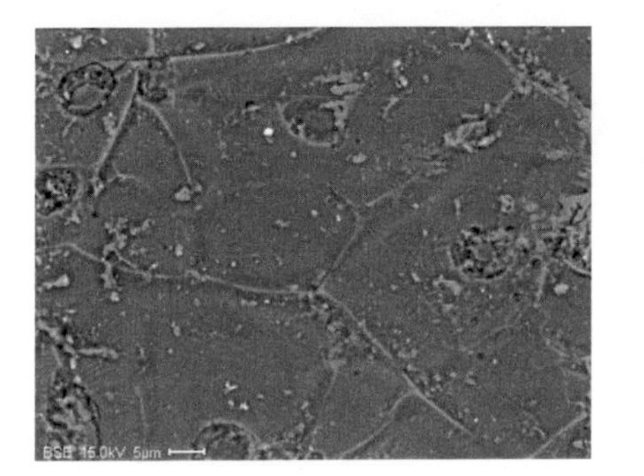

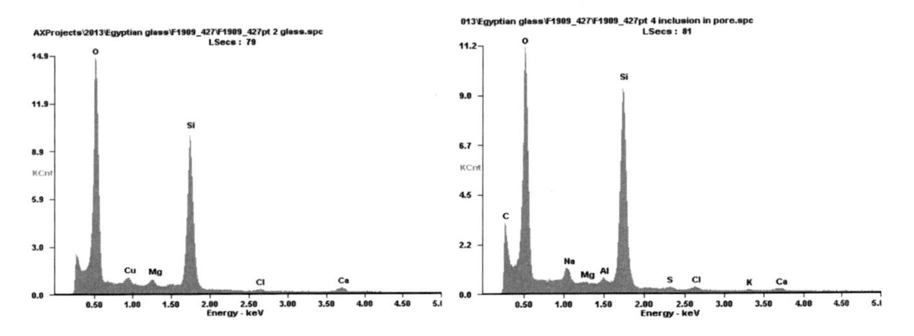

Figure 9. Backscattered electron image and EDS spectra of turquoise Egyptian glass from lotus opening of a kohl vial in the Freer Gallery, F1909.427. The image shows a microstructure of angular glass grains in a thin-walled matrix with partially filled pores both in the grains and at grain edges. The spectrum on the left is from the glass of a grain, that on the right from the interior of a pore. Spectra acquired at 15KV, amp time 25.6 for 100 seconds.

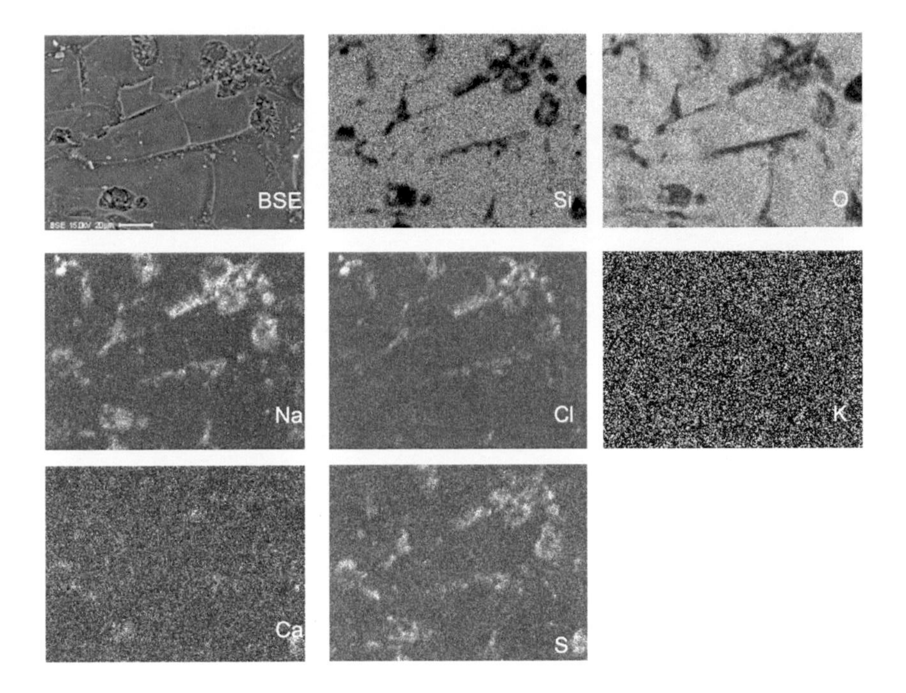

Figure 10. Backscattered electron image and EDS x-ray mapping of turquoise Egyptian glass, F1909.427. Maps acquired at 15KV, amp time 25.6 and 1310.7 seconds live time.

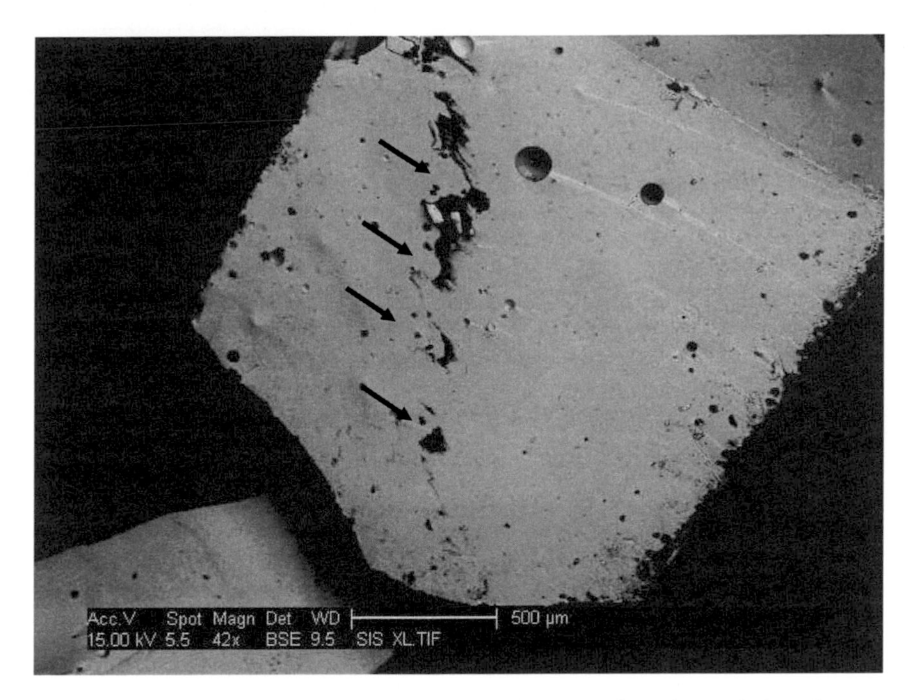

Figure 11. The backscattered electron image shows a fracture surface of a sample from the cobalt blue replica vessel depicted at the far left in Figure 2. Interior and exterior vessel walls are the two parallel faces of the fragment, and a wavy line of porosity (vertical in the image) proceeds diagonally between them. The porosity, that is neither aligned or perpendicular to vessel walls may be evidence of the interface between two powder particles. It provides evidence that viscous flow occurred as there are dips in the line of porosity (for example at arrows).

DISCUSSION

The CR images support the use of a powdered glass precursor. Both turquoise colored Egyptian glass and the turquoise colored replicates formed from powdered glass exhibit variation in wall thickness, an uneven interior and porosity of varying size and random location in the CR images. These features are not seen in vessels produced from molten glass. While the porosity is more easily seen in the processed images, it is necessary to exercise caution as the Photoshop filter used does not discriminate between pores and inclusions, but outlines both.

The XRF linescans show that compositional variations are present that are not solely due to colorants or inclusions in both turquoise (Figure 6) and black colored glass. The line scans of the colorant (Cu) and opacifier (Ca, Sb) ions increase and decrease in intensity at times similarly to, and at times differently from linescans of silicon (Figure 7) and the alkalies, Na and K. The variations that do not coincide with those of Cu, Ca and Sb point to differences in glass composition such as would occur from a powdered precursor. Backscattered electron images for turquoise colored Egyptian glass (F1909.427) and a turquoise colored replica both showed a structure of glass grains in a matrix with partially filled pores. The replica has much more rounded grains and more matrix material than the Freer Egyptian glass sample, perhaps due to a higher temperature furnace. Both have increased alumina in the pores, and both have increased sodium (visible for F1909.427 in the x-ray mapping in Figure 10) at the edges of the grains. Wet grinding or aging of the glass powder would produce extra sodium at the surface, alternatively the addition of a small amount of plant ash or other sodium source, to the glass powder would also form a high sodium containing glass at grain interfaces. The effect of such an addition would have been known to the Egyptians who used the method in efflorescence glazing during Egyptian faience production.

LA-ICPMS analysis underway at the Field Museum determined that F1909.427 was formed using a sodium plant ash alkali source so chlorine would be expected in small amounts in the glass (Bulk analysis for this glass is approximately 16.42% Na_2O, 4.74% MgO, 2.52% K_2O, 0.69% Al_2O_3, 0.4% Fe_2O_3). The presence of additional chlorine at grain interfaces for this glass is consistent with an addition of plant ash to the powdered glass [14]; however, given that this object had extensive fill material and overpaint, further study is necessary for confirmation.

A grain structure was not visible in the one cobalt blue replica examined to date (Figure 11) however the porosity in the sample provides evidence of glass movement, and the structure may have been lost due to one or more of the following: faster heating rate for a dark colored glass, subsequent marvering and heating steps, compositional differences, or the lack of opacifier particles that would have resulted in increased viscosity.

CONCLUSIONS

Microstructural evidence of powder technology was observed on a macroscopic scale using x-ray computed radiography for the Freer Egyptian glass. The radiography gave an overall view of the wall cross-section, the size, location and number of bubbles and inclusions in the glass. No striae or other differences in composition within the base glass were seen in the radiographs, and indeed, the decorative threads that did not contain lead (white and blue) were sometimes difficult to see. Therefore one cannot be certain that the radiographs reveal all of the heterogeneity in the glass that resulted from manufacturing. Because of this the radiographs of ancient glass alone do not provide definitive proof of a technique that uses powdered glass. However, the random location and great variation in size of the circular bubbles in the core-formed vessels is consistent with a process involving application of a powdered glass to a core with subsequent heating step(s) to melt and merge the glass into a continuous piece. Radiography of replica core-formed vessels made using powdered glass also showed this structure. Equally important, the size variation and great number of bubbles seen where a powder precursor was used were not seen in either blown vessels, or vessels formed from dipping the core in molten glass. No vessels made from winding glass threads were available for radiography.

Chemical variations that support but do not confirm a powder technology were seen in XRF linescans and in ICP-MS measurements taken at increasing depth.

SEM imaging confirmed the use of a powder technology for turquoise colored Egyptian glass. The microstructure was similar to that of a replica glass with a glass grains surrounded by a matrix and pores in grains and/or along grain interfaces. Pores contained material higher in aluminum, for both ancient glass and replica. In addition the pores in the ancient glass were higher in sodium and chlorine. The matrix was higher in sodium for both ancient glass and replica glass relative to the glass grains. In addition the ancient glass matrix contained higher chlorine and sulfur but less oxygen.

No evidence of a multigrain microstructure was visible in the cobalt blue glass replica in the SEM. Sintering or subsequent working of the glass can mask any evidence of a powder precursor. Evidence remains on a macroscale in the form of porosity, uneven wall thickness and rough or scraped vessel interior, and this was visible using x-ray radiography.

One fragmentary glass object in the Freer Gallery collection was identified that appears to have been formed by application of molten or low viscosity glass to a core.

ACKNOWLEDGEMENTS

We are grateful to Massumeh Farhad and Julian Raby, of the Freer|Sackler for granting permission to sample and to Dudley Giberson for a vessel replica.
The Leon Levy Foundation provided funding for this project.

REFERENCES

1. Nicholson, P. and Henderson, J. "Glass," in P. Nicholson and I. Shaw, eds. Ancient Egyptian Materials and Technology. Cambridge University Press: 195-224 (2000).
2. Giberson, Jr., D. "Core Vessel Technology: A new model," in: P. Vandiver, M. Goodway, J. Mass, eds. *Materials Issues in Art and Archaeology VI. Symposium held November 26-30, 2001 Boston, MA.* Materials Research Society: Warrendale: 571-8 (2002).
3. Smirniou, M. and Rehren, Th. "Direct Evidence of Primary Glass Production in Late Bronze Age Amarna, Egypt," *Archaeometry* 53.1: 58-80 (2011).
4. Shortland, A.J. "Who were the Glassmakers? Status, theory and method in mid-second millennium glass production," *Oxford Journal of Archaeology* 26.3: 261-74 (2007).
5. Hodgkinson, A. K. "High Stature Industries in the Capital and Royal Cities of the New Kingdom. In A. Hudecz, M. Petrik, eds. *Commerce and Economy in Ancient Egypt. Proceedings of the Third International Congress for Young Egyptologists 25-27 September 2009.* Budapest: 71-79 (2010).
6. Hodgkinson, A. K. "Mass-Production in New Kingdom Egypt: The Industries of Amarna and Piramesse," in D. Boatright *et al.* (eds.), *Current Research in Egyptology X: Proceedings of the Tenth Annual Symposium.* Oxford (2011).
7. McClellan, M. Core-formed Glass from Dated Contexts. PhD Dissertation, University of Pennsylvania, Philadelphia (University Microfilms International, Ann Arbor) (1984).

8. Stern, E. and B. Schlick-Nolte. *Early Glass of the Ancient World 1600 BCE-AD 50.* Ostfildern (1994).
9. Gunter, A. *A Collector's Journey. Charles Lang Freer and Egypt.* Scala Publications (2002).
10. Ettinghausen, R. *Ancient Glass in the Freer Gallery of Art.* Exhibition Catalogue (1962).
11. Trakosopoulou, E. "Glass Grave Goods from Acanthus," in G. Kordas, ed. *Hyalos – Vitrum - Glass. History, Technology and Conservation of Glass and Vitreous Materials in the Hellenic World.* Athens: 79-90 (2002).
12. Shortland, A.J. and H. Schroeder "Analysis of First Millennium BC Glass Vessels and Beads from the Pichvnari Necropolis, Georgia," *Archaeometry* 51: 947-965 (2008).
13. Nagel, Alexander; Blythe McCarthy, and Stacey Bowe. "Crafting Glass Vessels: Current Research on the Ancient Egyptian Glass Collections in the Freer Gallery of Art. Proceedings from a workshop by the New Archaeological Research Network for Integrating Approaches to Ancient Material Studies (2012): 659-671.
14. Rehren, T. and Pusch, E. B. "Glas für den Pharao – Glasherstellung in der Spätbronzezeit des Nahen Ostens.," in G. Wagner ed., *Fortschritte der Archäometrie*, 215-35. Heidelberg (2007).

Technical Art History: Pigment Identification, Reactivity and Transformation

Mater. Res. Soc. Symp. Proc. Vol. 1656 © 2014 Materials Research Society
DOI: 10.1557/opl.2014.823

Characterization of Bistre Pigment Samples by FTIR, SERS, Py-GC/MS and XRF

María L. Roldan[1], Silvia A. Centeno[1], Adriana Rizzo[1], Yana van Dyke[2]

[1]Department of Scientific Research. The Metropolitan Museum of Art, 1000 Fifth Avenue, New York, NY 10028, USA.
[2]Sherman Fairchild Center for Works of Art on Paper and Photograph Conservation. The Metropolitan Museum of Art, 1000 Fifth Avenue, New York, NY 10028, USA.

ABSTRACT

A combination of FTIR, normal Raman, SERS, Py-GC/MS and XRF was used to analyze commercial bistre samples to determine specific biomarkers that will allow for a rapid identification of the pigment in works of art. The results of the XRF analysis showed that potassium, calcium and iron are the main elements present. Characteristic bands belonging to phenolic components of lignin were observed in the FTIR spectra. The SERS analysis provided a fingerprint that may originate in the polymerization of the phenolic components catalyzed by the presence of the nanostructured silver surface under alkaline conditions. The Py-GC/MS analysis revealed the presence of lignin and cellulose biomarkers and a series of polycyclic aromatic hydrocarbons. The similarities observed between the commercial samples studied suggest that both originate in the same raw material, hardwood. The results demonstrate the potential of the multi-technique approach used for the characterization of this complex black-brown pigment.

INTRODUCTION

Bistre, also known as *bister*, is a black-brown pigment primarily used in watercolor and for washes in pen and ink drawings [1, 2]. Its use dates from the seventeenth century, although it became more common in the eighteenth century linked to the development of the watercolor technique. During the nineteenth century bistre was replaced by sepia [2]. Bistre was usually prepared from wood soot collected close to the flame. After collection, the soot was extracted with boiling water and the extract evaporated [1]. The final color of the pigment depends on the wood species and the preparation method, and it has been stated that beech (*Fagus* species) was preferred above other types of wood because of the brown hues that could be achieved [2, 3].

From a botanical point of view, beech is a hardwood. The lignin in hardwoods (*angiosperms)* consists of syringyl (S) and guaiacyl (G) units, while in softwoods (*gymnosperms*) lignins are composed mainly of guaiacyl units [4]. The S/G ratio of hardwoods is a distinctive feature for each wood species and greatly varies within species of the same genus [5]. In addition to lignin, cellulose and hemicellulose are the main building blocks of wood. These biopolymers have been extensively studied in wood science. Of particular interest for this work is the study of the products formed after wood is thermally treated. It is well known that when wood is burnt, smoke is released to the atmosphere accompanied by the deposition of tar and soot. Several chemical reactions, such as hydrolysis, oxidation, dehydration, and pyrolysis can take place when the temperature is raised and the biopolymers degrade [6]. The structural changes that occur during the pyrolysis of wood have been studied by different analytical techniques, especially Fourier transform infrared (FTIR) spectroscopy, Raman, and GC/MS [6-8].

The identification of bistre in works of art by non-invasive techniques such as Raman and X-ray fluorescence (XRF) is a highly desirable but problematic task due to intrinsic limitations of the techniques and the complexity of the material. Surface enhanced Raman scattering (SERS) offers the advantages of a high selectivity, small sample size and little sample preparation, while pyrolysis-gas chromatography-mass spectrometry (Py-GC/MS) provides a precise molecular composition and assists in the interpretation of the SERS spectra. In the present study, SERS was used in combination with XRF, normal Raman, FTIR, and Py-GC/MS to fully characterize two commercial samples of bistre and to determine specific biomarkers that will allow for rapid identification of the pigment in works of art. The information obtained is relevant to the understanding of the pigment-binder interactions and to the preservation of works containing bistre.

EXPERIMENTAL

Commercial bistre pigment samples were purchased as black-brown powders from Kremer Pigments Inc. (US) and from Zecchi (Italy). The rest of the chemicals, all analytical grade, were purchased from Sigma-Aldrich. Stock solutions of bistre were prepared by dissolving 1mg of the pigment in 1ml of a 0.5M NaOH solution because the material is sparingly soluble in water at neutral pH.

X-ray fluorescence (XRF) measurements were performed on the samples for equal live-times of 200 seconds at 40 kV and 500 μA, using a Bruker ARTAX 400 instrument equipped with a Rh tube, a 650 μm collimator, and a Si drift X-ray detector with a 10 mm^2 active area.

Fourier transform infrared (FTIR) spectroscopy analyses were carried out with a Vertex70 spectrometer coupled to a Hyperion 1000 microscope equipped with a cryogenic mercury-cadmium telluride (MCT) detector (all by Bruker Optics). The measurements were done in the transmission mode with the samples crushed between the windows of a diamond anvil cell (Spectra Tech). Each spectrum is the result of 128 scans and has a 4 cm^{-1}spectral resolution.

Normal Raman and surface enhanced Raman scattering (SERS) spectra were recorded with a Renishaw Raman System RM1000 equipped with edge filters and an electrically cooled CCD detector, and using a 514 nm excitation. For the SERS analyses, the laser power was set to 0.6 mW and a 20x objective was used; for the normal Raman measurements a 0.06 mW power and a 50x objective were used. The SERS spectra were acquired in a single scan with an integration time of 30 s and a spectral resolution of 2 cm^{-1}. The normal Raman spectra presented below are an average of 5 scans, each acquired with an integration time of 10 s and the same resolution as the SERS spectra. A hydroxylamine-reduced silver colloid prepared using a method reported in literature [9] was employed for the SERS measurements. The silver colloid suspensions were activated by adding 8 μL of 0.5 M KNO_3 to 200 μL of the colloid. Then, 2 μL of the stock solutions of the samples were added to the colloidal solutions and the pH was adjusted to 12 with 0.1M NaOH. Finally, 1 μL of the sample/Ag colloid system was placed in glass capillaries for the analyses.

Pyrolysis-gas chromatography-mass spectrometry (Py-GC/MS) analyses were performed with and without methylation with (tetramethyl) ammonium hydroxide (TMAH). Samples of approximately 100 μg were weighed out in an Ultralloy cup (Frontier Lab) with a UMX 2 (Mettler Toledo) ultramicrobalance. The samples were pyrolyzed at 550°C in the vertical micro-furnace of a double-shot 2020iD pyrolyzer (Frontier Lab). 550°C was considered an appropriate pyrolysis temperature, following the results of evolved gas analysis for both samples. The principle of evolved gas analysis is described elsewhere [7]. The micro-furnace is interfaced to the injector of an Agilent 6890 gas chromatograph, coupled with an Agilent 5973 Network Mass Selective Detector (Agilent technologies). For methylation, the samples were treated with 3 μL of TMAH solution 25% in methanol, prior to pyrolysis. The analysis was carried out in split mode 30/1 and a J&W DB-5MS (30m x 0.25mm x 0.25μm) was used for the chromatographic separation. The injector and the transfer line were kept at 320°C. He was used as carrier gas, with a 1.5ml/min constant flow. The oven program was set at 40°C for 1 min, ramped to 320°C at 10°C /min and isothermal for 10 min. The data was evaluated using the deconvolution software AMDIS, and the NIST/ Wiley mass spectral database; published references were also used for identification of the mass spectra.

RESULTS AND DISCUSSION

XRF analysis showed that K, Ca, and Fe are the main elements present in both samples, along with minor amounts of S, Mn, and Zn (Figure 1), in agreement with previous observations [10].

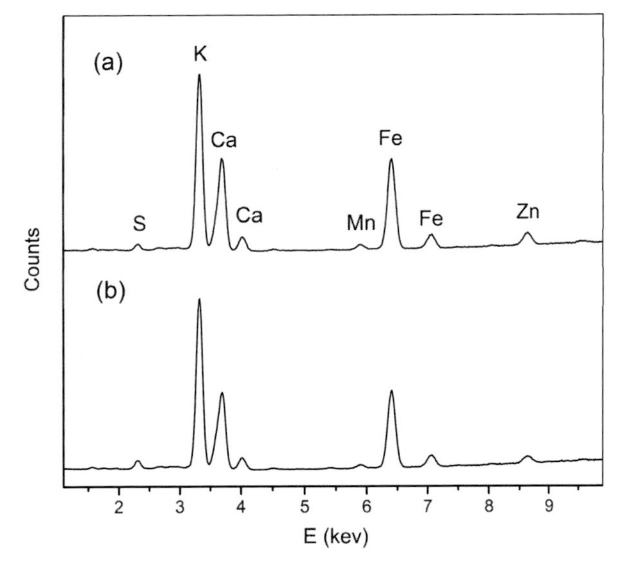

Figure 1. XRF spectra of bistre samples: Kremer (a), and Zecchi (b).

Figure 2 shows FTIR spectra acquired in the two samples studied in this work. The assignments of the features in these spectra were carried out based on two comprehensive IR studies of lignin from different wood species [11, 12]. The FTIR spectra obtained are dominated by bands originating in syringol and guaiacol components of lignin from *angiosperm* woods. The bands observed between 3000 cm^{-1} and 2800 cm^{-1} can be assigned to asymmetric and symmetric modes of methyl and methylene groups. The shoulder at 1690 cm^{-1} reveals the presence of conjugated carbonyl groups of aryl aldehydes and/or aryl ketones such as syringaldehyde or acetosyringone. The most intense band at 1595 cm^{-1} and the band at ca. 1517 cm^{-1} can be related to υ(C=C) in both syringol and guaiacol aromatic rings. A higher relative intensity of the former band respect to the latter suggests a higher content of syringyl units [11]. Bands at ca. 1456, 1427

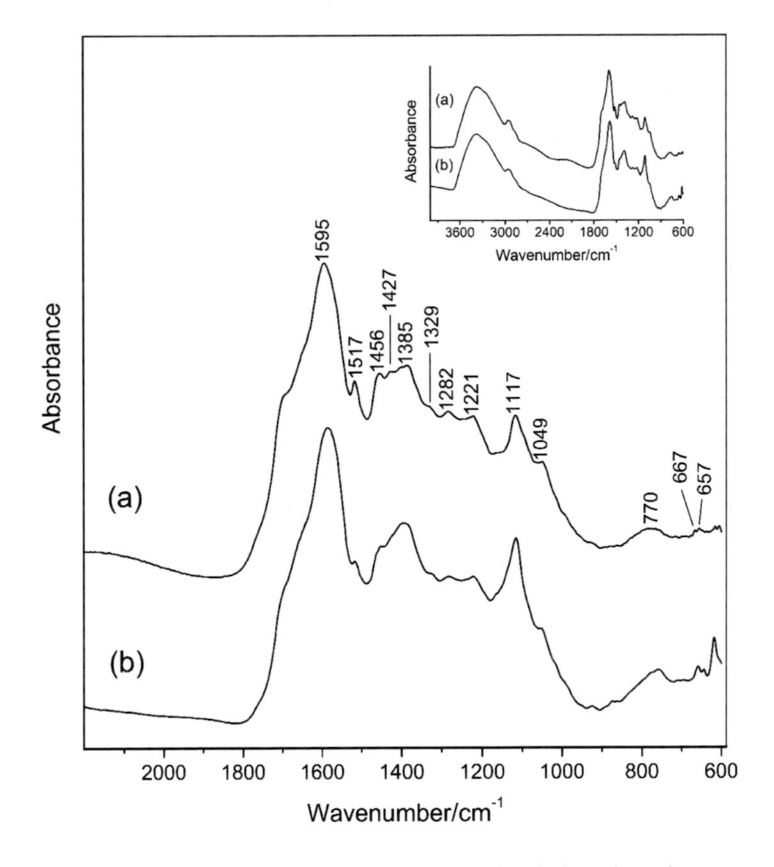

Figure 2. FTIR spectra of bistre samples: Kremer (a), and Zecchi (b). The inset shows the same spectra in the 4000-600 cm^{-1} range.

and 1385cm^{-1} may be associated with asymmetric C-H bending modes of methyl and methylene groups, aromatic stretching combined to in-plane C-H aromatic deformation, and symmetric C-H deformation of methyl groups. Bands characteristic of υ(C-O) in the 1300-1200 cm^{-1} range are used as main markers for the identification of syringyl and guaiacyl units [13]. It has been previously reported that the bands at ca. 1329 and 1117 cm^{-1} indicate the presence of syringol, and can be assigned, respectively, to υ(C-O) coupled to an aromatic ring stretching, and to the υ(C-O) of phenolic rings in syringol. The bands at ca. 1282 cm^{-1}, due to υ(C-O) coupled to an aromatic guaiacol-type ring stretching, and at ca.1049 cm^{-1}, that can be attributed to υ(C-O-C), confirm the presence of guaiacol. Also a contribution of the pyranose ring stretching in levoglucosan to the 1049 cm^{-1} feature has to be considered. The peak at ca.770 cm^{-1} in the spectrum of the sample from Kremer and at 758 cm^{-1} in the spectrum of the sample from Zecchi can be assigned to a ring deformation characteristic of 1,2,3- trisubstituted aromatic rings associated with syringyl units. The out-of-plane deformation mode of the OH group gives rise to features at ca. 667 and 657 cm^{-1}, and at 660 and 644 cm^{-1}, for the Kremer and Zecchi samples respectively.

The normal Raman spectra obtained (Figure 3) are characterized by two broad bands typical of amorphous carbonaceous materials at ca. 1603 and 1375 cm^{-1}, which are common to other complex brown-black pigments such as asphalt and van Dyke Brown [14]. By contrast, the signals in the SERS spectra (Figure 4) are significantly enhanced, revealing a unique fingerprint for bistre. It has been previously reported that phenols can undergo oxidative polymerization when they are in contact with a metal surface [15] and that this process can be accelerated by the

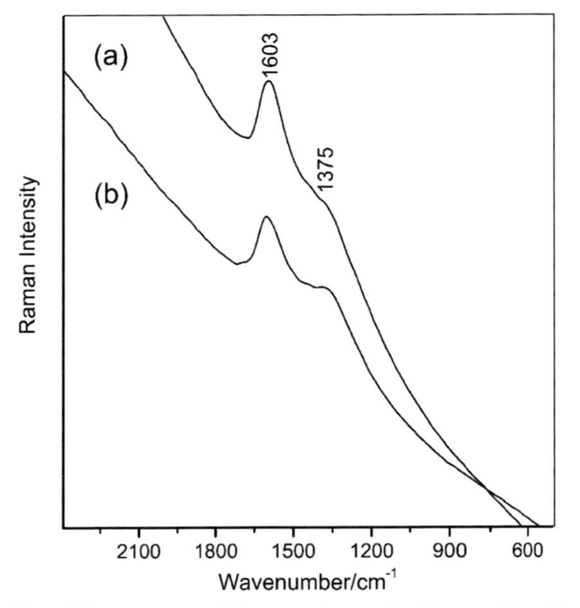

Figure 3. Normal Raman spectra of bistre powder samples: Kremer (a) and Zecchi (b).

presence of hydroxyl groups [16]. We expect that the polymerization products of the phenolic compounds produced during the combustion of wood and other products of the thermal degradation of the polymer solubilized at high pH contribute features to the SERS spectra.

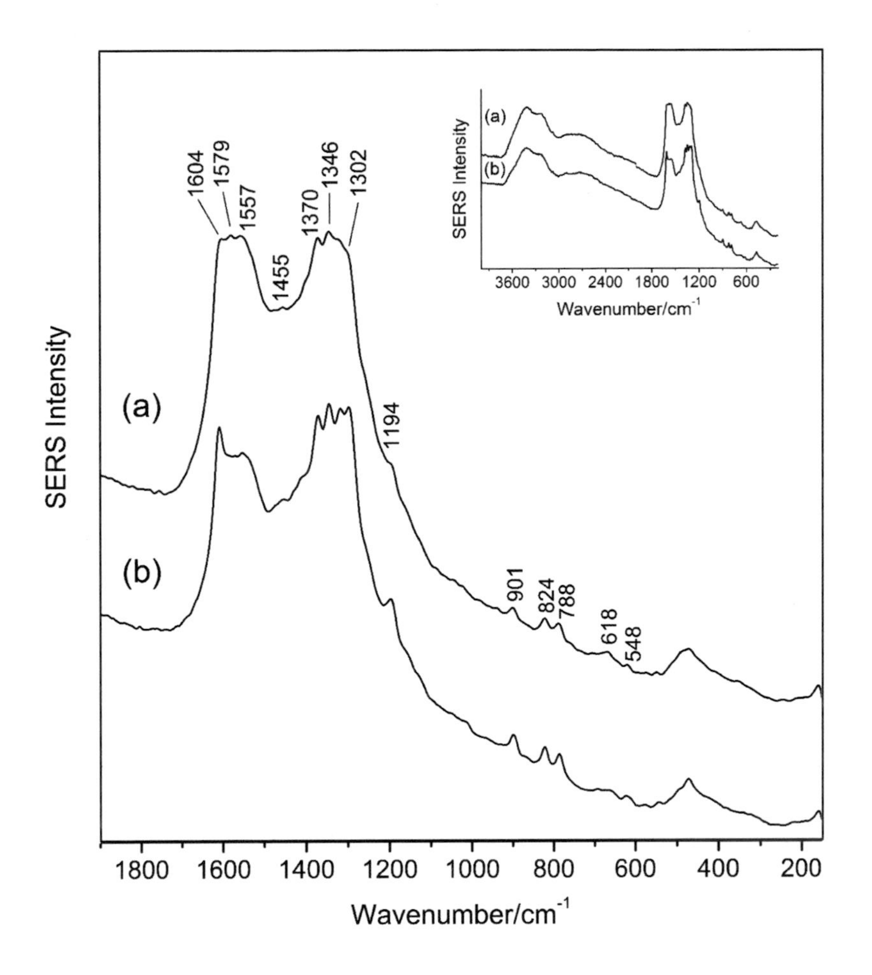

Figure 4. SERS spectra of bistre samples: Kremer (a) and Zecchi (b), acquired in a 1mg/ml of a 0.5M NaOH solution on a hydroxylamine-reduced silver colloid. λ_0=514 nm. The inset shows the same spectra in the 4000-200 cm^{-1} range.

In the range between 1600 and 1500 cm^{-1}, the three peaks observed at ca. 1604, 1579, and 1557 cm^{-1} can be due to aromatic ring stretching modes [15]. A band at ca. 1579 cm^{-1} has been assigned to highly substituted aromatic rings of polymerized phenols [15, 16]. Its presence in the SERS spectra obtained suggest that further polymerization reactions are taking place on the metal surface promoted by both the silver surface and the alkaline conditions. In these spectra, the aliphatic bending of methyl and methylene groups is observed at ca. 1455 cm^{-1}. The band at ca. 1370 cm^{-1} has been previously observed in the SERS spectra of several aromatic acids and has been attributed to COO^- groups [16]. This band could originate in the presence of syringic and *p*-hydroxybenzoic acids in the sample or as a consequence of the oxidation of aldehydic species to their corresponding acidic forms on the silver colloid surface at alkaline pH.

The bands at ca. 1346 and 1302 cm^{-1} can be attributed to υ(C-O) combined with ring stretching modes. The medium intensity band at ca. 1194 cm^{-1} can be related to υ(C-O) and C-H in-plane deformation modes. The three peaks at 901, 826, and 788 cm^{-1} can be assigned to aromatic ring deformations and C-H deformation modes. In the lower frequency region, bands can be observed at ca. 618 and 548 cm^{-1}. Similar bands have been reported for polymerized phenols and have been attributed to skeletal modes of the polymer formed on the surface [15]. As for the FTIR spectra, the SERS spectra of both pigments were found to be similar. However, peaks at ca. 1604 and 1302 cm^{-1} were more pronounced for the bistre sample from Zecchi possibly indicating a higher aromatic content as a result of the presence of higher amounts of polycyclic aromatic hydrocarbons (PAHs), as discussed for the Py-GC/MS measurements below. Further research will be necessary in order to support this assumption. It is noteworthy that the polymerization products and their specific modes of adsorption on the colloid can be used as biomarkers for the identification of bistre.

The similarity in composition of the two bistre samples was confirmed by Py-GC/MS analysis with and without methylation. The chromatograms are dominated by chemical components typical of lignin, cellulose, as well as by PAHs (Figure 5). In both samples, the presence of components derived from syringyl units of lignin, as well as the high S/G ratios, confirm the source as lignin from *angiosperms*, perhaps beech, as suggested by spectroscopic techniques [8]. Cellulose components, primarily levoglucosan, were detected but these are not diagnostic for the identification of the wood species [6, 17]. Several PAHs and their isomers are also present in both samples as a result of a high temperature being used in their manufacture. The bistre from Zecchi shows a relatively higher amount of polyaromatic hydrocarbons compared to the Kremer's, suggesting differences in the wood combustion processes for the two samples. Generally, more PAHs are released at the onset of wood combustion. The distribution of PAHs can be diagnostic for the differentiation between smoke from hardwood and softwood [18]. Because of the high number of PAHs co-eluting with their isomers, and sometimes other components, their accurate identification is difficult. However, among the PAHs which are well resolved and in high amounts are fluoranthene and pyrene (peaks 17 and 18), which are typically abundant in beech. Interestingly, Py-GC/MS indicates also the possible presence of retene, coeluting with another PAH at 20.8 min. This peak is more prominent in the Zecchi bistre, and is weak in the Kremer sample. Retene is generally abundant as a product of the pyrolysis of coniferous wood from *gymnosperm* species, which contain resin acids of the abietane-backbone. However, it has been found in low amounts with dehydroabietic acid (DHA) also in smoke generated from beech [18]. A trace peak for DHA methylester was detected in the Py-GC/MS chromatogram of the sample from Zecchi, as it would be typical for tars, and more distinctively in the chromatograms of both bistre samples obtained after methylation (not shown). Although some contribution of softwood such as pine could not be excluded, no peaks of oxidized pine wood were detected, indicating that the primary source of the bistre samples is hardwood.

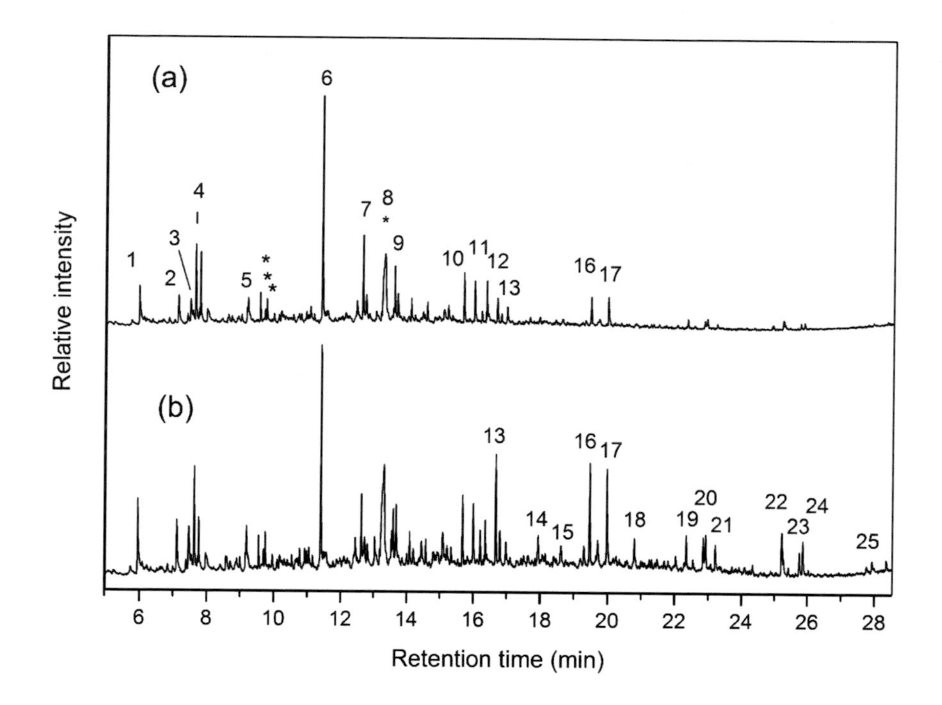

Figure 5. Chromatograms obtained from Py-GC/MS of bistre from Kremer (a) and bistre from Zecchi (b). The chromatograms have been normalized to show the relative distributions of the components. Components assigned (main m/z fragment; MW): **1**. Phenol (m/z 94). **2**. *o*-Cresol (m/z 108). **3.** *m*-Cresol (m/z 108). **4**. *p*-Guaiacol (m/z 109; 124). **5**. 4-Methyl guaiacol (m/z 138). **6.** Syringol (m/z 154). **7.** 4-Methyl syringol (m/z 168). **8**. Levoglucosan (m/z 60; 162). **9.** Unknown phenolic compound (m/z 167; 182). **10.** Syringaldehyde (m/z 182). **11**. Methoxyeugenol (m/z 194). **12.** Acetosyringone (m/z 181; 196). **13**. Syringil acetone? (m/z 167; 210). **14**. Anthracene/ phenanthrene (m/z 178). **15.** Methylanthracene/ methylphenanthrene (m/z 192). **16**. PAH (m/z 204). **17.** Fluoranthene (m/z 202). **18.** Pyrene (m/z 202). **19.** Substituted phenanthrene/ retene (m/z 219). **20.** Acepyrene? (m/z 226). **21**. Benz | a | anthracene /isochrysene (m/z 228). **22.** PAH (m/z 230). **23.** Benzo | b | fluoranthene and Benzo | k | fluoranthene (m/z 252). **24.** Benzo | e | pyrene (m/z 252). **25**. Benzo | e | pyrene. **26.** Indenopyrene (m/z 276). *Cellulose-derived components.

CONCLUSIONS

Commercial samples of bistre were analyzed by XRF, FTIR, normal Raman, SERS and Py-GC/MS. The assignments of the features in the vibrational spectra are supported by Py-GC/MS data that provided a precise molecular composition, particularly confirming the presence of lignin and PAHs from hardwood. The SERS spectra reported are suitable references for the identification of the pigment in works of art in a minimally invasive manner. The results show the potential of this multi-technique approach for the study of complex black-brown pigments.

ACKNOWLEDGMENTS

MLR thanks the Andrew W. Mellon Foundation for a Senior Fellowship at The Metropolitan Museum of Art.

REFERENCES

1. N. Eastaugh, V. Walsh, T. Chaplin, R. Siddall, *Pigment Compendium. A Dictionary of Historical Pigments*; Elsevier Butterworth-Heinemann: Amsterdam, 2004.
2. G. W. R. Ward, *The Grove Encyclopedia of Materials and Techniques in Art,* Oxford University Press, 2008.
3. R. White, National Gallery Technical Bulletin, **10**, 58-71 (1986).
4. A. M. Saariaho, A. S. Jaaskelainen, M. Nuopponen and T. Vuorinen, Applied spectroscopy, **57** (1), 58-66 (2003).
5. H. Pereira, J. Graça and J. C. Rodrigues, in J. R. Barnett, G. Jeronimidis (Eds.), Wood Quality and its Biological Basis; Blackwell CRC Press, USA and Canada, 2003, p.77.
6. B. R. T. Simoneit, J. J. Schauer, C. G. Nolte, D. R. Oros, V. O. Elias, M. P. Fraser, W. F. Rogge and G. R. Cass, Atmospheric Environment, **33** (2), 173-182 (1999).
7. S. Reale, A. Di Tullio, N. Spreti and F. De Angelis, Mass Spectrom Rev, **23** (2), 87-126 (2004).
8. A. Heigenmoser, F. Liebner, E. Windeisen and K. Richter, Journal of Analytical and Applied Pyrolysis, **100** (0), 117-126 (2013).
9. N. Leopold and B. Lendl, The Journal of Physical Chemistry B, **107** (24), 5723-5727 (2003).
10. J. Colbourne, in *Iron Gall Ink Meeting: Triennial Conservation Conference*, Newcastle, United Kingdom, 2000.
11. O. Faix, Holzforschung, **45** (s1), 21-27 (1991).
12. O. Faix, in Lin, S.Y., Dence, C.W. (Eds.), *Methods in Lignin Chemistry*, Springer, Berlin, pp. 233-241 (1992).
13. A. Antonović, V. Jambreković, J. Franjić, N. Španić, S. Pervan, J. Ištvanić, A. Bublić, Periodicum Biologorum, **112** (3), 327-332 (2010).
14. E. P. Tomasini, E. B. Halac, M. Reinoso, E. J. Di Liscia and M. S. Maier, Journal of Raman Spectroscopy, **43** (11), 1671-1675 (2012).
15. S. Sánchez-Cortés and J. V. García-Ramos, Journal of Colloid and Interface Science, **231** (1), 98-106 (2000).

16. S. Sánchez-Cortés, O. Francioso, J. V. García-Ramos, C. Ciavatta and C. Gessa, Colloids and Surfaces A: Physicochemical and Engineering Aspects, **176** (2-3), 177-184 (2001).
17. G. C. Galletti and P. Bocchini, Rapid Communications in Mass Spectrometry, **9** (9), 815-826 (1995).
18. M. A. Bari, G. Baumbach, B. Kuch and G. Scheffknecht, Atmospheric Environment, **43** (31), 4722-4732 (2009).

Mater. Res. Soc. Symp. Proc. Vol. 1656 © 2014 Materials Research Society
DOI: 10.1557/opl.2014.707

Analysis of Lead Carboxylates and Lead-Containing Pigments in Oil Paintings by Solid-State Nuclear Magnetic Resonance

Jaclyn Catalano[1,2], Yao Yao[2], Anna Murphy[2], Nicholas Zumbulyadis[3], Silvia A. Centeno[1], Cecil Dybowski[2]

[1]Department of Scientific Research, The Metropolitan Museum of Art, New York, NY 10028, U.S.A.
[2]Department of Chemistry and Biochemistry, University of Delaware, Newark, DE 19716, U.S.A.
[3]Independent Researcher

ABSTRACT

Soap formation in traditional oil paintings occurs when heavy-metal-containing pigments, such as lead white, $2Pb(CO_3)_2 \cdot Pb(OH)_2$, and lead-tin yellow type I, Pb_2SnO_4, react with fatty acids in the binding medium. These soaps may form aggregates that can be 100-200 μm in diameter, which swell and protrude through the paint surface, resulting in the degradation of the paint film and damage to the integrity of the artwork. In addition, soap formation has been reported to play a role in the increased transparency of paint films that allows the painting support, the preparatory drawing, and the artists' alterations to become visible to the naked eye. The factors that trigger soap formation and the mechanism(s) of the process are not yet well understood. To elucidate these issues, chemical and structural information is necessary which can be obtained by solid-state ^{207}Pb, ^{119}Sn, and ^{13}C nuclear magnetic resonance (NMR). In the present study, a combination of ^{207}Pb NMR pulse sequences was used to determine accurately the NMR parameters of lead-containing pigments and lead carboxylates known to be involved in soap formation, such as lead palmitate, lead stearate, and lead azelate. These results show that the local coordination environment of lead azelate is different from lead palmitate or lead stearate and therefore it is unlikely that lead azelate would be incorporated into an ordered structure containing lead palmitate and lead stearate. In addition, the chemical shifts of the pigments obtained are different from those of the soaps, demonstrating that ^{207}Pb NMR is useful in characterizing the components when present in a mixture, such as a paint film. The NMR methods discussed can also be applied to other Pb-containing cultural heritage materials, electronic and optoelectronic materials, superconducting materials, and environmentally contaminated materials.

INTRODUCTION

The formation of lead and other heavy-metal carboxylates, also called heavy-metal soaps, has been reported to be the cause of deterioration of hundreds of oil paintings dating from the fifteenth to the twentieth centuries [1-6]. Soaps form when heavy-metal-containing pigments, such as the commonly used lead white, $2PbCO_3 \cdot Pb(OH)_2$, and lead-tin yellow type I, Pb_2SnO_4, react with fatty acids that result from the hydrolysis of glycerides in the oil binding medium, or from protective coatings [1, 2, 4, 7, 8]. Soaps have been characterized and identified in samples from works of art by microanalytical techniques, such as FTIR, SIMS, GC-MS, DTMS, SEM-

EDX, micro-XRF, and Raman spectroscopy, inside aggregates or inclusions that can be as large as 100-200 μm in diameter and may break through the paint surface [1-6]. Soap formation has also been indicated as the cause of the increased transparency of paint films [5, 9-11]. Despite its widespread occurrence, the chemistry of soap formation is not yet fully understood.

Solid-state NMR (ssNMR) can provide structural and dynamical information on lead soaps that cannot be accessed by other methods. Other structural techniques such as solution NMR and X-ray crystallography are not possible means of study because lead soaps are very insoluble and single crystals are difficult to produce. The relationship between ^{207}Pb ssNMR parameters and solid-state structure has been reviewed by Fayon *et al.* [12], Dybowski and Neue [13], and more recently by Dmitrenko *et al.*[14] There is a strong dependence of the chemical-shift parameters on local structure, particularly on the coordination geometry around the lead atom. Hence, principal components of the ^{207}Pb chemical-shift tensor and the isotropic chemical shifts are indicative of the chemical identity of lead centers and, to the extent that different fatty acids lead to different coordination geometries, can be used to identify reaction pathways. We have previously characterized lead carboxylates known to be involved in soap formation [15]. In this proceeding, we report ^{207}Pb spectra and chemical-shift tensors for lead white and lead-tin yellow and show they can be distinguished from the lead carboxylates.

EXPERIMENTAL DETAILS

Lead palmitate was synthesized by a previously published procedure [15]. Basic lead carbonate (lead white) and lead nitrate were purchased from Sigma-Aldrich and lead-tin yellow type I 1010 light was purchased from Kremer Pigments (Germany).

^{207}Pb ssNMR spectra were recorded at 11.75 tesla (104.63 MHz lead-207 frequency) with a standard Bruker 4-mm probe. Approximately 100 mg of sample were packed in a 4-mm rotor. Mixtures of lead carbonate or lead white (95% by weight) and lead nitrate (5%) were prepared. Mixtures of lead palmitate and lead white were also prepared by percent weight of 50:50, 10:90, and 1:99. Solid lead nitrate was used as a secondary external reference for the ^{207}Pb spectra, the isotropic chemical shift being -3491 ppm relative to tetramethyllead (TML) at 298 K [16]. The spectrum of lead nitrate was recorded at different spinning speeds to compensate for the temperature increase due to spinning.

^{207}Pb Wideband Uniform Rate Smooth Truncation – Carr-Purcell Meiboom-Gill (WURST-CPMG) spectra of the samples were recorded using the parameters of MacGregor *et al.*[17] WURST pulse widths were 50 μs, with pulse shapes created via the shape tool in Topspin 3.1. Seventy-five Meiboom-Gill loops were acquired for the WURST-CPMG experiments, with a 200-μs echo, and a sweep range of 0.5 MHz in all cases. The recycle delay was 7 s. For lead-tin yellow, multiple WURST-CPMG spectra were collected at different carrier frequencies by shifting the carrier frequency a multiple of the spikelet separation (981.934 ppm) from spectrum to spectrum. The collected spectra were superimposed to form the final spectrum. For lead white, WURST-CPMG spectra covering the range from 4000 ppm to -5000 ppm were recorded with delays of 7 and 60 seconds.

For lead-tin yellow, ^{207}Pb spectra were acquired using direct excitation with spin-temperature alternation and magic-angle spinning (STA/MAS) at 10, 11, and 12 kHz to obtain the isotropic chemical shift. Spin-temperature alternation was used to minimize the effects of ringdown of the probe circuits [16]. General conditions for these experiments included a π pulse width of 8.5 μs, a delay of 1 ms, and a $\pi/2$ pulse width of 4.25 μs.

The analysis of the ^{207}Pb chemical-shift tensors was performed by fitting the WURST-CPMG envelope. The isotropic chemical shift acquired in the MAS experiment was fixed in the fitting procedure, because it could be measured accurately. Fits to the WURST-CPMG envelopes were aided by simulation of the powder pattern with the program WSOLIDS [18].

DISCUSSION OF RESULTS

Paintings are multilayered complex systems. To study paint samples by NMR, analysis of the individual components needs to be performed. The lead soaps of saturated monocarboxylic stearic (C18) and palmitic (C16) acids and of the dicarboxylic azelaic acid have been detected in samples removed from deteriorating oil paintings [2, 4, 19]. We have studied the spectra of lead carboxylates known to be involved in soap formation and determined the ^{207}Pb chemical-shift tensors by using both STA/MAS and WURST-CPMG experiments [15]. Lead has a large chemical-shift anisotropy that can spread the signal over thousands of ppm. The TA/MAS experiment maximizes the signal by condensing the broad spectrum into a spinning sideband pattern, where each peak is separated by an integer multiple of the spinning speed from the isotropic chemical shift. However, the STA/MAS experiment is unable to excite uniformly over a wide range. WURST-CPMG provides uniform excitation, through the use of adiabatic pulses, and CPMG loop design enables multiple echoes to be recorded during each acquisition, thereby greatly enhancing the signal-to–noise ratio. The STA/MAS spectrum of lead palmitate was acquired in 2 days compared to 4 hours by WURST-CPMG. The WURST-CPMG spikelet pattern that is produced is a result of the Fourier transformation of the echo train. As seen in Figures 1-4, from this pattern it is easy to determine the chemical-shift tensor components. By using both STA/MAS and WURST-CPMG, we are able to determine the chemical-shift tensors more accurately, because we use the isotropic chemical shift assigned from the STA/MAS experiments to refine the principal components extracted from the WURST-CPMG experiment.

Table 1. The Principal Components of the ^{207}Pb Chemical-Shift Tensors of Lead Azelate, Lead Stearate, Lead Palmitate and Lead Carbonate.[a]

Compound	δ_{11} (ppm)	δ_{22} (ppm)	δ_{33} (ppm)	δ_{iso} (ppm)	Ω (ppm)	κ
Lead Azelate[15]	-160±7	-604±11	-2800±8	-1188±3	2640±11	0.66 ±0.01
Lead Stearate[15]	-1810±3	-2007±6	-2555±3	-2124±4	745±4	0.47±0.03
Lead Palmitate[15]	-1820±3	-2013±5	-2560±3	-2131±3	740±4	0.48±0.02
Lead Carbonate[16]	-2311±2	-2481±2	-3075±8	-2622±3	764±8	0.55±0.03

[a]The fitting procedure used the isotropic shift, as determined from the MAS spectra of the material, with simulation of the values of δ_{11}, δ_{22}, and δ_{33} determined by the edges of the WURST-CPMG spectrum. Span (Ω) and skew (κ) were calculated. $\Omega = |\delta_{33} - \delta_{11}|$ and $\kappa = 3*(\delta_{iso} - \delta_{22})/\Omega$.

The chemical-shift tensors reported in Table 1 show that lead palmitate and lead stearate are very similar and indistinguishable by ^{207}Pb NMR spectroscopy, which is expected because the only structural difference between the two is the addition of two carbon groups on a large hydrocarbon chain. However, the lead NMR parameters of lead palmitate and lead stearate chemical-shift tensors are very different from those of lead azelate, which is reasonable since azelaic is a dicarboxylic acid. The difference in lead coordination geometry supports the hypothesis that it is unlikely that lead azelate would be incorporated into an ordered structure containing lead palmitate and lead stearate [4, 20].

Lead White

The ^{207}Pb spectrum of lead white, basic lead carbonate, was obtained with the WURST-CPMG experiment. (Figure 1) Only one lead species is observed under the current experimental conditions. The crystal structure of lead white shows layers of lead hydroxide and lead carbonate [21]. A fit to the spectrum in Figure 1 is consistent with the chemical-shift tensor of lead carbonate (Table 1) [16], and agrees with previous results by Verhoeven *et al.*[22]

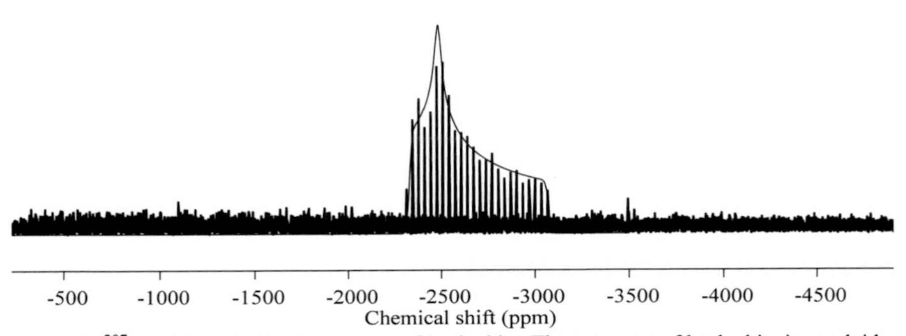

Figure 1. ^{207}Pb WURST-CPMG spectrum of lead white. The spectrum of lead white is overlaid with the chemical-shift tensor of lead carbonate as reported by Neue *et al.* [16]

Depending on how lead white is prepared, there usually is a residual amount of lead carbonate present [23]. To test if the observed signal is from the lead carbonate layers in lead white or a lead carbonate impurity, we mixed a known of amount of lead nitrate (5%) with lead white and with lead carbonate (95%). As shown in Figure 2, using the chemical-shift tensors of lead carbonate and lead nitrate, the fit for the lead white: lead nitrate mixture is 1:1 and lead carbonate: lead nitrate is 95:5. Thus, the lead species observed under our current experimental conditions is a lead carbonate impurity in lead white. The spectra of the lead species in lead white could be very broad, the T_1 could be very long compared to the recycle delay, or T_2 could be very short (which would interfere with the refocusing of the WURST-CPMG experiment), causing one not to detect the resonance of lead in this phase.

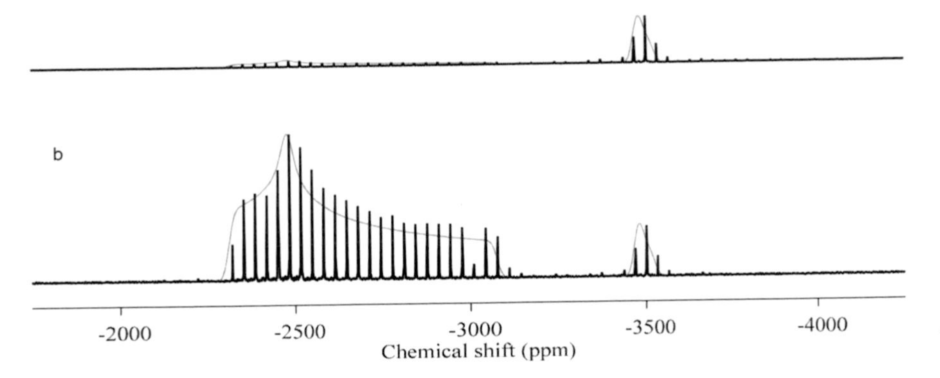

Figure 2. ^{207}Pb WURST-CPMG spectra of 5% lead nitrate in 95% (a) lead white or (b) lead carbonate. The fit of spectrum (a) has an area ratio of 1:1, whereas the fit of spectrum (b) has an area ratio of 95:5. These spectra demonstrate that, in the spectrum of lead white, only the lead carbonate impurity is observed.

Lead-Tin Yellow Type I

A preliminary fit of the ^{207}Pb spectrum of lead-tin yellow type I obtained with the WURST-CPMG sequence is shown in Figure 3. The fit consists of the overlap of the spectra of four lead species, two having similar chemical-shift tensors. The crystal structure [24] of Pb_2SnO_4 shows two lead species with similar coordination environments. The other two lead species were identified by the isotropic chemical shift of minium Pb(IV) site observed in the STA/MAS spectrum (not shown) [12, 25]. Minium is one of the starting materials for the synthesis of the pigment [26]. The principal components of the chemical-shift tensors for the two lead sites of minium are included in the fit of the spectrum in Figure 3. Future DFT calculations based on the crystal structure of lead-tin yellow type I will be used to refine the chemical-shift tensor components of the two main species in Pb_2SnO_4.

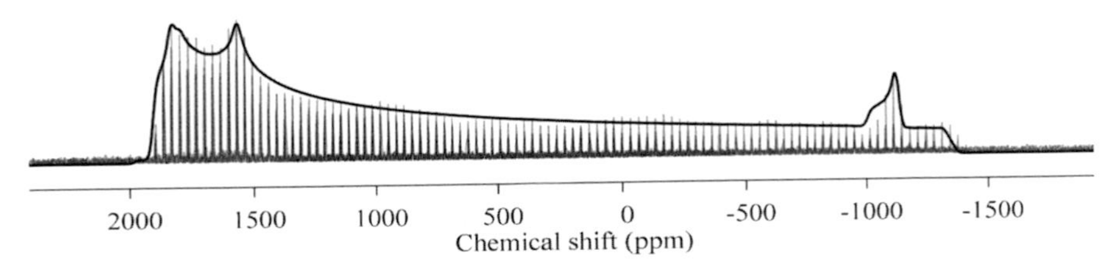
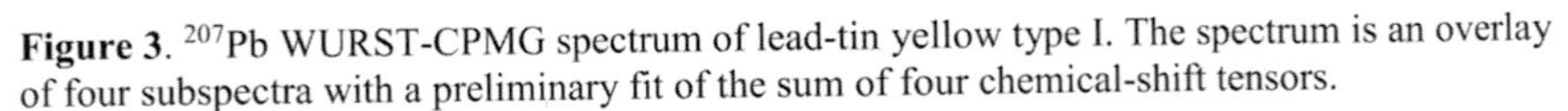

Figure 3. ^{207}Pb WURST-CPMG spectrum of lead-tin yellow type I. The spectrum is an overlay of four subspectra with a preliminary fit of the sum of four chemical-shift tensors.

Mixtures

^{207}Pb spectra of samples containing known amounts of lead white and lead palmitate (Figure 4), verify that one can distinguish these components in a mixture. The fits of the spectra to sums of two species with the known chemical-shift tensors are consistent with the assumption that only the lead carbonate impurity in lead white is detected. As shown in Figure 4a, we can observe 1% lead palmitate in a mixture under our current experimental conditions with a total data acquisition time of 24 hours, which sets the lower detection limit for this experiment. This fact is important when considering smaller sample sizes or more complex mixtures, where there is a smaller percentage of lead palmitate present. Because lead palmitate and lead stearate behave similarly, we expect the same to be true for lead stearate. However, under the same experimental conditions 1% lead azelate is not observable, as the chemical-shift tensor is broader and a longer experiment time is needed to obtain an acceptable spectrum.

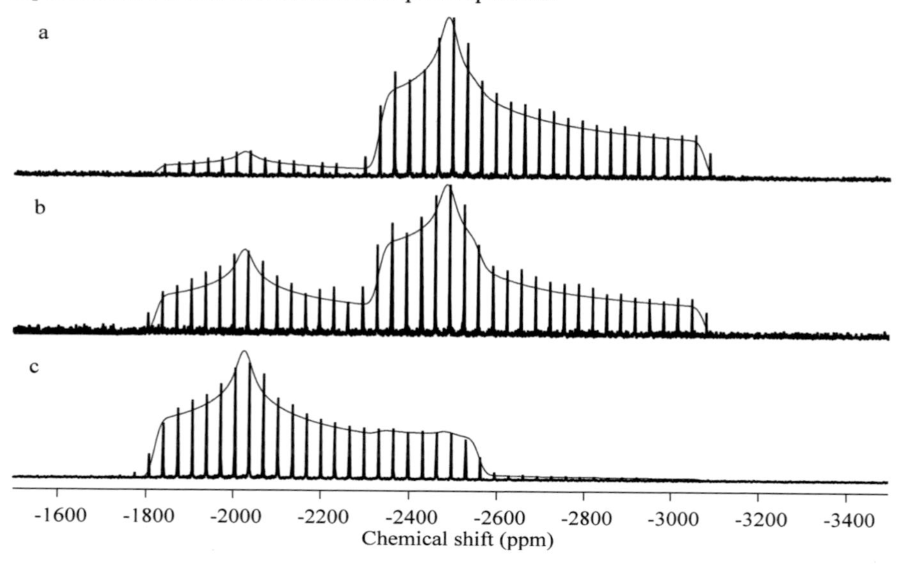

Figure 4. ^{207}Pb WURST-CPMG spectra of mixtures of lead white with lead palmitate; weight ratios of (a) 99:1, (b) 95:5, (c) 50:50. Spectrum (a) shows that 1% lead palmitate is observable under our current experimental conditions. The area ratios in these spectra only reflect the amount of lead palmitate relative to the lead carbonate impurity in lead white.

CONCLUSIONS

NMR chemical-shift tensor analysis of lead-containing pigments and lead carboxylates, known to be involved in soap formation, allows identification and quantification of the lead-containing components of a paint mixture. WURST-CPMG experiments enable shorter total experiment time than STA/MAS experiments, which is helpful when components are present in low quantities. Using WURST-CPMG experiments, we show that 1% lead palmitate is observable in 24 hours of experiment time. The next step in this study is to perform WURST-CPMG experiments on model paint samples subjected to accelerated aging. After optimization of the parameters for model paint samples along with the use of micro-sample holders, ssNMR may be considered a minimally invasive technique to study cultural heritage objects.

ACKNOWLEDGMENTS

CD acknowledges the support of the National Science Foundation under Grant CHE 1139192. SC acknowledges support of the National Science Foundation under Grant CHE 1139190.

REFERENCES

1. P. Noble, J. J. Boon and J. Wadum, ArtMatters **1**, 46-61 (2003).
2. K. Keune, PhD Thesis, University of Amsterdam, 2005.
3. M. Spring, C. Ricci, D. Peggie and S. Kazarian, Analytical and Bioanalytical Chemistry **392** (1), 37-45 (2008).
4. C. Higgitt, M. Spring and D. Saunders, National Gallery Technical Bulletin **24**, 75-95 (2003).
5. S. A. Centeno and D. Mahon, The Metropolitan Museum of Art Bulletin **67** (1), 12-19 (2009).
6. C. Hale, J. Arslanoglu and S. A. Centeno, in *Studying Old Master Paintings. Technology and Practice*, edited by M. Spring (Archetype Publications and The National Gallery, London, 2011), pp. 59-64.
7. J. Van der Weerd, PhD Thesis, University of Amsterdam, 2002.
8. J. J. Boon, J. van der Weerd, K. Keune, P. Noble and J. Wadum, in *ICOM-CC 13th Triennial Meeting* (Rio de Janeiro, 2002), Vol. 1, pp. 410-406.
9. A. Eibner, *Malmaterialienkunde als Grundlage der Maltechnik.* (Verlag von Julius Springer, Berlin, 1909), pp. 121.
10. P. Noble, A. van Loon and J. J. Boon, in *ICOM Committee for Conservation 14th Triennnial Meeting* (James and James, The Hague, 2005), Vol. 1, pp. 496-503.
11. A. van Loon, PhD Thesis, University of Amsterdam, 2008.
12. F. Fayon, I. Farnan, C. Bessada, J. Coutures, D. Massiot and J. P. Coutures, Journal of the American Chemical Society **119** (29), 6837-6843 (1997).
13. C. Dybowski and G. Neue, Progress in Nuclear Magnetic Resonance Spectroscopy **41** (3-4), 153-170 (2002).
14. O. Dmitrenko, S. Bai, P. A. Beckmann, S. van Bramer, A. J. Vega and C. Dybowski, Journal of Physical Chemistry A **112** (14), 3046-3052 (2008).
15. J. Catalano, Y. Yao, A. Murphy, N. Zumbulyadis, S. A. Centeno and C. Dybowski, Applied Spectroscopy **68** (3), 280-286 (2014).
16. G. Neue, C. Dybowski, M. L. Smith, M. A. Hepp and D. L. Perry, Solid State Nuclear Magnetic Resonance **6** (3), 241-250 (1996).
17. A. W. MacGregor, L. A. O'Dell and R. W. Schurko, Journal of Magnetic Resonance **208** (1), 103-113 (2011).
18. K. Eichele, HBA 3.1 and WSOLIDS, University of Tubingen, 2013.
19. K. Keune and J. J. Boon, Stud. Conserv. **52** (3), 161-176 (2007).
20. M. J. Plater, B. De Silva, T. Gelbrich, M. B. Hursthouse, C. L. Higgitt and D. R. Saunders, Polyhedron **22** (24), 3171-3179 (2003).

21. P. Martinetto, M. Anne, E. Dooryhée, P. Walter and G. Tsoucaris, Acta Crystallographica Section C **58** (6), i82-i84 (2002).
22. M. A. Verhoeven, Carlyle, L., Reedijk, J., Haasnoot, J.G., in *Reporting Higlights of the De mayerne Programme.*, edited by J. J. a. F. Boon, E.S.B. (The Hague: Netherlands Organisation for Scientific Reserach (NWO). 2006), pp. 33-42.
23. B. Keisch, in *Studies in the History of Art* (1971-1972), Vol. 4, pp. 121-133.
24. J. R. Gavarri, J. P. Vigouroux, G. Calvarin and A. W. Hewat, Journal of Solid State Chemistry **36** (1), 81-90 (1981).
25. S. P. Gabuda, S. G. Kozlova, V. V. Terskikh, C. Dybowski, G. Neue and D. L. Perry, Solid State Nuclear Magnetic Resonance **15** (2), 103-107 (1999).
26. R. J. H. Clark, L. Cridland, B. M. Kariuki, K. D. M. Harris and R. Withnall, Journal of the Chemical Society. Dalton Transactions. (16), 2577-2582 (1995).

Mater. Res. Soc. Symp. Proc. Vol. 1656 © 2014 Materials Research Society
DOI: 10.1557/opl.2014.826

Fine Pore Structure Characterization in Two Gessoes Using Focused Ion Bean Scanning

Michael Doutre[1], Ashley Freeman[1], Brad Diak[2], Alison Murray[1], George Bevan[1], and Laura Fuster-López[3]
[1] Art Conservation Program, Queen's University, Kingston, Ontario, K7L 3N6
[2] Mechanical and Materials Engineering, Queen's University, Kingston, Ontario, K7L 3N6
[3] Instituto Universitario de Restauracion del Patrimonio, Universidad Politecnica de Valencia, Valencia, Spain, 46022

ABSTRACT

The movement of fluids through a porous medium is a function of the material type and the size and morphology of the voids. In the conservation of painted artworks, the movement of materials (for example, cleaning solutions) is a major factor in how a work reacts to treatments and a large influence on how the work will change with time. Of particular importance in the conservation of painted surfaces is the preparatory layer. This is a highly active transport medium because it is generally highly porous and a comparatively large and uniform component of a painted surface. In this work, a gesso film of calcium carbonate and rabbit skin glue, typical of the preparatory layer of many painted works, and an acrylic-based gesso film were imaged by focused ion beam (FIB) scanning. The gessoes were milled and scanned with gallium ions serially, in sequential planes orthogonal to the plane of the film. This yielded quantifiable measurements of the fine internal structures at a resolution far higher than previously reported for this type of material. This enabled a greater understanding of the geometry of the internal surfaces, increasing the understanding of the mechanics of capillary flow and diffusive behavior in this extremely common and significant material.

INTRODUCTION

The structure of a painting generally comprises many different materials organized as layers on a support, including a size for the support layer, the preparatory layer, and the pictorial layer. Traditionally, the preparatory layer, also referred to as the gesso or ground layer, consists of a white pigment, such as chalk or gypsum, bound with an adhesive, such as hide glue. More recently, modern commercially available materials have been used as the adhesive to bind different pigments, such as an acrylic adhering titanium dioxide particles. While the chemical and physical characteristics of the preparatory layer were the central focus of prior research, more recent studies have examined the diffusive characteristics [1,2,3] of aqueous treatments and the internal morphology of the preparatory layer. The resulting information may help to determine how the gesso layer interacts with the paint layer; how the gesso layer may facilitate movement of mobile applied materials, such as paint and cleaning solutions; and how physical damage may occur and spread. Another application is that the gesso forms a reactive layer to environmental changes in temperature and relative humidity.

Numerous techniques can visualize the internal morphology of artworks. Computed tomography (CT) is a non-destructive technique for characterizing various works of art and samples from paintings. Using synchrotron x-ray microtomography, Ferreira *et al.* [4] were able to observe the morphology and internal structure of samples from two different paintings by Swiss artist Cuno Amiet. After a visual examination of the first painting, *Winter in Oschwand,* various factors, including a uniform thickness of the preparatory layer and a smoothness of the pictorial layer, suggested that a commercially prepared canvas was used. The second painting, *Portrait of Max Leu,* did not display the same visual characteristics as the commercially prepared canvas, suggesting that the artist applied the preparatory layer after the canvas was stretched. The samples collected from these paintings were scanned at the Swiss Light Source (SLS) in Villigen (MS-X02DA-TOMCAT beam line) and the resulting image data cubes generated three-dimensional computer models enabling a qualitative assessment of the porosity. The commercially prepared gesso sample displayed a smooth top surface with small deviations on the surface, while the artist-prepared ground was rough and inhomogeneous. Also, two different layers were observed in the commercially prepared ground while only one layer was noted in the artist-prepared ground. Overall, a better understanding of the samples was possible after examining the different orthoslices, as aggregates or voids found in one slice were missing in another; however, no quantitative data regarding the pore distribution were presented.

Lin *et al.* [5] used focused ion beam (FIB) milling coupled with scanning electron microscopy (SEM) to examine the spatial distribution of pigment particles in dried paint samples. The milling executed by FIB-SEM allowed thin layers of the paint to be removed from the sample; imaging showed that the pigments did not change position. Serial cross-section images were collected during this process and used to produce reconstructions in three dimensions (3D) of the paint samples. With the use of a Euclidean distance map and ultimate eroded points (regional maxima), the dispersal of the pigment particles within the paint samples could be quantified in 3D.

In the work presented here, a Micrion 2500 FIB instrument was used to examine the internal structures of two gessoes. The images captured during this process were used to make three-dimensional models and measurements of the internal morphology of the materials in order to gain a better understanding of the functionality, variability and durability of the preparatory layer.

EXPERIMENTAL

Sample Preparation

Two gessoes were examined. The first was a laboratory-prepared gesso consisting of 57g of International Gilders' Supplies 9µm calcium carbonate ($CaCO_3$) to 50mL of a 10% (w/w) aqueous solution Lefranc & Bourgeois rabbit skin glue (RSG). The second material was Golden Acrylic Gesso, a commercially prepared gesso consisting mainly of titanium dioxide (TiO_2) and calcium carbonate ($CaCO_3$) bound in a methylmethacrylate-butylacrylate copolymer. Both gessoes were cast on Mylar in wells of 3M® electrical tape to a dried thickness of approximately 120 µm. Castings were then created by drawing down the wet material with the long side of a microscope slide. Once the casting formed a surface film, the tape was removed and the casting dried for at least one week at ambient conditions. Thus one side of the casting was in direct contact with the Mylar sheet (film-substrate interface), while the other side was in contact with the air (film-air interface). The dried casting was peeled from the Mylar and slivers, ranging from 120 – 150 µm, were cut from it using a glass blade. The film-substrate interface side of the sample was affixed to a FIB sample stage with carbon tape. Copper tape was then placed over the remaining visible carbon tape (Figure 1).

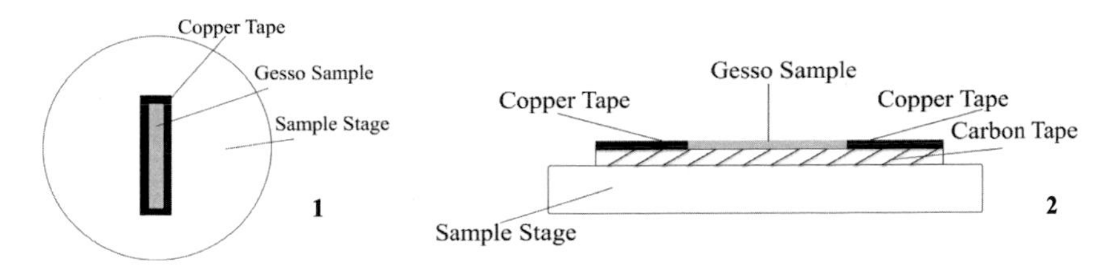

Figure 1. Schematics of the FIB sample preparation process: Top-view (1) and cross-section (2) (not drawn to scale).

Rough Milling

Deposition, milling and imaging was carried out in a Micrion 2500 Focused Ion Beam (FIB) instrument using a gallium ion source operating at 30kV. Before milling tungsten was deposited on the surface of the sample in three sequential rectangles producing a protective layer approximately 90 μm x 40 μm along the region of interest. Trench milling was started using an aperture of 150 μm at the region of interest. As milling continued small areas of the sample became charged. To dissipate this accumulated charge, gold was deposited over the surface in a sputtering chamber. Rough milling of the sample was continued until a small rectangular trench was cut out (Figure 2).

Figure 2. A SEM image (1) and schematic of the trench milled in the Golden Acrylic Gesso sample (2).

Sequential Polishing

Once the trench was formed, sequential polishing of one wall was conducted, orthogonal to both the bottom and sides of the casting well. For this process the aperture was set to 75μm giving a source current of 99pA. Eight successive images were captured for the RSG gesso and eleven images were captured for the Golden Acrylic Gesso, in secondary electron mode (Figure 3), after each 200 nm milling step.

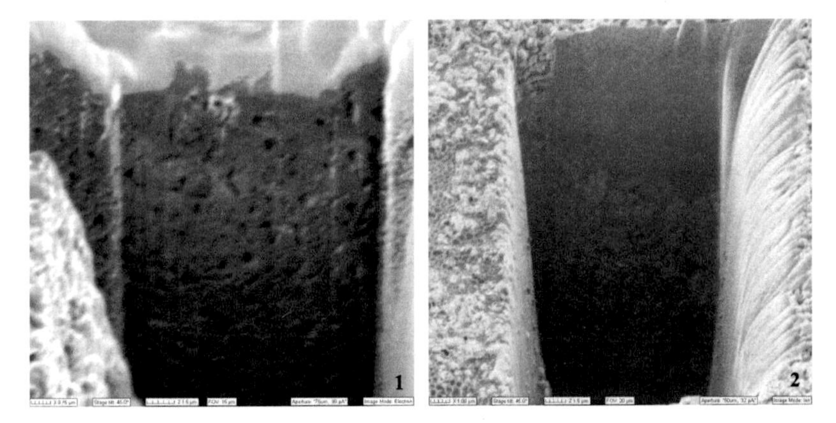

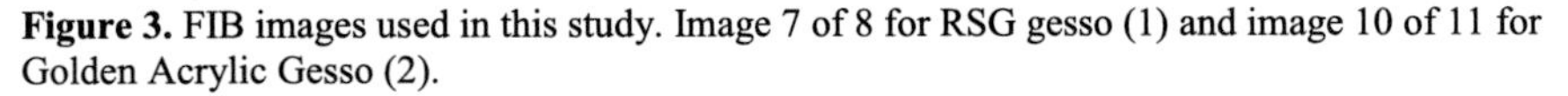

Figure 3. FIB images used in this study. Image 7 of 8 for RSG gesso (1) and image 10 of 11 for Golden Acrylic Gesso (2).

Image Processing

The collected images were processed in ImageJ [6] and were stacked in sequential order. The stack was then stretched in its y-axis by sin (45°) to compensate for the tilt of the stage, aligned using a Rigid Body alignment function [7], and cropped to remove the trench edges. To remove the tone gradient created by charge buildup on the surface of the film, each image was duplicated, and the duplicate median was filtered with a kernel size much larger than the features being examined ($32px^2$). The filtered duplicates were then subtracted from each corresponding original image. The stack was then binarized with Otsu thresholding (Figure 4).

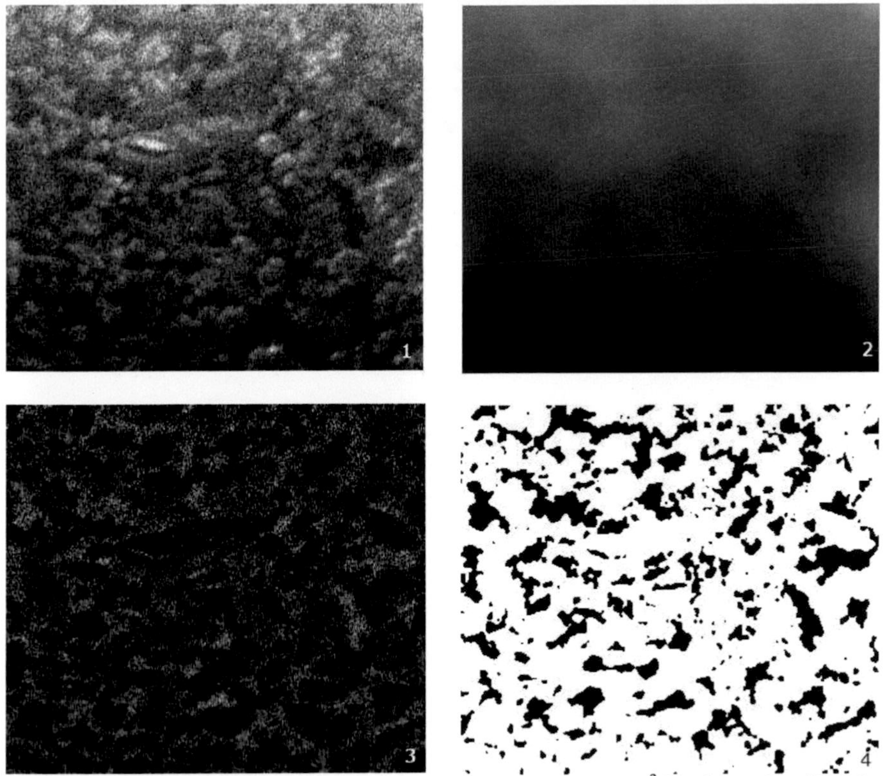

Figure 4. The Golden Acrylic Gesso raw image stack cropped to 7.7μm^2 (1), the median filtered stack (2), the stack resulting from the subtracting the inverse of the filtered stack from the inverse of the original (3), and the Otsu binarized stack (4).

RESULTS AND DISCUSSION

Using this technique the RSG gesso was found to have a closed network of round pores ranging between 0.022 and 2.10 μm^2 and a mean area of 0.10(±0.21) μm^2. These rounded pores are homogeneously distributed throughout the film with an average separation of 1.45μm, covering 6.7% of the examined area (Figure 5). As the orientation of void space within the RSG gesso was of interest and visual examination of the fairly round pores offered no obvious directional bias, the orientation was examined by fitting ellipses to the pores and examining the direction of the major axis [8]. Using this method, a bias to 50° off vertical to one side of the casting was observed (Figure 6), likely produced by the casting method used. Based on its density, the bulk porosity of this material is thought to be approximately 30%, while only 6.7% of the examined area was found to be voids. It is likely that most of the porosity of this material may be in voids smaller than what could be imaged with the ~20nm resolution achieved in this work.

Figure 5. Otsu binarized stack of RSG gesso showing rounded pores in black and $CaCO_3$ grains and agglomerates in white. Image stack cropped to 7.7μm^2.

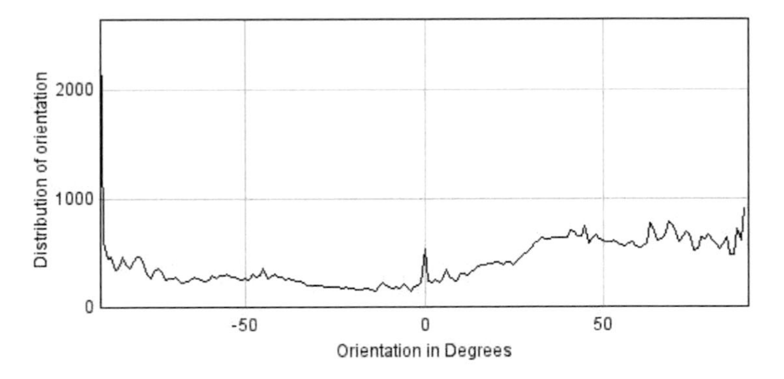

Figure 6. Orientations of the largest dimension of each pore within the RSG gesso. The sharp peaks at zero and 90 are artifacts of the analysis method.

In contrast, the Golden Acrylic Gesso was found to have a fairly open network of pores ranging from 0.0023 μm^2 up to 50.0 μm^2 with a mean area of 0.23(±1.86) μm^2. The high deviation being the result of the porosity existing largely as connected networks. The total porosity of the Golden Acrylic Gesso was 45% in the examined area (Figure 7), roughly in line with the predicted bulk porosity (~40%) of the material, based on the density of the film.

Directional analysis of the Golden Acrylic Gesso offered no indication of any directional bias to the orientation of the pores (Figure 8).

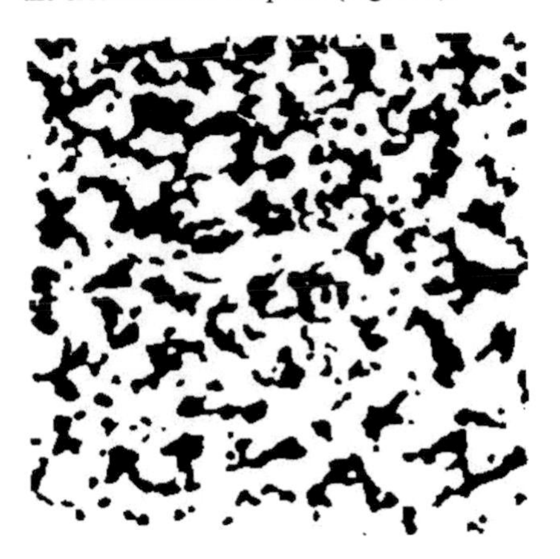

Figure 7. Otsu binarized stack of Golden Acrylic Gesso showing rounded pores in black and TiO_2 and acrylic grains and agglomerates in white. Image stack cropped to $7.7\mu m^2$.

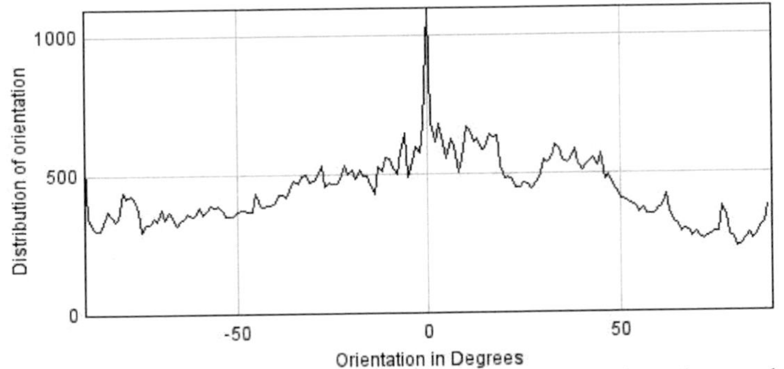

Figure 8. Orientation of the primary direction of pore segments (areas between branches in the pore network) within the acrylic gesso. The narrow peak at zero degrees is an artifact of the analysis method.

As the imaging method for both these materials used sequential slices, the images used in this study can be constructed into 3D images of the void spaces, as shown in Figure 9 for the RSG gesso. For this study, the 3D models were used to examine the pore morphology qualitatively, in the plane orthogonal to the imaged planes. No discrepancies were observed between the size and shape of the pores measured from the directly imaged planes versus the orthogonal plane which was constructed using the sequential slices; however, the lower resolution out of the imaged plane (200nm orthogonal to the imaged planes compared to 20nm within the imaged plane) means only the largest pores could be examined (greater than 0.6μm, for the plane orthogonal to the imaged plane).

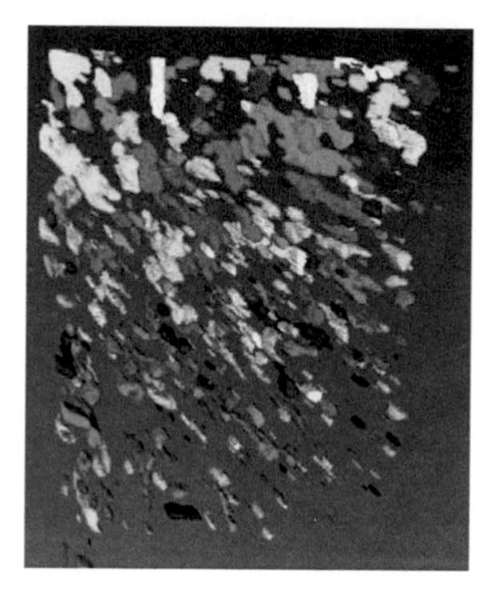

Figure 9. A three-dimensional model of the void space within the RSG gesso.

CONCLUSIONS

Sequential FIB imaging offers great possibilities as a method to study the fine structures of paint films in three dimensions, at higher resolutions than easily achievable with other techniques, such as x-ray micro-computed tomography. Gessoes, such as those examined in this work, are of particular significance in art conservation. These materials are usually the largest single component of the paint layers in a painting on canvas and are commonly used to fill losses in the pictorial layer in preparation for in-painting. In measuring the size and morphology of these pores, the eventual goal of this research is to relate these properties to the transport of fluids within these materials and more broadly painted artworks. To understand thoroughly how these materials interact with fluids such as cleaning solutions, varnishes, in-paints, and ambient moisture, the characterization of the pore structure, such as presented here, is essential.

The internal structure of two very different types of materials, traditional gesso (rabbit skin glue and $CaCO_3$) and an acrylic gesso from Golden Artist Colors (acrylic copolymer adhering TiO_2 particles) were examined. Exactly the same steps were applied to both materials: preparation, milling, polishing, and image processing. The results gave the average pore size and the distribution of pore sizes, how much of the material was made of pores, whether the pore system was connected or not, the orientations of the pores, and three-dimensional models. Despite serving the same purpose in many applications, the two materials examined were found to differ greatly in their internal morphology. The RSG gesso was found to have a closed network of round pores, with the casting method likely to play a role in the morphology of the pore network. The Golden Acrylic Gesso was found to be an open network of large systems of pores that owe their morphology to the material's formulation, with the casting method likely to have a much smaller effect on the final pore structure.

ACKNOWLEDGMENTS

This work was supported by the Natural Sciences and Engineering Research Council of Canada and Queen's University.

REFERENCES

1. P. Whitmore, H. Morris, and V. Colaluca, *Penetration of Liquid Water through Waterborne Acrylic Coatings*, Modern Paints Uncovered, London (2007).
2. K. Ulrich, S. Centeno, J. Arslanoglu, and E. Del Federico, *Absorption and Diffusion Measurements of Water in Acrylic Paint Films by Single-sided NMR*, Progress in Organic Coatings 71: 283–289.9 (2011).
3. M. Doutre, A. Freeman, H.F. Shurvell, and A. Murray, *Gessoes: Porosity and the Effects of Capillary Action*, Presented at the 41st AIC Annual Meeting, Indianapolis, Indiana (2013).
4. E. Ferreira, J. Boon, J. van der Horst, N. Scherrer, F. Marone and M. Stampanoni, *3D Synchrotron X-ray Microtomography of Paint Samples*, Proc. of SPIE Vol. 7391 (2009).
5. J. Lin, W. Heeschen, J. Reffner, and J. Hook, *Three-Dimensional Characterization of Pigment Dispersion in Dried Paint Films Using Focused Ion Beam–Scanning Electron Microscopy*, Microsc. Microanal. 18, 266–271 (2012).
6. W.S. Rasband, ImageJ, U. S. National Institutes of Health, Bethesda, Maryland, USA, http://imagej.nih.gov/ij/ (1997-2012).
7. P. Thévenaz, U.E. Ruttimann, and M. Unser, *A Pyramid Approach to Subpixel Registration Based on Intensity*, IEEE Transactions on Image Processing vol. 7, no. 1, pp. 27-41 (1998).
8. D. Sage, OrientationJ, Biomedical Image Group, École Polytechnique Fédérale de Lausanne Lausanne, Switzerland, http://bigwww.epfl.ch/demo/orientation/ (2012).

Mater. Res. Soc. Symp. Proc. Vol. 1656 © 2014 Materials Research Society
DOI: 10.1557/opl.2014.828

Effects of Humidity on Gessoes for Easel Paintings

Michael Doutre[1], Alison Murray[1], and Laura Fuster-López[2]

[1] Art Conservation Program, Queen's University, Kingston, Ontario, K7L 3N6
[2] Instituto Universitario de Restauracion del Patrimonio, Universidad Politecnica de Valencia, Valencia, Spain, 46022

ABSTRACT

Gessoes are widely used in easel painting as grounds or preparatory layers; in art conservation, gessoes are employed as infill materials to level a loss in the paint surface in preparation for inpainting. The goal of this investigation was to establish the relationship between the mechanical behavior of various gessoes when exposed to different relative humidities (25%, 50%, and 100%) and to compare modern commercial gesso products with a traditional gesso. The materials included two commercial artists' acrylic gessoes (composed of largely titanium dioxide and aqueous dispersions of acrylic polymers), two commercial spackling compounds frequently used in the conservation of easel paintings, and a traditional gesso (calcium carbonate and rabbit skin glue). Uniaxial tensile testing was used to characterize the elastic modulus, strain at failure, and ultimate tensile strength (UTS) of the materials. By understanding the physical limits of these materials under different conditions, damage to artworks and the failure of conservation treatments containing these types of materials may be prevented or reduced.

INTRODUCTION

Various materials are used to fill losses in a painting and thereby maintain the artist's original intent. The additional material does, however, make the entire painting system more complex. Any fill material should react to changes in the environment, for example changes in relative humidity (RH), in a manner that does not compromise the surrounding original material. A complete characterization of gessoes at different relative humidities is therefore important.

Historically, gessoes were made of a white pigment, such as chalk or gypsum, and a binder, often rabbit skin glue. The 1950's saw the emergence of the widespread application of commercial emulsified polymers and acrylic artists' materials. Since then, interest has grown in the use of modern polymeric materials in art conservation, both as a response to the growing number of art works made using modern materials that require conservation and because modern materials are used in treatments. Compared with traditional materials, these new polymers have greater reversibility and their chemical and physical properties can be better controlled [1]. Often the properties of the traditional and modern materials differ widely.

Identifying and understanding where and how the forces are distributed among the various layers is key for properly handling an artwork. Any materials applied during conservation must not contribute additional stress. Fuster-López described the effect of various relative humidity situations on paintings [2]. For paintings executed on canvas supports, at standard RH levels, the primary forces are concentrated in the size layer, which is in the layer closest to the fabric support and far from the paint layer. The size layer, which is essentially a layer of adhesive applied to the fabric support, generates tension as it locks the matrix of the fabric support. Fillers need to supply the structure of the painting with the appropriate strength to avoid mechanical damage and present good adhesion and cohesion between the size and pictorial layers. When considering the effect of different RH levels, understanding how the entire painting reacts as a system is critical. At RH levels exceeding 85%, the size and paint layers become very flexible and plastic, and no longer offer support for the painting structure. As a result the canvas becomes the primary support for the entire painting and the bond between the canvas size and the ground is nearly lost. Under low RH levels, the dramatic shrinkage that occurs in the size layer generates very high stresses in paintings. The size and the paint layers will develop high stresses and possible cracking, and the size layer will become the primary support for the painting.

The mechanical properties of various materials were tested at 25%, 50%, and 75% relative humidity. Included were two commercial artist acrylic gessoes, two acrylic-based drywall spackling compounds often used in conservation treatments, and one lab-made rabbit skin glue (RSG) gesso.

EXPERIMENTAL

Four commercial gessoes tested were used as purchased, without modifications: Becker's Latexspackle, Flugger Spackle, Golden Acrylic Gesso, and Liquitex Acrylic Gesso. The rabbit skin glue (RSG) gesso consisted of 67% $CaCO_3$ in a 10% (w/w) aqueous solution rabbit skin glue. The brand of rabbit skin glue (RSG) was Lefranc & Bourgeois and a 10% solution was chosen because previous research had used this concentration for making fills [3,4].

The samples for all tests were cast on Mylar film in wells approximately 6cm by 30cm that were created by outlining the well with layers of electrical tape. The wells were filled with the liquid gesso materials and brought level with the tape using a glass slide. Care was taken to prevent bubbles from forming in the material.

The castings were prepared for tensile testing by cutting strips 1.27cm wide by approximately 6cm long. The thickness of each strip was measured at three points along its length and the average value was used in determining the cross section of the sample.
The stress-strain behavior was determined for the samples at 25%, 50%, and 75% relative humidity. The chambers were controlled to maintain a stable RH within 2% of the desired level. The tests were carried out at the ambient laboratory temperature, which ranged from 21-24°C.

Preliminary tests showed some samples were brittle and others were elastic. The brittle samples, Becker's, Flugger and RSG, were analyzed as described by Hagan [5]. The apparatus consisted of a manual tensiometer coupled to a Vishay P3500 strain indicator. Hagan automated the tensiometer by attaching a stepper motor to the driveshaft that elongated the sample holder. The motor was controlled by a BasicX-24 microcontroller. The samples were elongated at 17%/min. The elastic samples, Golden and Liquitex, were run on a different environmentally controlled uni-axial tensiometer, which had a linear actuator (Haydon Kerk, Inc., USA). The load cell (Transducer Techniques, USA) was used to measure the force being applied to the test sample, rather than the cantilever-style strain gauge used in the other system. The samples were elongated at 100%/min. The environmental chamber included a thermoelectric temperature system (TE Technology, USA).

RESULTS

Table 1 summarizes the ultimate yield strength, elastic modulus, and strain at failure for all five materials. Seven replicates were tested for each result. The values for the acrylic materials (Golden and Liquitex) were found to be similar in magnitude and behavior to previously reported values for these types of acrylic paints [6, 7].

Table 1. Summary of test results. All values are the mean of seven replicates.

Ultimate Tensile Strength (MPa)				**Standard Deviation**		
	25%	50%	75%	25%	50%	75%
Liquitex	7.22	5.15	4.01	0.64	1.04	0.64
Golden	10.69	9.71	5.42	1.61	1.05	1.03
Flugger	2.14	1.44	0.44	0.21	0.07	0.08
Becker's	1.94	1.65	0.74	0.30	0.19	0.08
RSG	11.54	10.61	8.56	1.34	1.56	1.45
Elastic Modulus (MPa)				**Standard Deviation**		
	25%	50%	75%	25%	50%	75%
Liquitex	412.2	57.3	50.1	20.3	6.0	36.9
Golden	303.3	93.9	42.9	63.3	18.8	12.0
Flugger	388.0	244.4	35.7	56.3	23.5	6.7
Becker's	416.3	205.0	71.7	89.8	25.2	18.7
RSG	1511	1730	1995	134	145	127
Strain at Failure (mm/mm)				**Standard Deviation**		
	25%	50%	75%	25%	50%	75%
Liquitex	0.033	0.233	0.128	0.018	0.032	0.052
Golden	0.200	0.632	0.535	0.125	0.181	0.222
Flugger	0.016	0.038	0.064	0.004	0.005	0.013
Beckers	0.019	0.029	0.041	0.003	0.005	0.008
RSG	0.030	0.027	0.024	0.003	0.002	0.003

Figure 1 shows that the ultimate tensile strength was inversely related to humidity ($p=0.001$). On average, the ultimate strength of the gessoes changed by approximately 1% (1.0±0.4%) for each percentage point of deviation from 50% relative humidity. Similarly, the elastic modulus (Figure 2) of all the materials decreased with humidity (p=0.045) except in the case of the RSG gesso, which displayed the opposite trend, with its elastic modulus increasing with humidity. The artists' acrylic materials (Golden and Liquitex) in particular stiffened greatly at 25% RH, displaying a large increase in their elastic moduli and a large decrease in their strain at failure at the lower humidity (Figures 2 and 3).

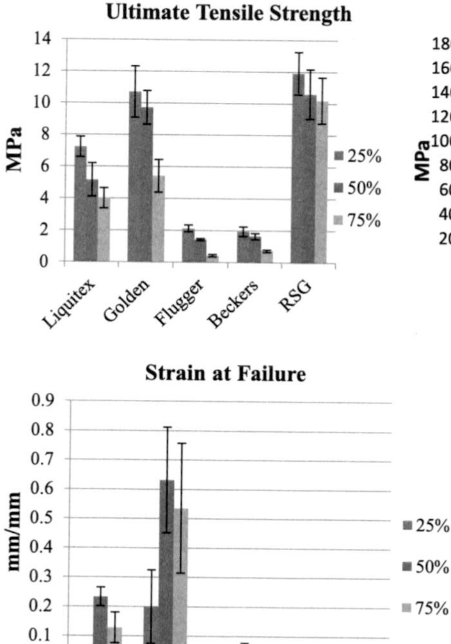

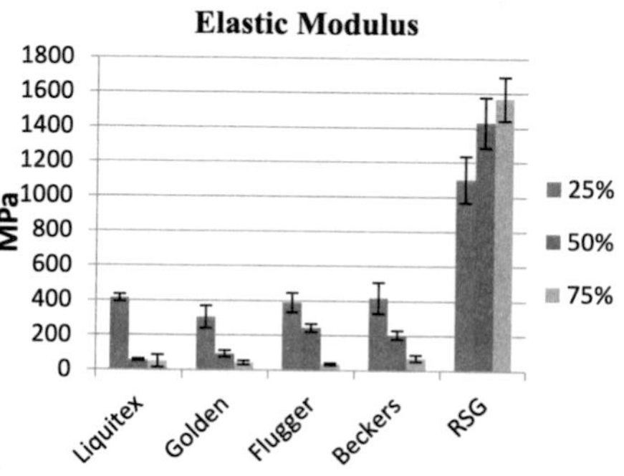

Figures 1, 2, 3. Graphical presentation of the ultimate tensile strength, elastic modulus, and strain at failure for all materials examined. The 25% RH is the left column, 50% is the middle, and 75% is the right. Error bars show standard deviation. As the Golden and Liquitex materials are far more elastic than the other materials, these samples were run at a far higher elongation rate, relative to the other materials. This will have greatly increased the measured elastic moduli and ultimate strength values and decreased the strain at failure.

CONCLUSIONS

The mechanical properties of these materials are dependent on relative humidity, but the degree of change and even direction of change varied widely depending on the material. Overall the traditional gesso made of rabbit skin glue displayed the lowest degree of change with humidity. In lower humidity conditions, the artists' acrylic materials stiffened considerably and broke at far lower strains than at relative humidities at or above 50%. In choosing appropriate fillers for areas of loss in paintings, the entire painted system needs to be taken into consideration in order to ensure that the product is compatible with the rest of the system. The RH levels studied in this work were quite extreme in the context of easel paintings. Future research could focus on how changes in the mechanical properties progress between the RH levels examined in this project, particularly for RH levels below 50%, in order to determine the range of conditions in which artworks containing these materials can be safely handled.

ACKNOWLEDGMENTS

This work was supported by the Natural Sciences and Engineering Research Council of Canada and Queen's University. The authors would like to thank to all those who supported this project, in particular Marion Mecklenburg and the Smithsonian Center for Materials Research and Education (SCMRE) for long term loan of some of the tensile testers used in this project. Kaslyne O'Connor is to be thanked for her Macgyver-esque humidity-control ideas.

REFERENCES

1. E. Jablonski, T. Learner, J. Hayes, and M. Golden, *Conservation Concerns for Acrylic Emulsion Paints: A Literature Review*, Tate Papers (2004).
2. L. Fuster-López, *Filling Materials for Canvas Paintings: Technical Evolution and Physico-Mechanical Analysis*, Ph. D. diss., Universidad Politecnica de Valencia, Spain (2006).
3. H. Smith, *An Investigation of the Suitability of Modostuc® as a Fill Material.* Unpublished Report. Queen's University, Kingston (2004).
4. L. Fuster-López, M.F. Mecklenburg, M. Castell-Agustí, and V. Guerola-Blay, *Filling Materials for Easel Paintings: When the Ground Reintegration Becomes a Structural Concern*, ICOM-CC Working Group Paintings: Scientific Study, Conservation, and Restoration, p.180-186 (2008).
5. E. Hagan, *A Comparison of Age, Climate, and Aqueous Immersion Effects on the Mechanical Properties of Artist's Acrylic Paints,* Master's thesis, Queens University, Canada (2004).
6. E. Hagan, and C. Young, *Cold Temperature Effects of Modern Paints Used for Priming Flexible Supports*, ICOMCC (2009).
7. J.D. Erlebacher, M.F. Mecklenburg, and S.J. Tumosa, *The Mechanical Properties of Artists' Acrylic Paints with Changing Temperature and Relative Humidity*, Polymer Preprints 33(2): 646–7 (1992).

Alteration, Technology and Interpretation of Archaeological Ceramics, Glazes and Glasses

Mater. Res. Soc. Symp. Proc. Vol. 1656 © 2014 Materials Research Society
DOI: 10.1557/opl.2014.829

Role of Weathering Layers on the Alteration Kinetics of Medieval Stained Glass in an Atmospheric Medium

Aurélie Verney-Carron[1], Anne Michelin[1], Lucile Gentaz[1], Tiziana Lombardo[1], Anne Chabas[1], Mandana Saheb[1], Patrick Ausset[1], Claudine Loisel[2]

[1]LISA UMR7583 CNRS/UPEC/UPD, 61 avenue du Général de Gaulle, 94010 Créteil, France.
[2]LRMH, USR 3224 CNRS, 29 rue de Paris, 77420 Champs-sur-Marne, France.

ABSTRACT

In order to model and predict the alteration of medieval potash-containing stained glass, it is necessary to understand the mechanisms of alteration layer formation at the glass surface and its role on the evolution of alteration kinetics. Moreover, the alteration layers observed on stained glasses are particular, as they are often fractured and heterogeneous in terms of thickness, with the appearance of pits and the detachment of scales. Contrary to silicate glasses altered in aqueous environment where the gel layer has a protective role, cracks and scales are harmful to the durability of stained glasses altered in air. In order to address these mechanistic issues, a program of experiments in the laboratory and in the field were performed. The fracturing was shown to be caused by the growth of the alteration layers and amplified by the alternation of humid and dry periods changing the density of hydrated layers. The pitting is initiated by defects at the glass surface and increased in external atmospheric medium as these defects fix the precipitated salts. However, despite fracturing and pitting, the development of an altered layer imposes a diffusive transport of the solution between the external medium and the bulk glass.

INTRODUCTION

Alteration of silicate glasses has geochemical consequences as the release of elements from natural glass contributes to the composition of rivers and oceans [1]. High-level activity nuclear wastes are also vitrified, and the assessment of their durability is essential to constrain the radionucleides release in the geological disposal [2]. Likewise, archaeological glass objects or stained glass windows are subject to alteration whether it is in buried soils, in seawater (shipwrecks), in museums or in contact with the atmosphere and they are required to be preserved as cultural heritage.

That is why numerous studies deal with the understanding of the glass alteration mechanisms, the measurements of the kinetics of these processes and the development of predictive models of alteration. Mechanisms of glass alteration are common to all silicate glasses in contact with aqueous solutions. Three mechanisms can be distinguished: (1) interdiffusion or ion exchange between glass alkali elements and hydrogenated species in solution (H_3O^+, H_2O) that leads to a selective leaching and to an increase of pH of the solution [3-5], (2) dissolution of the glassy network by hydrolysis of iono-covalent bonds (Si-O-Si, Si-O-Al) [6], and (3) secondary phase precipitation. These alteration mechanisms lead to the formation of alteration products. Interdiffusion and/or local reactions of hydrolysis/condensation cause the formation of a hydrated glass or a gel [7] for all kinds of glasses and alteration media. Secondary phases can also precipitate at the glass – solution interface. They mainly consist in phyllosilicates and carbonates in aqueous medium [8-9] or in salts (sulfates, nitrates, carbonates, phosphates) in atmospheric medium [10-12].

Alteration kinetics of the different mechanisms have been studied on ancient samples and in laboratory. The alteration rate evolves with time, and several kinetic regimes have been identified [13]. Kinetic parameters have been experimentally determined for several glasses: for basaltic glass [14, 15], for archaeological glass [16] and for nuclear glass [13]. They are specific to each glass composition.

Geochemical glass alteration models are based on a conceptual description of these mechanisms (diffusion, dissolution) and the experimental parameterization of associated kinetics. Speciation of elements in solution evolving as alteration progresses is calculated and leads to the precipitation of secondary products [13, 16, 17]. However, these models required relatively constant alteration conditions. This is in contrast with the atmospheric environment characterized by the alternation of dry and humid or rainy periods, causing greater complexity of processes. Indeed, climate plays a major role, especially water in the form of rainfall or gaseous condensation (corresponding to wet and dry deposition, respectively). Besides, over the last 200 years, the rising energy needs have been responsible for an increase in the emissions of gases (SO_2, CO_2, NO_x) and particulate matter. Especially, gases cause an acidic deposit onto glass which accelerates their dissolution and the formation of salts which can in turn improve alteration [18].

Even if alteration mechanisms are the same in atmospheric and in aqueous environment [19], stained glass weathered in atmosphere presents specific features, such as pits, cracks, and scaling [10, 20-22]. Their formation is not always understood and in the prospect of modeling the glass alteration in atmosphere, the assessment of their role on the glass alteration kinetics is required. The presence of cracks involves that the circulation of solution up to the interface between bulk and altered glass can be fast relative to the diffusion of solution through the porosity of alteration layers. The departure of scales (loss of material) also leads to the appearance of new bulk glass surfaces and can delay the establishment of an alteration rate drop. Moreover, local rainfall events favors an intense leaching of the surface, and evaporation results in very thin water films whose composition rapidly evolves in terms of composition and pH.

The observation of ancient glasses gives values of an apparent alteration rate (alteration thickness divided by exposure time). The mean rate of alteration (r_{alt}) ranges from 0.06 to 0.28 µm/year for medieval potash glasses [10, 20-22] and can be variable on a same piece of stained glass.

Thus, the objective of this paper is not only to understand the processes responsible for specific features observed on weathered stained glass windows, such as fracturing and pits, but also to assess their influence on alteration kinetics. For this, experiments are performed in controlled conditions on synthetic stained glass samples.

EXPERIMENTAL METHODOLOGY

Various experiments were performed on a synthetic glass that models ancient potash stained glass. Its composition obtained by SEM-EDS (Scanning Electron Microscopy - Energy Dispersive X-ray Spectroscopy) is (in wt.%): $SiO_2 = 55.5 \pm 0.7$, $K_2O = 21.1 \pm 0.6$, $CaO = 15.8 \pm$

0.3, $MgO = 3.2 \pm 0.1$, $Al_2O_3 = 2.4 \pm 0.1$ and $P_2O_5 = 2.0 \pm 0.3$ (standard deviation is calculated for various samples and zones). The glass samples (2 x 2 x 0.3 cm) are polished to obtain a micrometer scale roughness surface and have a geometrical surface of 10.4 cm^2.

Experiment 1 (1Unsh3 – 1Unsh6 – 1Sh36) consists in exposure of three glass samples to the urban atmosphere on top of Saint-Eustache tower in Paris: two were exposed unsheltered from the rainfall for 3 and 6 months (1Unsh3 and 1Unsh6) and one was placed in a sheltered condition during 36 months (1Sh36) [12].

Experiment 2 (2UnshDC – 2UnshDA) aimed at determining if the alternation of rainfall events and drying periods is responsible for fracturing of altered layers. For this, experiments in the laboratory were carried out on glass samples in dynamic conditions. The dynamic system (Figure 1) allows renewing a solution continuously thanks to a peristaltic pump (flowrate = 0.7 mL/min) and PharMed® tubes connected to a small Savillex® cell (60 mL) containing glass sample and placed in an oven at 50°C. The leachates are collected at regular intervals. The alteration solution is a synthetic rainwater whose composition is close to rains in Paris that we analyzed (Table 1). In the first experimental run, the solution is constantly renewed during 1 month (2UnshDC). In the second one, the time of leaching is also 1 month but the sample undergoes drying phases (3 days at ambient temperature after each 4-days alteration periods) (2UnshDA).

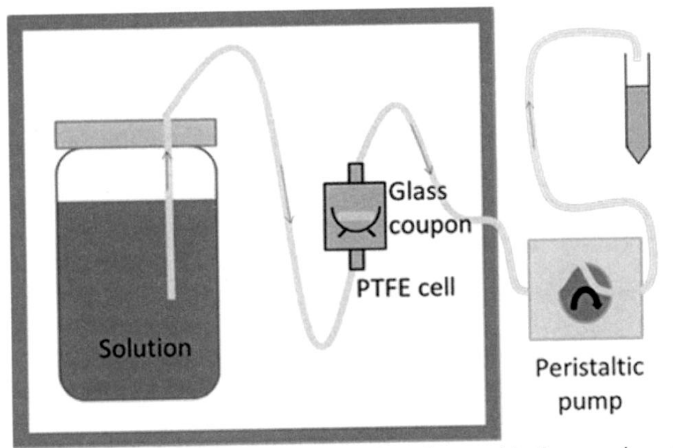

Figure 1. Dynamic system of glass alteration used in the experiments

Table 1. Composition of synthetic rainwater used in the experiment and similar to rain collected in Paris

Cations	Na^+	NH_4^+	Ca^{2+}	
C (mol/L)	$1,20 \cdot 10^{-4}$	$6,75 \cdot 10^{-5}$	$5,90 \cdot 10^{-5}$	
Anions	Cl^-	NO_3^-	SO_4^{2-}	HCO_3^-
C (mol/L)	$1,29 \cdot 10^{-4}$	$4,80 \cdot 10^{-5}$	$3,55 \cdot 10^{-5}$	$3,98 \cdot 10^{-6}$

Experiment 3 (3UnshC2) consists in exposing a glass sample placed in a PTFE reactor at 20°C containing 100 mL of synthetic rainwater during 635 days and maintained in static condition. Leachates are collected at regular intervals during the first month (1, 2, 7, 10, 14, 21 days).

Experiment 4 (4Ch1) aimed at reproducing a dry deposition in the Interactions between Materials and Environment Chamber (CIME). This device is a modular system for simulating varied climatic and environmental contexts, with the ability to integrate the climatic parameters (temperature, relative humidity) and gaseous pollutants (SO_2, NO_2). One glass sample was weathered for 37 days at 20°C with a relative humidity ranging from 80 to 92%. Five injections of 400 ppb of SO_2 and two of 80 ppb of NO_2 per day were carried out.

For experiments 1 to 3, the collected glass samples were dried and cut in two pieces using a diamond knife. One piece was directly gold-coated. The other piece was embedded in epoxy resin to prepare a cross-section, polished using SiC paper (grade < 1200) and diamond suspensions (9−3−1 μm) and then gold-coated. Samples were observed by SEM-EDS (JEOL JSM-6301F SEM linked with a Link ISIS300 EDS detector) at 20 kV. The sample of experiment 4 (4Ch1) was observed without coating by using a Hitachi TM3030 Tabletop SEM at 5 kV.

All the leachates of experiments 2 and 3 were acidified with pure HNO_3 (1% of the volume) and analyzed by Inductively Coupled Plasma – Atomic Emission Spectroscopy (ICP-AES) (Si, Al, Ca, Mg, K, P). The uncertainty ranged from 3% to 5% depending on the elements considered. The concentrations C_i (where i is the element analyzed) in $g \cdot L^{-1}$ are corrected for the dilution factor and are used to calculate the normalized mass loss of the glass (NL_i) in $g \cdot m^{-2}$:

$$NL_i = \frac{C_i \times V}{S \times x_i} \quad (1)$$

with V the volume of solution in the reactor, S the reactive surface area and x_i the mass fraction of i in the glass.

RESULTS AND DISCUSSION

Morphology of the altered layer: observations and processes

By comparing the morphology of altered samples exposed in unsheltered and sheltered conditions (1Unsh3, 1Unsh6 and 1Sh36) with similar alteration thicknesses by SEM-EDS, it is possible to highlight the role of the environment. The sample 1Unsh3 displays a zonation with dark altered zones and light zones corresponding to bulk glass (Figures 2a and 2b). Several cracks and some precipitated phases are present in the dark zones. The sample 1Unsh6 shows the same characteristics with an increasing alternation of dark altered zones (poor in K and Ca) and light pristine glass (Figures 2c, 2d and 3). The altered parts are much more densely fractured and some scaling is observed. The secondary phases on the surface probably correspond to calcium carbonate minerals ($CaCO_3$) and silica (SiO_2). The sample 1Sh36 also displays a zonation in the form of waves corresponding to K and Ca enriched or poor zones and caused by the weathering (Figures 2e and 2f). Some cracks are visible, as well as sylvite (KCl), calcium carbonate and sulfate (probably syngenite, $CaK_2(SO_4)_2 \cdot H_2O$) precipitation.

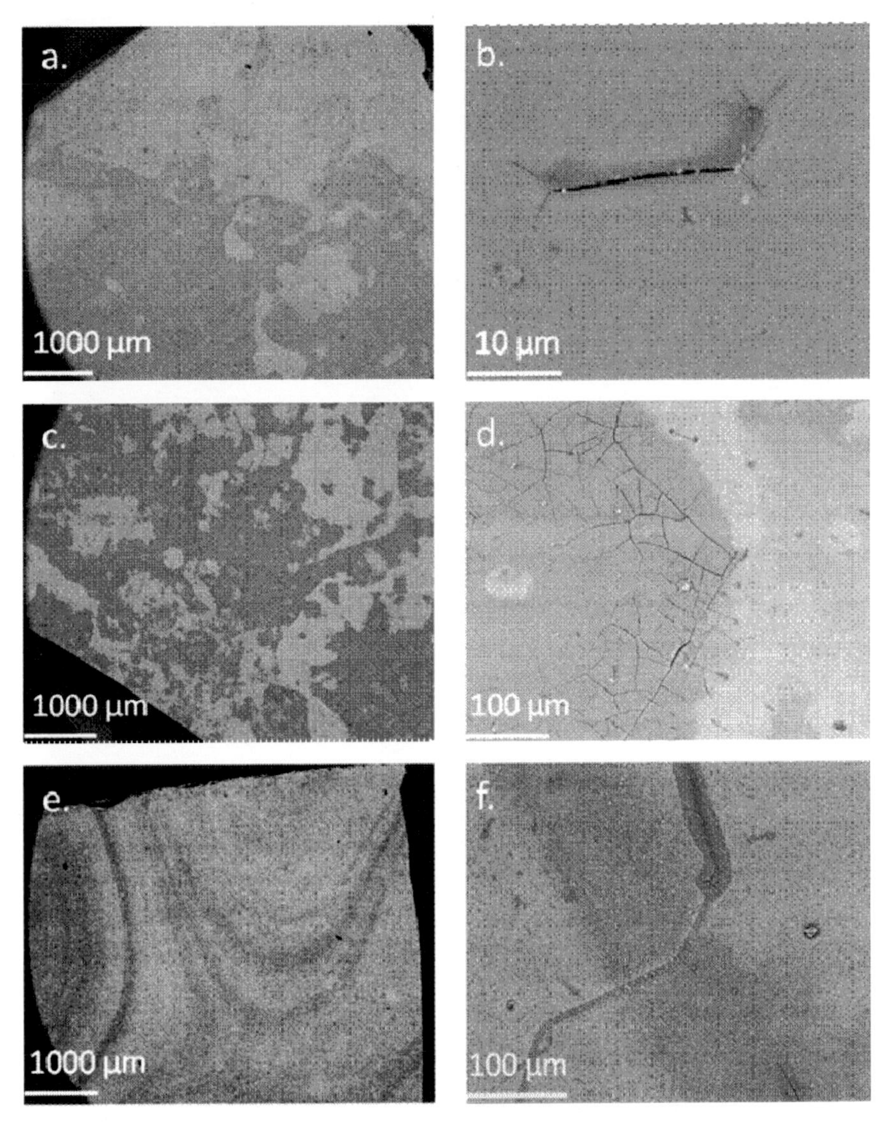

Figure 2. SEM photographs of the samples surface: 1Unsh3 (a,b), 1Unsh6 (c,d) and 1Sh36 (e,f). (a), (c) and (e) display a large view of the weathered surface of each sample, whereas (b), (d) and (f) zoom in on cracks present at the surface.

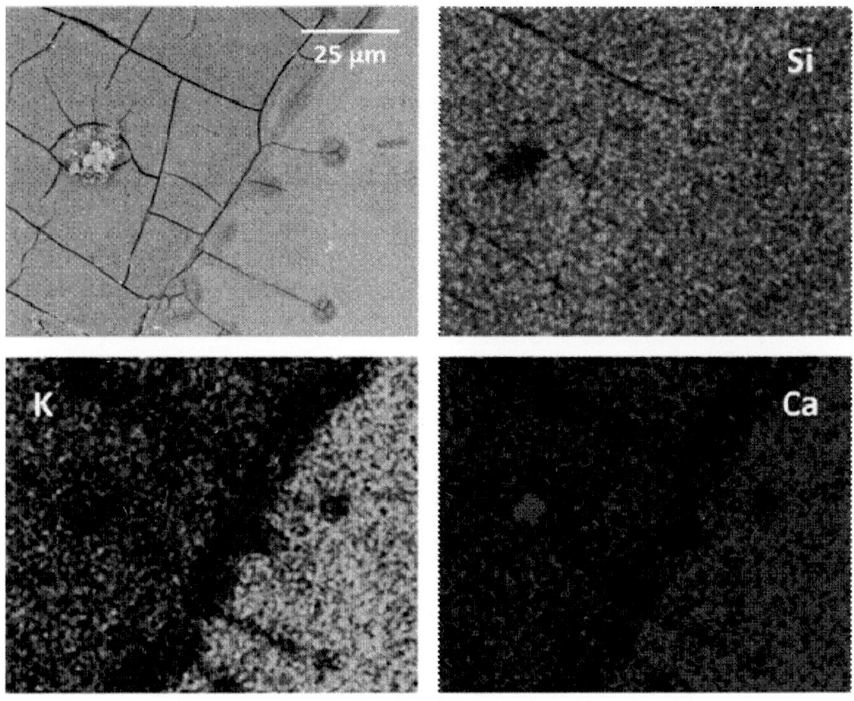

Figure 3. X-ray cartography of the sample surface 1Unsh6.

Cross-sections allowed altered layers to be analyzed and thicknesses to be measured. The altered layer is confirmed to be poor in K, Ca and Mg. 1Unsh3 displays an average alteration thickness of 750 nm (varying from 250 to 1000 nm), 1Unsh6 of 1700 nm (range of 500 nm to more than 2000 nm), and 1Sh36 of 1000 nm (range of 400 to 3200 nm).

The comparison of these samples shows that the morphology of altered layers, formed by an interdiffusion process, changes as a function of the sheltered (zonation) / unsheltered (scaling) exposure. The apparent alteration kinetics are also different, increasing by an order of magnitude between sheltered and unsheltered samples depending on the duration of exposure. Therefore fracturing seems to be induced more by the growth of the alteration layer (with an evolving density) than the environmental (rainfall, variations of relative humidity) conditions.

SEM observations of samples from experiment 2 (Figures 4a and 4b) show that the use of a dynamic system is too aggressive to reproduce realistic rainfall conditions. The glass is dissolved at a high rate (see below), leading to a very thin remaining hydrated layer (650 nm on 2UnshDC and 2UnshDA) poor in K, Ca and Mg, and local thick precipitation (between 1 and 18 µm on 2UnshDC) of Si, Al, Mg rich phases (probably phyllosilicates [16]), calcium phosphates and carbonates, caused by local hydrodynamic effects favoring saturation. Cracks are also present at the surface. However, despite the loss of glass by dissolution, the alternation of drying phases leads to an increase of the fracturing process on the remaining hydrated glass layers.

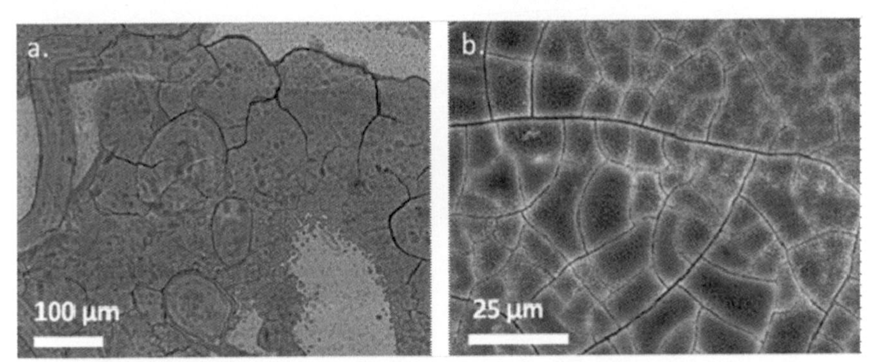

Figure 4. SEM photographs of the samples surface of 2UnshDC (a) and 2UnshDA (b).

The sample in experiment 3 (3UnshC2) was altered for 635 days in synthetic rainwater. Before its exposure, only face 1 of the sample was polished at the micrometer scale (Figure 5a) after cutting. Although this sample was in constant contact with solution, surfaces, especially the non-polished face 2, display numerous cracks and iridescence (Figure 5b) caused by the alteration.

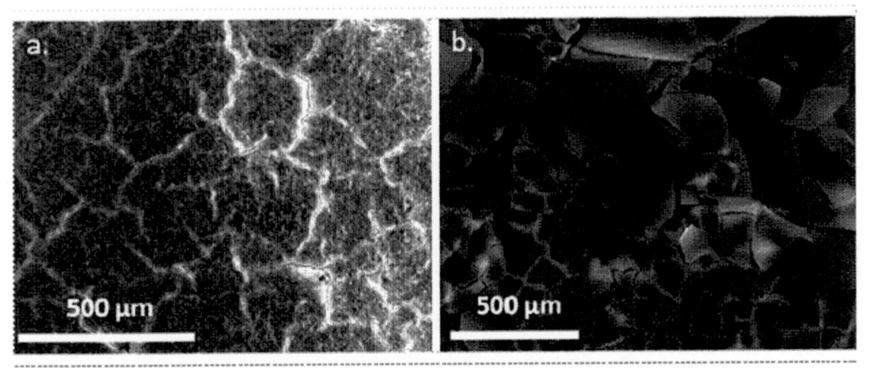

Figure 5. SEM photographs of the sample surface 3UnshC2: initially polished face 1 (a) and non-polished face 2 (b).

The difference of roughness between face 1 (Figures 6a and 6c), face 2 (Figure 6b) and lateral borders (Figure 6d) on sample 3UnshC2 induces different alteration morphologies. The alteration layer displays a relatively constant thickness on face 1 except for some local points, contrary to damaged face 2 and lateral borders. This shows that roughness or little cracks present on the initial surface involve very local chemical evolution of the solution. The release of alkalis and the consumption of protons by interdiffusion increase pH that favors the fast dissolution of glass. Moreover, the local saturation leads to the secondary phase precipitation.

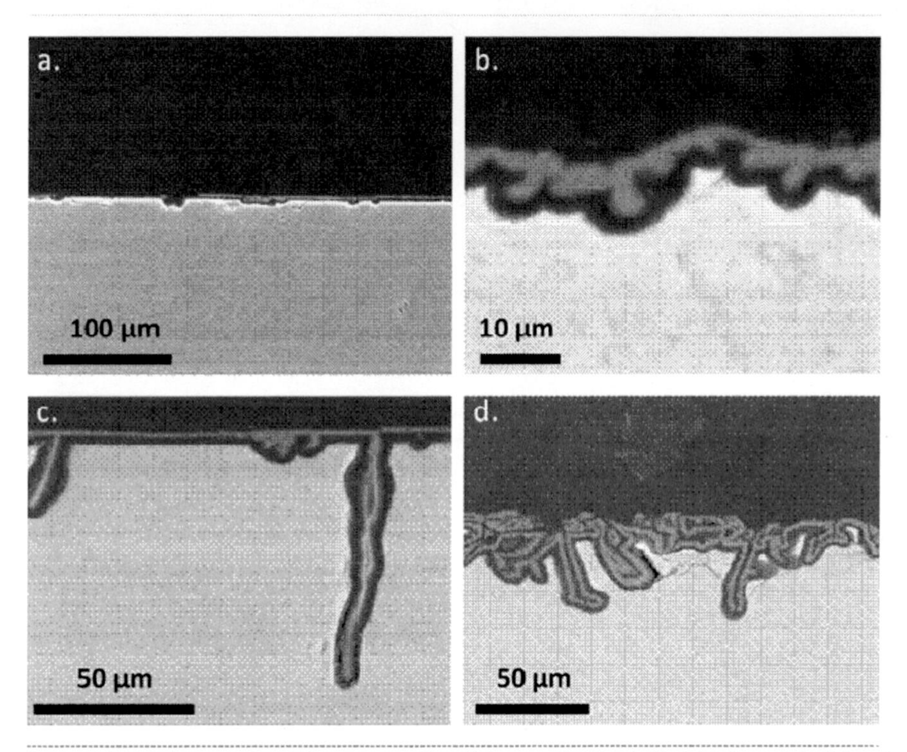

Figure 6. SEM photographs of the cross-section of the sample 3UnshC2: initially polished face 1 (a,c) and face 2 (b) and non-polished lateral border (d).

Finally, the SEM observations of the sample 4Ch1 weathered in the atmospheric chamber (CIME) show that sulfates (probably gypsum and syngenite) are formed on the surface (Figure 7). The size of these salts ranges from a few microns to 25 µm. Around the salt, a leaching zone shows that the alteration begins preferentially at the contact of salts. This in agreement with other's experiments [18], which have demonstrated that the presence of salts extend the time of wetness of the glass surface and can form saline solutions in the case of deliquescent salts leading to a strong leaching.

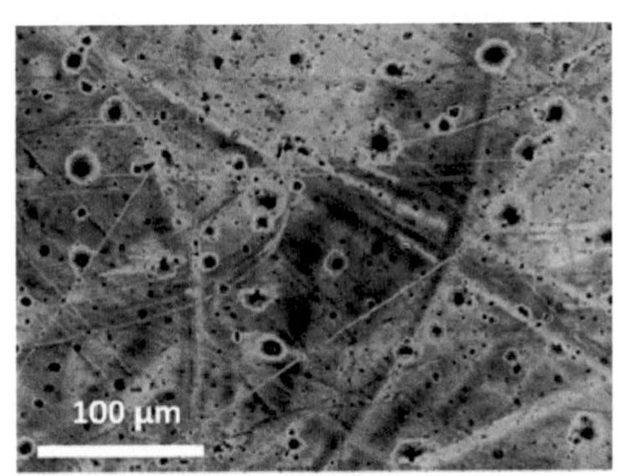

Figure 7. SEM photograph of the sample surface 4Ch1 (backscattered mode).

In summary, fracturing of the leached layer is caused by its growth with time, even in an aqueous medium, and its intensity is improved by the alternation of drying periods. The initial roughness results in a preferential leaching in isolated zones during rainfall events and becomes the predominant location for the fixation of salts in dry deposition conditions.

Evolution of the kinetics of alteration

Exposure conditions of stained glass windows are not well-known, but are never totally sheltered. The apparent alteration rate measured for medieval potash stained glass (after around 650 years of exposure) varies between 0.06 and 0.28 $\mu m \cdot year^{-1}$ in the altered zone (Table 2). This range could be underestimated if altered glass is lost but in some cases the initial surface is visible when it remains grisaille layer for example [22]. In order to constrain the evolution of alteration kinetic during several centuries, the kinetic data deduced from the different experiments are summarized in Table 2 and extrapolated over 650 years. The samples exposed in real conditions unsheltered from rain at 12°C for short durations have an apparent alteration rate of 3-3.4 $\mu m \cdot year^{-1}$. By assuming a linear rate and extrapolating it over 650 years, it gives an expected alteration thickness of 2000 µm, which is one order of magnitude higher than what has been observed on ancient samples. Apparent short-term alteration rate in sheltered condition (0.33 $\mu m \cdot year^{-1}$) is more in agreement with ancient samples.

Table 2. Alteration kinetic data deduced from the literature and the different experiments (r is the dissolution rate, D the diffusion coefficient, r_{app} the apparent alteration rate (measured by dividing the thickness by the time) and e the alteration thickness)

Source	Exposure condition	Temp. (°C)	r ($\mu m \cdot year^{-1}$)	D ($m^2 \cdot s^{-1}$)	r_{app} ($\mu m \cdot year^{-1}$)	e after 650 years (µm)
[10,22]	Unsh./Sh.	12			0.06-0.28	40-180
1Unsh3/6	Unsh.	12			3-3.4	2000
1Sh36	Sh.	12			0.33	220
2UnshDC/DA	Rainwater	50	407			
3UnshC2	Rainwater	20	21	$9.6 \cdot 10^{-19}$		
Calculation	Rainwater	12	9.8			6400

The analyses of leachates in experiments 2 and 3 show that the dissolution is congruent for Si, Al, K, Ca, Mg and P and the dissolution rate (r_0) is equal to 0.15 g·m^{-2}·d^{-1} at 20°C (i.e. 21 µm·year^{-1} by dividing by the glass density ρ = 2.6 g·cm^{-3}) (Figure 8a) and 2.9 g·m^{-2}·d^{-1} at 50°C (i.e. 407 µm·year^{-1}) (Figure 8b). There is no difference of kinetics between 2UnshDC and 2UnshDA as the dissolution of glass prevents the formation of a significantly thick altered layer. The activation energy (*Ea*) can be deduced by using an Arrhenius' law:

$$r_0 = k \cdot \exp(-Ea/RT) \quad (2)$$

with *k* a kinetic constant, *R* the ideal gas constant and *T* the temperature.

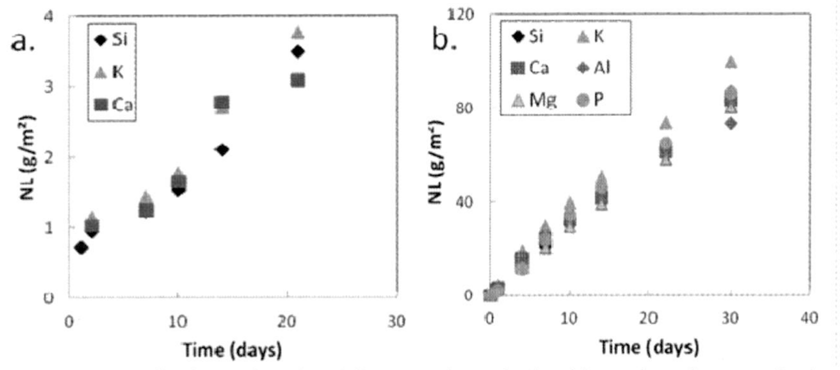

Figure 8. Normalized mass loss (Eq. (1)) versus time calculated from glass elements: for the experiment 3UnshC2 at 20°C (a) and 2UnshDA at 50°C (b).

The activation energy is found to be 75 kJ·mol^{-1}, which is typical of the dissolution of silicate glasses [16] and this gives an expected dissolution rate of 0.07 g·m^{-2}·d^{-1} or 9.8 µm·year^{-1} at 12°C (the average annual temperature in North of France) and an expected thickness of dissolution of 6.4 mm after 650 years. However, by considering that rainfall events occur 6% of time (548h·y^{-1} in average between 1961 and 1990 at Paris Montsouris, Meteofrance data), the expected thickness of dissolved glass would be 384 µm. This value is conservative as it assumes that rainfall events are strong enough to lead to a constant renewal of the solution and to a maximal dissolution rate, whereas there are only 16 days per year (over 187 rainy days) where precipitation is higher than 10 mm per day (Meteofrance data). However, this value of dissolved glass is too high compared to the observed glass loss on ancient samples.

Moreover, the solution analyses of experiment 2 and 3 (Figure 8) also lead to the calculation of a diffusion coefficient (*D*) in saturated conditions by using the second Fick's law:

$$e = NL_K/\rho = 2\sqrt{D \cdot t/\pi} \quad (3)$$

with *e* the alteration thickness, ρ the density of glass and t the time.

The diffusion coefficients are equal to $9.6 \cdot 10^{-19}$ $m^2 \cdot s^{-1}$ at 20°C and $2.4 \cdot 10^{-17}$ $m^2 \cdot s^{-1}$ at 50°C. By using Eq. (2) for the diffusion coefficient, the activation energy of the process is found to be 85 $kJ \cdot mol^{-1}$ that is in agreement with the literature data [16] and the diffusion coefficient at 12°C is $3.5 \cdot 10^{-19}$ $m^2 \cdot s^{-1}$, which is lower (more than one order of magnitude) than the values determined for ancient potash medieval glasses [23]. Nevertheless, by using this value, the expected alteration thickness is 1.9 μm after 3 months, 2.7 μm after 6 months and 96 μm after 650 years, consistent with measured values on 1Unsh3 and 1Unsh6, as well as ancient samples, respectively.

In summary, two hypotheses can be proposed to explain the long-term evolution of alteration kinetics. Either the glass is altered at a high dissolution rate during local events, or the alteration rate of glass is initially high and decreases with time. The first hypothesis can be questioned as the high short-term apparent alteration rate required events of too long duration and leads to huge altered thickness for long times. On the contrary, the second hypothesis based on a diffusion-controlled kinetics is consistent with short and long term exposures. This means that the alteration layer imposes a global diffusive transport, and the presence of cracks does not induce a significant preferential access of solution to the interface bulk – altered glass.

CONCLUSIONS

Medieval potash-containing stained glasses display characteristic altered layers with pits, cracks and scales. Fracturing is caused by alternation of humid and dry periods changing the density of hydrated layer, as well as by the growth of alteration layers during a nearly constant contact with an aqueous solution. The pitting alteration is found to be initiated by roughness and increased by the precipitation of salts at the surface in atmospheric medium. Both phenomena, fracturing and pitting, lead to the progression of the alteration at the bottom of pits often crossed by a crack. Moreover, the development of the altered layer leads to a diffusive transport through the altered glass, and thus, a progressive drop in the alteration rate.

ACKNOWLEDGMENTS

The authors would like to thank the French Ministry of Culture and Communication and ADEME for their financial support. They sincerely acknowledge an anonymous reviewer for its thorough reading.

REFERENCES

1. H. Staudigel and S.R. Hart, *Geochim. Cosmochim. Acta* **47**, 337-350 (1983).
2. C. Poinssot and S. Gin, *J. Nucl. Mater.* **420**, 182-192 (2012).
3. Z. Boksay, G. Bouquet and S. Dobos, *Phys. Chem. Glasses* **9**, 69-71 (1968).
4. R.H. Doremus, *J. Non-Cryst. Solids* **19**, 137-144 (1975).
5. R. Newton, "Deterioration of Glass," *Conservation of Glass*, eds R. Newton, S. Davison (Buttherworth Heinemann, 1989) pp. 135-164.
6. B.C. Bunker, *J. Non-Cryst. Solids* **179**, 300-308 (1994).

7. N. Valle, A. Verney-Carron, J. Sterpenich, G. Libourel, E. Deloule and P. Jollivet, *Geochim. Cosmochim. Acta* **74**, 3412-3431 (2010).
8. J. L. Crovisier, T. Advocat and J.L. Dussossoy, *J. Nucl. Mater.* **321**, 91-109 (2003).
9. A. Verney-Carron, S. Gin and G. Libourel, *Geochim. Cosmochim. Acta* **72**, 5372-5385 (2008).
10. J. Sterpenich J. and G. Libourel, *Chem. Geol.* **174**, 181-193 (2001).
11. I. Munier, R.A. Lefèvre and R. Losno, *Glass Technol.* **43**, 114-124 (2002a).
12. L. Gentaz, T. Lombardo, C. Loisel, A. Chabas and M. Vallotto, *Environ. Sci. Pollut. Res.* **18**, 291-300 (2011).
13. P. Frugier, S. Gin, Y. Minet, T. Chave, B. Bonin, N. Godon, J.E. Lartigue, P. Jollivet, A. Ayral, L. De Windt and G. Santarini, *J. Nucl. Mater.* **380**, 8-21(2008).
14. C. Guy and J. Schott, *Chem. Geol.* **78**, 181-204 (1989).
15. I. Techer, T. Advocat, J. Lancelot and J.M. Liotard, J. *Nucl. Mat.* **282**, 40-46 (2000).
16. A. Verney-Carron, S. Gin, P. Frugier and G. Libourel, *Geochim. Cosmochim. Acta* **74**, 2291-2315 (2010).
17. P. Frugier, T. Chave, S. Gin and J.E. Lartigue, *J. Nucl. Mater.* **392**, 552-567 (2009).
18. L. Gentaz, T. Lombardo, A. Chabas, C. Loisel and A. Verney-Carron, *Atm. Env.* **55**, 459-466 (2012).
19. I. Munier, R.A. Lefèvre, F. Geotti-Bianchini and M. Verità, *Glass Technol.* **43**, 225-237 (2002b).
20. N. Carmona, M.A. Villegas and J.M. Fernández Navarro, *J. Mater. Sci.* **41**, 2339-2346 (2006a).
21. N. Carmona, L. Laiz, J.M. Gonzalez, M. Garcia-Heras, M.A. Villegas, C. Saiz-Jimenez. *Int. Biodeter. Biodegrad.* **58**, 155-161 (2006b).
22. T. Lombardo, C. Loisel, L. Gentaz, A. Chabas, M. Verità and I. Pallot-Frossard, *Corr. Engin. Sci. Technol.* **45**, 420-424 (2010).
23. J. Sterpenich and G. Libourel, *J. Non-Cryst. Solids* **352**, 5446–5451(2006).

Mater. Res. Soc. Symp. Proc. Vol. 1656 © 2014 Materials Research Society
DOI: 10.1557/opl.2014.663

Study of Cloisonné enamel glaze of decorative components from Fuwangge in the Forbidden City by means of LA-ICP-MS and micro-Raman Spectroscopy

Hongying Duan[1] ,Liang Qu[1*],Xiaolin Cheng[2] ,Yan Su[3], Aiguo Shen[3] and Shiwei Wang[4]
[1] Conservation Department, the Palace Museum, Beijing, 100009, China
[2] Center for Conservation, National Museum of China, Beijing, 100006, China
[3] College of Chemistry and Molecular Sciences, Wuhan University, Wuhan, 430072, China
[4] Architecture Department, the Palace Museum, Beijing, 100009, China
* Corresponding author, email: lionat528@gmail.com

ABSTRACT

Two Cloisonné enamel architectural components from Fuwangge in the Forbidden City that were produced from Yangzhou (one production center) in Qing Dynasty (1616-1911 A.D.) were chosen and analyzed. A combination of Laser ablation-inductively coupled plasma-mass spectrometry (LA-ICP-MS) and micro-Raman spectroscopy was successfully used to analyze eight colors in enamel glazes (yellow, white, pink, turquoise, yellow green, deep blue, red and deep green). Chemical composition results reveal that the enamel glaze matrix belongs to lead-potash-lime glass (PbO-K_2O-CaO-SiO_2). Based on Raman spectroscopy, lead-tin yellow types II, cassiterite, lead arsenate, fluorite and hematite were found as opacifiers and/or colorants. In addition, a detailed discussion of raw materials, such as fluorite and borax, might provide valuable information to trace manufacturing technology and provenance.

INTRODUCTION

The enamel technique is the application of colored glass to metal substrate for decoration. This technique is known as early as the Aegean culture sphere in Cyprus in the thirteenth century B.C [1]. It is widely accepted that in the Yuan Dynasty (1271-1368 A.D.) Chinese artisans took the art of enamel from the western countries and developed it (falang, Cloisonné enamel). During Emperor Qianlong reign of Qing Dynasty (1736-1795A.D.), because it was deeply appreciated by the emperor, the technique of Cloisonné enamel had great development and Cloisonné enamel was widely used in royal artifacts, including decorative, sacrificial, daily use and architectural elements. There were several famous manufacture centers in Qing Dynasty (1616-1911 A.D.): Beijing (The Royal Workshop), Yangzhou, Suzhou and Guangzhou. According to literature record, batches of Cloisonné enamel components made in Yangzhou were adopted for architectural interior furnishing of Fuwangge [2]. Fuwangge, located northeast of the Forbidden City (see Figure 1), was built in the 36th year of Emperor Qianlong reign (1771 A.D.). These enamel components represented the highest quality handiworks of the country at that time [3]. In this work Cloisonné enamel samples from Fuwangge were analyzed to study raw materials and manufacturing technology of Yangzhou Cloisonné enamel in the Qing Dynasty, and to provide scientific evidence for copy and supplement work of enamel components. This will be helpful in our future work to explore the difference among different manufacture centers.

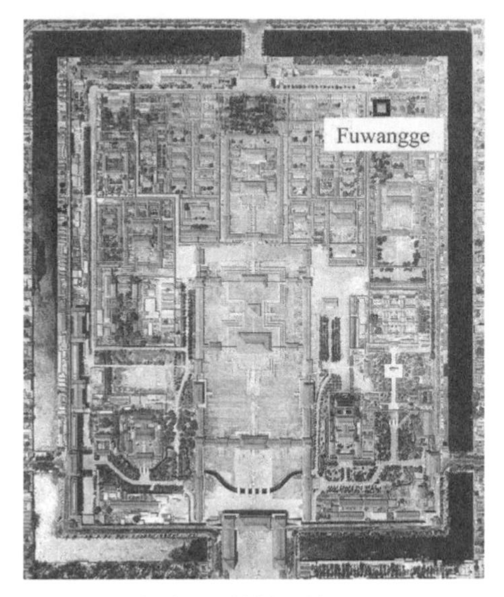

Figure 1. The location of Fuwangge in the Forbidden City

Although enamel glaze is similar to ancient glass, their research, especially analytical study, is far less popular than ancient glass. Concerning Western enamel glaze studies, most focus on chemical composition analysis through both non-invasive and destructive analytical methods, such as electron probe microanalysis (EPMA)[4], proton-induced X-ray emission (PIXE)[5], scanning electron microscopy-energy dispersive spectroscopy (SEM-EDS)[6], energy dispersive X-ray fluorescence (EDXRF)[7,8]. On basis of chemical composition, the flux, opacifier, and colorant are discussed, which is helpful to understanding raw materials, technology, period of manufacture and provenance studies.

Very limited analytical studies have been conducted on Chinese Cloisonné enamel glaze. To date, to the best of our knowledge, only four works have been published. Two used a destructive sampling for SEM/EDS, X-ray diffraction, EPMA, and atomic emission spectroscopy analysis [9,10]; the other two are non-invasive analyses using PIXE and Raman spectrometry [11,12]. In our case, to ensure the integrity and protect the value of the artifacts, only non-invasive or micro-destructive methods for chemical composition and molecular structure analysis should be applied to the Cloisonné enamel components which have defined manufacture time and origin.

Laser ablation-inductively coupled plasma-mass spectrometry (LA-ICP-MS) possesses many advantages such as low limits of detection, rapid analysis, determination of 50–60 major, minor and trace elements simultaneously and minimal damage to the object [13]. The direct sampling method on solid samples, allows on site analysis and micro-destruction to samples, which is optimal for the analysis of valuable silicate relics such as ceramics [14] and glass [15]. However, LA-ICP-MS analysis does not provide a molecular fingerprint for the area under observation. As demonstrated by the investigation of various glass [16], ceramics [17], and enamels [18], micro-Raman spectroscopy is an ideal tool to study the colorant and opacifer in objects made from glass because it is non-invasive, has high spatial and spectral resolution, and can be used on samples of non-uniform shape. This provides a valuable way to trace their possible raw material, manufacturing techniques and origins.

In the present study, the combination of LA-ICP-MS and micro-Raman spectroscopy were successfully used to analyze Cloisonné enamel decorative components from Fuwangge in the Forbidden City. The emphasis was focused on the enamel glass matrix and the identification of opacifiers and colorants. Based on the results obtained, raw materials and manufacturing techniques are discussed briefly.

CLOISONNE ENAMEL SAMPLES

Through investigation of Cloisonné enamel components in Fuwangge in the Forbidden City, two Cloisonné enamel architectural elements were chosen. Sample FWGFL-1 contains turquoise and dark blue, two common color enamel glazes. Sample FWGFL-2 covered eight kinds color Cloisonné enamel glazes for a representative and comprehensive survey of materials. A general description of the samples is listed in Table 1, with photographs in Figure 2 and Figure 3. In both artifacts all areas are opaque or translucent, indicating that certain opacifiers were used. Many holes were found on Cloisonné enamel glazes. Wax, or wax with colored pigment were used to fill those holes.

Table 1. Description of samples

Sample		Decoration	Date	Color
FWGFL-1	FWGFL-1-D01	*kui* dragon pattern	Qianlong reign of Qing Dynasty (1736-1795 AD)	dB (dark blue)
	FWGFL-1-D02			Tu (turquoise)
FWGFL-2	FWGFL-2-D03	Scrolling lotus and bat pattern		dB
	FWGFL-2-D04			Tu
	FWGFL-2-D05			W (white)
	FWGFL-2-D06			dG(deep green)
	FWGFL-2-D07			yG (yellow green)
	FWGFL-2-D08			Y (yellow)
	FWGFL-2-D09			R (red)
	FWGFL-2-D10			P (pink)

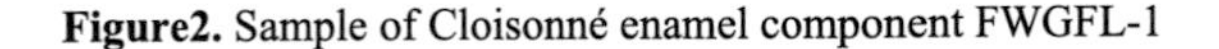

Figure2. Sample of Cloisonné enamel component FWGFL-1

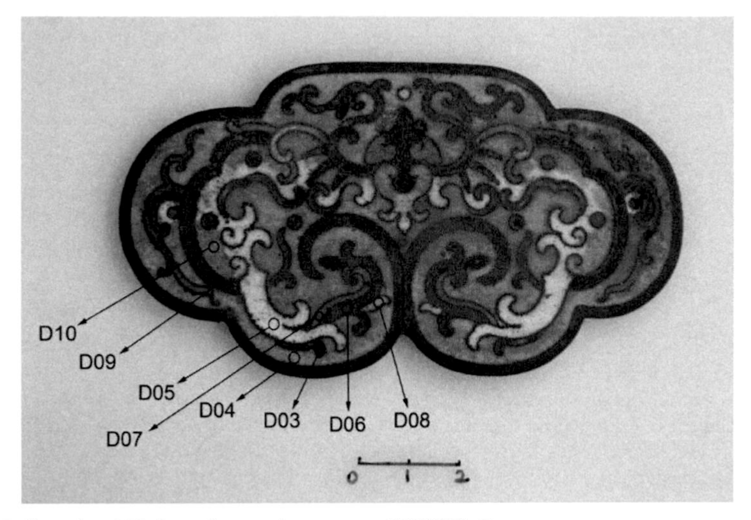

Figure3. Sample of Cloisonné enamel component FWGFL-2

EXPERIMENTAL

The analyses carried out involved ICP-MS (Finnigan Element II, Germany) and laser ablation system (New Wave UP213, United States) for direct introduction of solid samples. The single point analysis mode with a laser beam diameter of 40μm, operating at 80% of the laser energy (0.11 mJ) and at a pulse frequency of 10 Hz was used. A pre-ablation time of 40 s was set up in order to eliminate the transient part of the signal and be sure that a possible surface contamination or corrosion does not affect the results of the analysis. Blank count rates were measured for 20 s prior to ablation. For each glass sample, an average of three measurements corrected from the blank was considered for the calculation of concentrations. Two different standard reference materials, NIST SRM 612 (United States) and KL2-G (Germany) were used and ^{29}Si isotope was used for internal standardization. Concentrations for all components, including silica, were calculated assuming that the sum of their concentrations in weight percent equal to 100%.

The Raman microscopy used in this study was Nicolet Almega spectrometer system configured with an Olympus BX-50 microscope and an integrated motorized stage. Objectives of 10 × and 50 × were used. The excitation wavelengths, 532 nm was used in correspondence to light source of frequency doubled Nd:YVO DPSS laser. The laser power on the sample surface was set to a value 7 mW.

RESULTS AND DISCUSSION

Enamel glaze matrix and raw material

In our previous work, FWGFL-1 was sampled and measured by many elemental analysis methods to evaluate and confirm the feasibility of LA-ICP-MS, which provided almost all elements composition results in Cloisonné enamel glaze except Fluorine [19]. Table 2 lists major, minor and some trace elemental compositions of two objects analyzed by LA-ICP-MS.

Table 2. Chemical composition of analyzed samples using LA-ICP-MS (wt%)

Content	D01 dB	D02 Tu	D03 dB	D04 Tu	D05 W	D06 dG	D07 yG	D08 Y	D09 R	D10 P
B_2O_3	5.48	4.20	5.62	3.91	5.93	4.58	1.86	0.02	5.25	6.55
Na_2O	2.25	1.42	2.25	1.87	2.77	2.03	0.91	0.13	2.44	5.52
MgO	0.10	0.03	0.10	0.04	0.05	0.08	0.67	0.09	0.07	0.14
Al_2O_3	0.63	0.20	0.65	0.23	0.29	0.53	0.45	0.78	0.91	0.51
SiO_2	46.25	40.93	46.30	40.02	33.00	38.55	28.72	26.87	34.14	51.43
P_2O_5	0.05	0.01	0.05	0.01	0.01	0.01	0.01	0.01	0.02	0.01
K_2O	18.10	15.96	18.78	19.36	17.78	14.90	9.32	4.10	14.12	20.23
CaO	10.61	7.63	10.77	9.23	13.97	7.32	2.66	0.96	2.26	4.75
TiO_2	0.03	0.01	0.03	0.01	0.01	0.02	0.02	0.02	0.03	0.03
MnO	0.03	0.00	0.03	0.00	0.00	0.08	0.00	0.00	0.00	0.01
Fe_2O_3(t)	0.75	0.09	0.77	0.09	0.09	3.43	0.09	1.54	3.62	0.24
PbO	9.37	25.01	8.06	22.38	21.36	24.14	48.68	57.45	32.59	6.63
Co	0.34	0.00	0.36	0.00	0.00	0.01	0.00	0.00	0.00	0.00
Ni	0.16	0.00	0.17	0.00	0.00	0.00	0.00	0.00	0.00	0.00
Cu	0.03	1.35	0.05	1.49	0.01	2.83	1.08	0.12	2.28	0.01
Zn	0.01	0.00	0.02	0.00	0.00	0.02	0.00	0.06	0.25	0.00
Sn	0.00	0.00	0.00	0.00	0.00	0.03	5.41	7.79	1.91	0.01
Sb	0.01	0.01	0.00	0.01	0.01	0.01	0.02	0.01	0.02	0.00
As	5.44	3.11	5.60	1.32	4.66	1.36	0.02	0.01	0.04	3.87
Bi	0.24	0.00	0.25	0.00	0.00	0.00	0.03	0.00	0.00	0.00
Au*	0.0	3.4	0.0	0.0	0.4	0.4	0.4	0.5	0.2	58.8

* unit of Au content is μg/g

The chemical composition of dark blue and turquoise areas in samples FWGFL-1 and FWGFL-2 is very close (see table 2), suggesting the raw materials and manufacturing technology of Fuwangge Cloisonné enamel glaze was very uniform. The eight different color enamel glazes all contained lead oxide with minimum and maximum values of 7% and 57%, and most contents were above 20%. The other major components were silica, potassium oxide and calcium oxide, and total content of these four components ranged from 83% to 91%. The average contents of sodium and potassium oxide were 2.2% and 15.3% respectively, illustrating dominant alkali was potassium oxide. Therefore, enamel glaze matrix belongs to lead-potash-lime glass (PbO-K_2O-CaO-SiO_2).

According to the published data, Western enamel glazes are soda-lime-based glass or lead-soda-lime-based glasses [18]. Some painted Limoges enamels are potash-lime-lead glass [18]. However, most Chinese Cloisonné enamels analyzed (fifteenth, sixteenth, eighteenth and nineteenth centuries) have a similar composition and are lead-based potash-lime type glasses [9, 11]. The result of our analyzed Cloisonné enamel components from Fuwangge, assigned to eighteenth century production, also supported this result. The difference in alkali between Western enamel (sodium) and Chinese Cloisonné enamel (potassium) may be due to different abundance of raw materials. This situation was also the same in ancient glass of China and the West. Although the origin and development progress of Chinese potassium glass was complex, K_2O-CaO-SiO_2 glass dominated from the Yuan to the Qing Dynasty (thirteen to nineteenth century) [20]. In late Yuan Dynasty, KNO_3 had already been utilized and added into glass as flux in Boshan, which was the biggest ancient glass manufacture center in the Ming and Qing Dynasties [21, 22]. The use of KNO_3 in glass recipes as flux was very popular in the Qing Dynasty [23]. Chinese Cloisonné enamel manufacture technology originates from the West and achieved the peak in Emperor Qianlong reign of Qing Dynasty. After such a long period development, native enamel glaze raw materials were very consistent [24]. Therefore, from the point of technology maturity and raw material resources, it was quite reasonable that native potassium glass recipe was applied to Cloisonné enamel.

It would be easier to study Cloisonné enamel raw materials if ancient artisans recorded enamel glaze recipes. Unfortunately they did not have this habit. Wangshixiang found one artisan manuscript about size, material of one Cloisonné enamel tower made in the 37th Qianlong reign. He published that valuable raw material information which was recorded in local mineral names [25]. In combination with our chemical composition, the possible chemical substances of those minerals are: Mayashi (Quartz, SiO_2), Dingfen ($PbCO_3$), Pengxiao (KNO_3), Pishuang (arsenic trioxide, As_2O_3), Pengsha (borax, $Na_2B_4O_7 \cdot 10H_2O$) and Zishi (fluorite, CaF_2).

Table 3. The contents of B_2O_3 and Na_2O and their mole ratio in Cloisonné enamel glaze samples

Sample	D01	D02	D03	D04	D05	D06	D07	D08	D09	D10
B_2O_3	5.48	4.20	5.62	3.91	5.93	4.58	1.86	0.02	5.25	6.55
Na_2O	2.25	1.42	2.25	1.87	2.77	2.03	0.91	0.13	2.44	5.52
n_B/n_{Na}	2.17	2.63	2.22	1.86	1.90	2.01	1.81	0.11	1.92	1.06
Mean $n_{Na}/n_B{}^*$	1.95									

* Result of D08 was not calculated for low content of boron trioxide and sodium oxide which may be impurities from other raw materials

As shown in table 3, mean mole ratio of sodium to boron is 1.95, very close to 2 which is the ratio of sodium to boron in borax ($Na_2B_4O_7 \cdot 10H_2O$). Therefore it could be deduced that sodium and boron were both from borax. Boron is difficult to detect in many analytical methods; therefore the presence of boron in enamel glazes has not been quantitatively determined. According to Bertran [26], nineteenth century enamel starting compositions contained significant amount of borax. The use of borax before the nineteenth century has to be questioned. Besides our study, boron was also qualitatively detected in Cloisonné enamel of Qing Dynasty [10], which confirms the use of borax in China Cloisonné enamel before the nineteenth-century. In one alchemy book of Tang Dynasty (616-907 A.D.), one glass recipe recorded the addition of boron into PbO-K_2O-SiO_2 matrix glass. This is the earliest document about the use of boron in glass in China. But this recipe failed to spread and soon disappeared [27]. It is not until the Qing

Dynasty that boron was found again in glass [28, 29]. In addition to glass, contemporary enamel decorative porcelain (Falangcai), imitation of Cloisonné enamel, where glass is applied to a porcelain body, were found to contain boron [30], demonstrating boron was one characteristic of enamel glaze. It mainly functioned as flux, but also can adjust the dilatation coefficient of enamel glaze to avoid scaling from the metal body [31].

From the data in table 2, it is difficult to determine zishi raw materials because element fluorine cannot be detected by LA-ICP-MS. In our previous work, enamel glaze powders of sample FWGFL-1 had been measured through EMPA and a potentiometric titration method. The results of fluorine content are listed in table 4. It was found that certain content of fluorine exists in the enamel glaze. In the Raman spectrum, the small peak around 319 cm^{-1}, which was detected in white, turquoise, deep blue and deep green colored (Figure 4c,e-g) enamel glazes, corresponds to fluorite (CaF_2). CaF_2 was observed by TEM in sixteen century Chinese Cloisonné enamel [9]. CaF_2 was added as opacifier to make enamel glaze opaque, just as it does in glass. The period where CaF_2 was used as a glass opacifier in China was much earlier than in west. It can be traced back to the Tang Dynasty through the analysis of opal glass. CaF_2 was also found in blue glass excavated in Boshan glass factory (workshop) site of the fourteenth century [32]. According to Turner, it was not until the nineteenth century that CaF_2 opacifier was used in west [33]. Nowdays CaF_2 is a popular opacifier in modern western metal enamels.

Table 4. Analytical results of fluorine content of FWGFL-1 (wt/%)

Sample	Glaze color	F (%)	
		EMPA	Potentiometric titration
FWGFL-1-D01	dB	6.47	5.13
FWGFL-1-D02	Tu	4.41	4.09

Opacifiers and colorants

Under the microscope, crystalline materials were observed which represent pigments, opacifiers or residues of raw materials distributed heterogeneously within the enamel glaze matrix. Figure 4 shows the representative Raman spectra of eight color enamel glazes obtained by focusing the laser beam directly on the micro-crystalline materials dispersed in the vitreous matrices. It should be noted that quality of some spectra was poor which were related to their poor optical quality as they had bubbles, stains and wax on the surface.

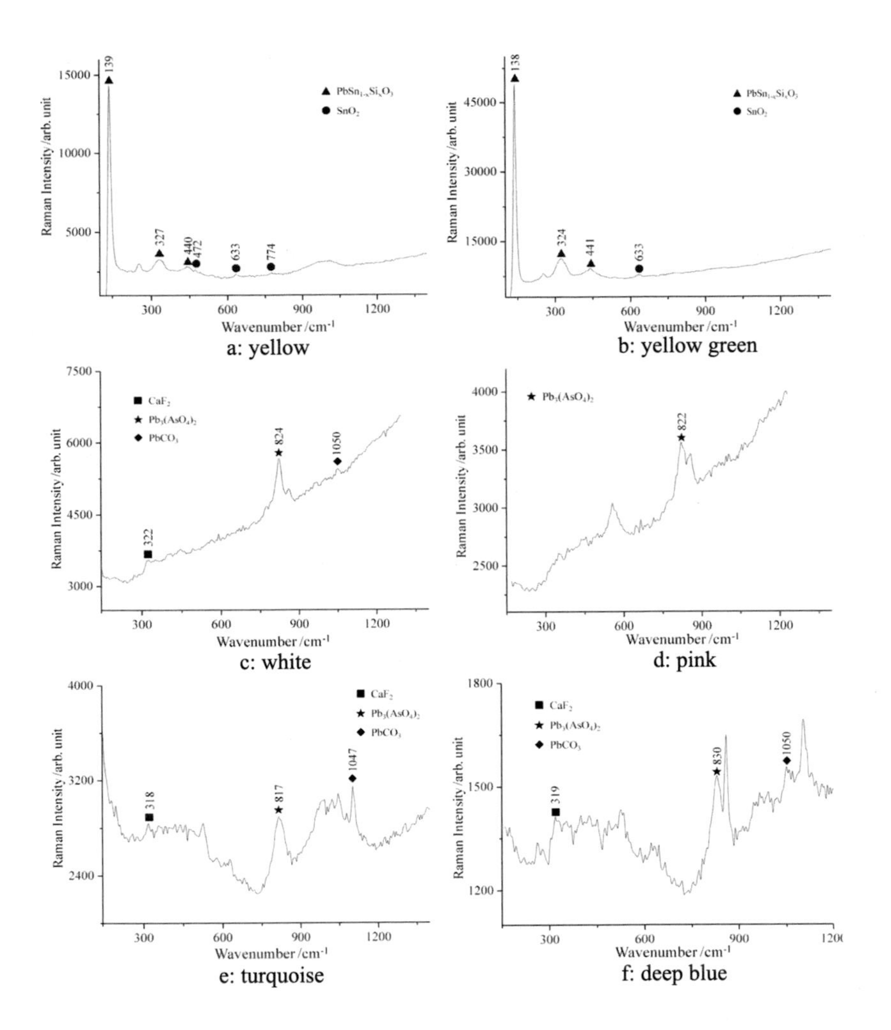
a: yellow
b: yellow green
c: white
d: pink
e: turquoise
f: deep blue

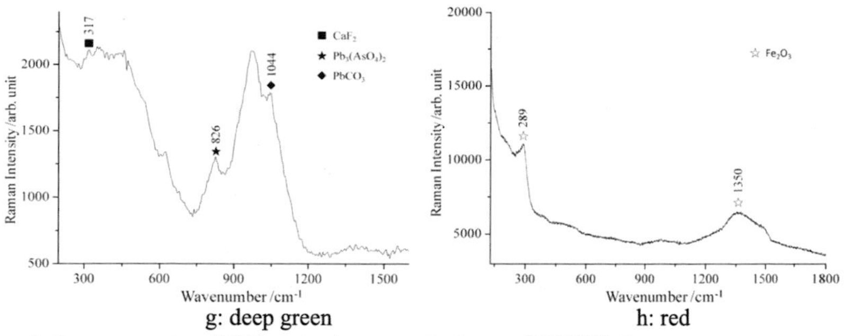

Figure 4. Raman spectra of different color enamel glazes of FWGFL-2

As shown in Figure 4a (yellow) and 4b (yellow green), typical Raman spectra with bands at 138–139, 324–327, and 440–441cm^{-1} clearly matched that of lead-tin yellow types II ($PbSn_{1-x}Si_xO_3$) [34,16]. In addition to $PbSn_{1-x}Si_xO_3$, small amounts of SnO_2 (cassiterite) also existed. The Raman bands at 472, 633, and 774 cm^{-1} coincide with the typical Raman bands for cassiterite reported previously [34, 35]. According to the classification of opacifiers by Freestone [36], SnO_2 (cassiterite) belongs to Type A (colorless or white color opacifier) and lead-tin yellow types II ($PbSn_{1-x}Si_xO_3$) belongs to Type B (colorful opacifier with opaque and coloring function in the process of glass production). In yellow and yellow green enamel glaze, tin-based opacifiers and colorants including $PbSn_{1-x}Si_xO_3$ and SnO_2 were found. The yellow green color was due to the joint effect of lead-tin yellow types II particles and Cu^{2+} ions (1.08 wt%).

In white, pink, turquoise, deep blue and deep green enamel glazes, 1.4-5.6% wt% As_2O_3 was detected (Table 2). The Raman spectra of these glazes with bands at 817- 830 cm^{-1} (Figure 4c-g) confirms that arsenate (AsO_4^{2-}) was used as opacifier and colorant (only white) in those enamel glazes. As an opacifier, arsenate was found in glass, enamel and porcelain. Because Raman bands of many compound containing arsenate are close to each other, Raman bands around 825 cm-1 has been regarded as Ca/Pb arsenate [37,38] or lead arsenate [18,9,39]. In China, it was documented that the material "Bolibai" which contains lead arsenate was used in enamel decorative porcelain and Famille Rose porcelain in the Qing Dynasty as mild opacifier [30,40]. Cross-influence and link of raw materials and technical knowledge existed among these crafts [9], indicating lead arsenate was the opacifier in China Cloisonné enamel. The small peaks, at 317-322 and 1044-1050 cm-1, which were detected in white, turquoise, deep blue and green enamel glaze, corresponded to fluorite (CaF_2) and lead carbonate ($PbCO_3$). As discussed above, the former was functioned as an opacifier and the latter was an unmelted raw material in the enamel glaze.

In spectra of Figure 4d-g, there was no Raman signature of a pigment but only that of opacifier and glass matrix. In combination with chemical composition in table 2, it could be concluded that the coloring agents in those four color enamel glazes were metal oxides dissolved into glass matrix or a dispersion of metal nanoprecipitates, which have no distinctive Raman signals (except high metal ions content [11,37,38]). The content of Au in pink colored enamel glaze is 59 μg/g, which was far higher than others glazes. It is speculated that pink color was attributed to the presence of gold particles, which is consistent with Widauer [9]. The technology of Au colorant originated from the West in early Qing Dynasty and was applied to Cloisonné enamel, enamel decorative porcelain and glass [17, 23, 30]. As shown in Figure 5, there is no obvious boundary between the white and pink glaze, revealing that the pink glaze was obtained by a two layer technique: a white layer that is coated by Au containing colored glaze. Copper in its oxidized state (Cu^{2+}) is responsible for turquoise and deep green (in high lead lime-alkali glazes [9,41]) color in glazes and a deep blue color was obtained by Co^{2+} (0.36 wt%).

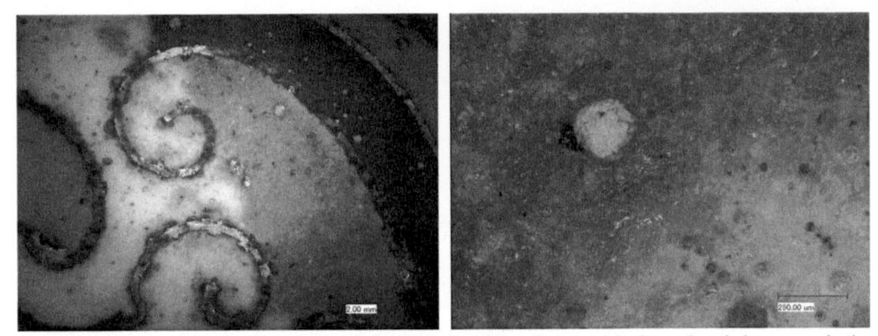

Figure 5. Microscope image(left:20×, right:200×) of two layer technique in pink enamel glaze

For red color enamel glaze, hematite (F_2O_3) was detected with its characteristic peak at 289 cm^{-1} and a broad band near 1350 cm^{-1}. No Raman signals of cuprite (Cu_2O) which was very popular in Celtic enamels and glass were found [34, 42, 43]. Although the content of Sn in red enamel glaze reached 1.91 wt%, no peaks of Sn-containing opacifier was found in Figure 4h, which may be due that no suitable spots were chosen and analyzed. It was reported that Sn-containing glazes had no Raman signal, demonstrating that actually tin remains dissolved in the glass and does not contribute to the opacification of the glaze [44]. More attention should be paid on opacifier in red enamel glaze in our next work.

Tin based opacifiers, one of the most common types opacifier used in glass objects, were first applied to glass production in Europe from the second to first centuries B.C. [45]. From then to the fourth century A.D., tin based opacifiers and antimony based opacifiers coexisted in glass production, and a gradual transition was made from antimony based opacifiers to tin based opacifiers. After that, tin based opacifiers were widely used in the making of glasses, glazes, mosaics and enamels in Europe and western Asia. A new type arsenate opacifier was developed in Italy (Venice) to make lattimo glass during the sixteenth century [46] and became a commonly used opacifying agent in the eighteenth and nineteenth centuries. However, tin oxide still remained in use, either alone or mixed with lead arsenate. In our study, tin based opacifier (in yellow and yellow green color) and arsenate opacifier (in white, pink, turquoise, deep blue and deep green) were both found. From their use situation, we concluded that arsenate opacifier was more common than tin based opacifier in eighteen century China Cloisonné enamel.

CONCLUSIONS

This study demonstrates the potential of a combination of LA-ICP-MS and micro-Raman spectroscopy for the analysis of Chinese Cloisonné enamel. LA-ICP-MS, as a micro-destructive method, was first adopted to obtain enamel glaze chemical composition by providing sixty major, minor and trace elemental composition results, including boron which is difficult to detect by many analysis methods. Raman spectroscopy was a useful and powerful tool for micro-crystalline identification in the glass matrix to obtain opacifier and colorant information. This combination can also be used in porcelain glaze, mosaic and glass.

The enamel glaze matrix belongs to lead-potash-lime glass (PbO-K_2O-CaO-SiO_2). The main alkali was potassium instead of sodium, which was the dominant alkali in the West. This situation was also the same in ancient glass of China and the West, which may be caused by different abundance raw materials. Fluorite (CaF_2) and borax ($Na_2B_4O_7 \cdot 10H_2O$) were demonstrated as enamel glaze raw materials, which was consistent with historical documents. Lead-tin yellow types II ($PbSn_{1-x}Si_xO_3$), cassiterite (SnO_2), lead arsenate ($Pb_3(AsO_4)_2$), fluorite (CaF_2) and hematite (Fe_2O_3) were found as opacifiers and/or colorants in enamel glazes. Transition metal ions, such as Co^{2+} and Cu^{2+}, dissolved into glass matrix played colorant role in deep blue, turquoise and deep green enamel glazes. A two-layer technique was used in pink enamel glaze by coating a gold containing colored layer on a white ground layer.

Limited by the number of analyzed samples, the present work is preliminary. In our future work, for the purpose of better understanding of raw materials and production technology of Chinese Cloisonné enamel, more attention will be focused on analysis and comparison of additional enamel samples, not limited to architectural decorative objects, from different manufacturing centers and in different periods.

ACKNOWLEDGMENTS

The authors would like to thank Dr. Qinghui Li and Dr. Yong Lei for discussions and important view of the manuscript. This research is supported by the National Science Foundation of China (No.51102051).

REFERENCES

1. H. Brinker, A. Lutz, in *Chinese cloisonnes. The pierre uldry collection,*(The Asian Society Publishers, New York, 1989) p.23.
2. SX. Zhang, in *From Palace to local - The technological exchange of 17th and 18th centuries*, (Forbidden City Press, Beijing, 2010) p.163.
3. Y. Li and Shiwei Wang, *Forbidden City*. 207, 10-23 (2012).
4. E. Kashchieva, S. Tsaneva, Y. Dimitriev and R. Kirov, *Journal of Non-Crystalline Solids*. **323**, 137-142 (2003).
5. M. Weldon, J. Carlson, S. Reedy and C. P. Swann, *Nuclear Instruments and Methods in Physics Research*. **B109/110**, 653-657 (1996).
6. J. Chamón, P. C. Gutierrez, J. Barrio, A. Climent-Font and M. Arroyo, *Appl Phys*. **A 99**, 377-381 (2010).
7. C.P. Stapleton, I. C. Freestone and S. G. E Bowman, *J. Archaeol. Sci.* **26**, 913–92 (1999).
8. S. Rohrs and H. Stege, *X-Ray Spectrometry*. **33**, 396-401(2004).
9. J. Henderson, M. Tregear and N. Wood, *Archaeometry*. **31** (2), 133-146 (1989).
10. JM. Miao, *Journal of Palace Museum*, **111**, 139-155 (2004).
11. B. Kırmızı, P. Colomban and B. Quette, *J. Raman Spectrosc*. **41**, 780-790 (2010).
12. I. Biron, B. Quette, *Techne*. **6**, 35 (1997).
13. B. Gratuze, *J. Archaeol. Sci.* **26** (8), 869–881 (1999).
14. S. L. Eckert, W. D. James, *J. Archaeol. Sci.* **38**, 2155-2170 (2011).

15.L. Dussubieuxa, P. Robertshawb and M. Glascockc, *International Journal of Mass Spectrometry*, **284**, 152–161 (2009).
16.QH. Li, S. Liu, BM. Su, HX. Zhao, Q. Fu and JQ. Dong, *Microscopy Research and Technique*. **76**(2), 133-140 (2012).
17.L. Zhao, H. Li, D. Mu, JH. Wang and JM, in *Corpus of the International Symposium on Ancient Ceramics*, edited by HJ. Luo and XM. Zheng (Shanghai Science and Technology Publishers, Shanghai, 2009). p.424-433.
18.B. Kirmizi, P. Colomban and M. Blanc, *J. Raman Spectrosc*. **41**, 1240-1247 (2010).
19.L. Qu, Thesis,University of Science and Technology Beijing,2013.
20.FX. Gan, in *Development of Ancient Chinese Glass Technology*, (Shanghai Science and Technology Publishers, Shanghai, 2005). p.236.
21. JL.Yi, XJ. Tu, *J chin Ceram Soc*. **12** (4) ,404-410 (1984).
22.F. Li, QH. Li, FX. Gan, B. Zhang and HS. Chen. *J chin Ceram Soc*.**33**(5), 584 (2005).
23.BD. Yang, *Journal of Palace Museum*, **48**, 17-18 (1990).
24.JF. Li, *Journal of Palace Museum*, **66**, 12-27(1994).
25.SX. Wang, *Journal of Palace Museum*, **22**, 61-73(1983).
26.H. Bertran, *Nouveau Manuel Complet de la Peinture sur Verre, sur Porcelaine et sur Email* , Edited by L. Mulot (Encyclop′edie-Roret, Paris, 1913). p.405.
27.KH. Zhao, *Studies in the History of Natural Sciences*, 10 (2) ,145-146 (1991).
28.P. England, J. C. Y.Watt and L. V. Zelst, *Scientific Research in Early Chinese Glass*, Edited by R. H. Brill and J. H. Martin (The Corning Museum of Glass, New York, 1991). p.103.
29.BD. Yang, *Journal of Palace Museum*, **34**, 3-17 (1986).
30.FK.Zhang and ZG.Zhang, *J chin Ceram Soc*. **8** (4), 339-350 (1980).
31.ZH. Xin, *Ceramics Science & Art*, **9**, 71-73 (2008) .
32.JL. Yi and XJ. Tu, *Scientific Research in Early Chinese Glass*, Edited by R. H. Brill and J. H. Martin (The Corning Museum of Glass, New York, 1991). p.99.
33.W.E.S.Turner. *Proc. Chem. Soc*, **4**, 93 (1961).
34.N. Welter, U. Schussler and W. Kiefer, *J. Raman Spectrosc*. **38**, 113-121 (2006).
35.HX.Zhao, QH. Li, S. Liu and FX.Gan, *J. Raman Spectrosc*. **44**, 643-649 (2013).
36.I. Freestone, in *Science and the Past*, Edited by S. Bowman (University of Toronto Press, Toronto,1991) p.37-56.
37.P. Ricciardi, P. Colomban, A. Tournie and V.Milande, *J. Raman Spectrosc*. **40**, 604-617 (2009).
38.A. Tournie, L. C. Prinsloo and P. Colomban, *J. Raman Spectrosc*. **43**, 532-542 (2012).
39.V. V. Linden, O. Schalm, J. Houbraken, M. Thomas, E. Meesdom, A. Devos, R. V. Dooren, H.Nieuwdorp, E. Janssen and K.Janssens, *J. Raman Spectrosc*. **39**, 112-121 (2010).
40.JM. Miao, BR. Yang and JH. Wang, in *Corpus of the International Symposium on Ancient Ceramics*, edited by HJ. Luo and XM. Zheng (Shanghai Science and Technology Publishers, Shanghai, 2009). p.441-446.
41.N. Wood, *Chinese Glaze*, (A & C Black, London,1999). p.63.
42.N. Brun and M. Pernot, Archaometry. **34** (2), 235-252(1992).
43.C.P. Stapleton, I. C. Freestone and S. G. E Bowman, *J. Archaeol. Sci*. **26**, 913–92 (1999).
44.P. Colomban and C. Truong, *J. Raman Spectrosc*. **35**, 195 (2004).
45.J. Henderson, *Oxford Journal of Archaeology*, **4**. 267-291 (1985).
46.L. C. Prinsloo, P. Colomban and A. Tournie, 4th International Conference on the Application of Raman Spectroscopy in Art and Archaeology, Modena, 3-7 sept. 2007.

Mater. Res. Soc. Symp. Proc. Vol. 1656 © 2014 Materials Research Society
DOI: 10.1557/opl.2014.813

Technological Behavior in the Southwest: Pueblo I Lead Glaze Paints from the Upper San Juan Region

Brunella Santarelli[1], David Killick[2] and Sheila Goff[3]

[1]Department of Materials Science and Engineering, University of Arizona, Tucson, AZ 85721 U.S.A.
[2]School of Anthropology, University of Arizona, Tucson, AZ 85721 U.S.A.
[3]History Colorado, Denver, CO 80203 U.S.A.

ABSTRACT

Although widely employed in Eurasia, lead glazes were produced in only two small regions of the Americas prior to European contact, both in the Southwest. Southwestern glaze paints are unique in that they developed as decorative elements instead of as protective surface coatings. The first independent invention of glaze paints was in the Upper San Juan region of southwestern Colorado during the early Pueblo I period (ca. 700-850 CE). Despite recent interest in the later Pueblo IV glaze paints of New Mexico (ca. 1275-1700 CE), there have been no technological analyses of the Pueblo I glaze paints. This research project presents the first analysis and technological reconstruction of the Pueblo I glaze paints. It is in the production of the glaze paints that the potters were innovating and experimenting with materials. These early glaze paints have the potential to provide important information regarding both technology of production as well as the relationships and interactions of potters during this period in the Upper San Juan region. Preliminary results reveal a pattern of traits that involves raw materials, processing, properties and performance of the final product suggesting the existence of a patterned technological behavior.

INTRODUCTION

In the prehistoric American Southwest changes in ceramic traditions have long been associated with social transformations; one of these major changes was the introduction of glaze paints. Southwestern glazes are unique in that they were produced as paints, and thus functioned as a purely decorative element. The first invention of glaze paints was in the Upper San Juan River drainage of southwestern Colorado during the early Pueblo I period, from ca. 700-850 CE [1,2], these glaze paints had a limited temporal and spatial distribution. Lead glazed paints were reinvented along the Mogollon Rim of eastern Arizona during the 13th century, from there they spread into the Little Colorado and Zuni regions of New Mexico, and then to the Rio Grande Valley where the technology reached its peak during the Pueblo IV period (ca. 1275-1400 CE), and eventually disappeared after the Pueblo Revolt of 1680 CE. Studies of these Pueblo IV glaze paints have increased our understanding not only of pottery production, but also of the social networks and communities of the potters who made them [3,4]. Despite recent interest in the Pueblo IV glaze paints, there have been no technological analyses of the Upper San Juan glaze paints. These early glaze paints have the potential to provide important information regarding both technology of production as well as the relationships and interactions of potters during this period in the Upper San Juan region. This paper presents the preliminary results of a technological analysis of Pueblo I glaze-painted ceramics, using both composition and microstructural data to reconstruct the technology of production.

Glaze-painted ceramics from the Upper San Juan are designated as Rosa black-on-white, a type that has been recognized in the ceramic literature since the 1930s. Rosa black-on-whites are distinguished from other contemporary ceramics, such as Chapin black-on-white and Piedra black-on-white of the Northern San Juan (Mesa Verde), by temper, paint type and decorative style. The most distinctive characteristic of the Rosa black-on-white ceramics is the presence of a thick glaze, which ranges in color from yellow/green to black, and was made using a ground lead ore (Figure 1). The ore used was most likely galena (PbS), which has been found in Pueblo I contexts [2,5]. This glaze paint is occasionally associated with a washy organic paint, which suggests that ground lead ore was either mixed with or applied over an organic solution [1]. The Rosa black-on-white ceramics are formed either by the coil and scrape method or by using a basket as a mold. They are smoothed and unslipped, and fired in a reducing or neutral atmosphere. The shapes and designs on Rosa black-on-white ceramics are fairly conservative and similar to other types. The culturally significant quality of these ceramics is the paint; it is in paint production that potters were innovating and experimenting with materials.

Samples for this research project were selected from sites excavated by the Animas-La Plata (ALP) project in Ridges Basin, a valley south of Durango, adjacent to the Animas River, and Blue Mesa, to the east of Ridges Basin. A large number of the sites investigated fell within the Pueblo I period, between 700-850 CE. The Pueblo I period in the San Juan drainage is characterized by the aggregation of dispersed hamlets into villages, the migration of populations, and the resulting interactions between groups with different culture histories [6]. Settlements consisted of a common plaza surrounded by clusters of smaller room blocks. The ALP sites are located between the Piedra tradition to the west (Mesa Verde) and the Rosa tradition to the east, two groups with distinct culture histories and settlement patterns [7]. Analysis of attributes such as architecture, site layout and mortuary practices in the project area suggests that variation is due to immigrants with different culture histories coming into the area, settling into villages and experimenting with organizing into communities for one to two generations [8]. These variations have been interpreted as strategies for signaling the diverse cultural identities in this landscape. The early Pueblo I occupation in this area dates to 750-825 CE, after which the area was rapidly abandoned following several incidents of violence. This area was not reoccupied. The short occupation and rapid and complete abandonment of Ridges Basin and Blue Mesa has been generally attributed to a lack of community integration of the diverse cultural groups settled in the area [8].

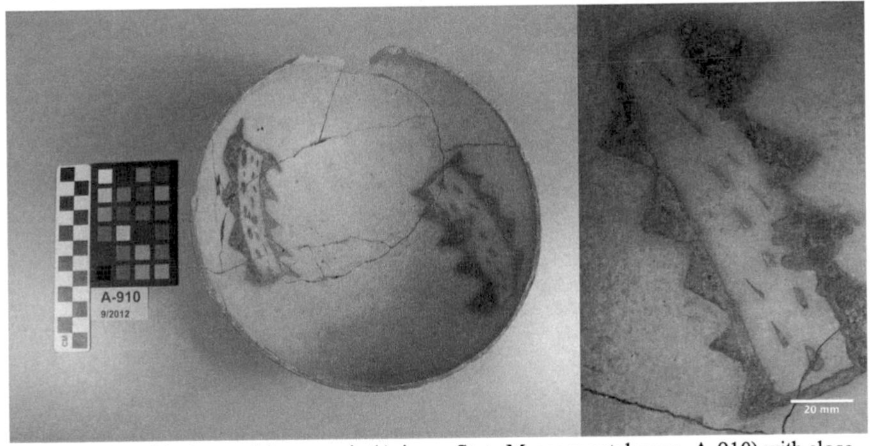

Figure 1 Rosa black-on-white ceramic (Arizona State Museum catalog no. A-910) with close-up of glaze paint

SAMPLE SELECTION AND METHODOLOGY

The sample set selected for this project consists of materials collected during the Animas-La Plata (ALP) project carried out by SWCA Environmental Consultants between 2002 and 2005 in the area around Durango, Colorado [9]. This project was selected for sampling because of the extensive presence of glaze-painted Rosa black- on-white ceramics: approximately 70% of the painted ceramics were glaze-painted [5]. The excavations were tightly controlled, which provides both provenience information and context for the ceramics. There is also chronometric data available for the excavations, which dates the occupation of this area to 750-825 CE [9].

For this study, 110 glaze-painted sherds were selected from the ALP collections. Sherds were collected from nine sites in the five settlement clusters identified by the project; site number, cluster and site overview are listed in Table I. Sherds were selected from sites that demonstrate the diversity of the inhabitants that made up the newly formed Pueblo I communities in the Upper San Juan region. The criteria for sample selection were established after a thorough review of the literature and discussion with project archaeologists. Samples were selected from sites with:

- evidence of pottery production,
- anomalies suggesting differences in culture history or ethnicity,
- features suggesting the sites were gathering places where people may have shared ideas, or
- evidence of possible relationships to sites in other clusters.

Table I ALP sites selected for study

Site No.	Cluster	Comments
5LP239	Eastern	Single family habitation site, evidence of pottery production (polishing stones and a kiln)
5LP240	Eastern	Single family habitation site, evidence of pottery production (two kilns), similarities to sites in North-Central cluster
5LP184	Western	Multiple habitation site, similarities to Sacred Ridge, unusual burials
5LP246	Western	Multiple habitation site, similarities to Sacred Ridge, evidence of specialization in faunal processing
5LP185	North-Central	Earliest occupied habitation site, later used as a cemetery
5LP236	North-Central	Sequentially occupied single family habitation site, burned remains found on floor of feature
5LP237	North-Central	Multiple habitation site, unusual burials, galena found at the site
5LP245	Sacred Ridge	Only real village in Ridges Basin, unique for time period due to size, layout and architectural forms, possible site of ritual activity
5LP2026	Blue Mesa	Multiple habitation site, occupation dates overlap with the end of the occupation in Ridges Basin

The sherds were photographed and visually inspected under low magnification to document color, degree of vitrification and presence or absence of an underlying organic paint in decoration. An Olympus InnovX x-ray fluorescence (XRF) unit was used as a quick, semi-quantitative way to analyze the paints and confirm the presence of lead. A sub-sample of 34 sherds was selected for electron microprobe analysis to collect quantitative compositional data and examine the microstructure of the glaze paints. A small piece of the ceramic was removed using a low speed saw to ensure the preservation of the fragile glaze. The samples were embedded in epoxy, polished, carbon coated and analyzed in the Cameca SX100 microprobe, housed in the Department of Lunar and Planetary Sciences at the University of Arizona. Compositional data was collected with analysis settings of 20kV and 40nA, with a 1μm beam. The microprobe was calibrated for the following oxides: Na_2O, Al_2O_3, SiO_2, K_2O, FeO, ZnO, PbO, CaO, MgO, and TiO_2. Other oxides, including CuO and MnO, were also analyzed, but concentrations were below the detection limit and are not reported. A set of of 3-5 data points was collected for each glaze paint analyzed, the points were normalized and averaged (analytical totals were between 97-100%).

RESULTS AND DISCUSSION

Examination of the glaze paints under low magnification showed extensive variability in the visual appearance of the paint. The glaze paints ranged from fully vitrified, thick yellow glazes, to matte black paints with no evidence of vitrification. XRF analysis of the glaze paints confirmed the presence of lead even in the non-vitrified paints, which suggests that the visual differences are simple variability as opposed to an intentional distinction. The sub-sample of 34 sherds for electron microprobe analysis was selected to represent this observed variability in visual appearance of the glaze paints.

Compositions of the glaze paints collected using wavelength dispersive spectroscopy (WDS) are given in Table II. The three main components of the glaze paints are SiO_2, Al_2O_3 and PbO; minor components include K_2O, FeO and ZnO. The composition of PbO in the glaze paints ranges from 10wt% to 60wt%, with the majority of the samples falling between 30wt% and 55wt%. The average concentration of PbO in the analyzed glaze paints is significantly higher than those reported for the later Pueblo IV glazes from eastern Arizona and the Zuni region of New Mexico [10,11]. Compositional analysis also showed no evidence of added colorants, such as MnO and CuO, both of which are commonly seen in the Pueblo IV Rio Grande glaze paints [12].

The results of this preliminary compositional analysis suggest a different technology of glaze paint production during the Pueblo I period than that recorded for the Pueblo IV period in eastern Arizona and New Mexico. The distribution of PbO concentrations measured seem to suggest a preference for a high-lead recipe, although at this point it is not possible to determine whether distinct recipes were being used. The common use of high lead paints does, however, suggest intentionality in glaze paint production. The variability seen in visual properties can be attributed to the different compositions measured, as paints with low lead contents would explain the matte paints if there was not enough flux present to lead to vitrification. The variability can also be attributed to lack of control of the firing technology. Glazes are complex materials that are affected by numerous variables during the production sequence; variables include not only the glaze paint recipe, but also how the materials are processed and fired. If the Pueblo I potters could not control their firing environment, it is likely that the wide variability in visual appearance is due to inconsistent firing regimes, which would result in attempted glaze paints that did not vitrify. Only three features that could be potential pottery kilns were identified during the ALP project, one in 5LP239, and two in 5LP240; both sites are located in the Eastern cluster. These kilns consist of sandstone slabs lining the sides and bottom of a pit, with a layer of charcoal over the bottom slabs [13]. One of the kilns had evidence of a layer of dirt added to produce a reducing atmosphere during firing. The kilns found by the ALP project are unique in that they were located directly in a habitation site, and they were much smaller than other recorded Pueblo I kilns from nearby areas [14]. The use of kilns indicates the potters' awareness of the importance of the firing stage in ceramic production, however, since kilns were only found in 2 sites out of the 73 excavated by the ALP project, it is impossible to know with certainty the extent to which this firing technology was used in the production of glaze-painted ceramics throughout Ridges Basin and Blue Mesa.

Table II Composition of Glaze Paints

Sample ID	Oxide Concentration (wt%)*									
	Na_2O	Al_2O_3	SiO_2	K_2O	FeO	PbO	CaO	MgO	ZnO	TiO_2
184.102.1.11	0.75	13.06	53.48	2.62	3.28	20.95	2.59	2.11	0.33	0.83
184.115.1.9	0.55	14.29	56.00	9.13	4.31	10.40	2.83	2.03	0.09	0.37
184.115.4.11	0.19	9.57	40.04	0.55	1.73	44.64	1.69	0.73	0.36	0.49
184.115.1.11	0.25	4.83	38.00	0.62	3.48	50.84	0.70	0.57	0.50	0.21
185.161.14.21	0.40	3.82	35.45	0.44	1.10	50.54	1.74	0.61	5.68	0.21
185.161.14.18	1.52	17.90	46.47	4.51	1.89	23.64	1.99	0.78	0.94	0.35
185.161.14.33	0.82	14.55	48.41	2.02	4.11	26.11	1.94	1.01	0.48	0.56
236.44.3.17	0.17	5.13	30.05	0.39	2.03	60.32	1.06	0.59	0.02	0.24
236.44.3.18	0.63	9.58	44.34	1.20	3.54	35.37	2.59	1.08	1.24	0.43
236.44.8.2	0.48	14.95	33.89	1.41	2.10	44.70	1.13	0.70	0.22	0.42
236.59.1.4	0.35	19.35	37.77	2.11	1.48	36.24	1.39	0.88	0.09	0.34
237.73.1.32	0.19	7.09	43.39	0.81	1.61	44.59	0.83	0.74	0.45	0.31
237.59.1.60	0.31	6.27	41.78	1.01	2.81	40.99	2.54	1.04	2.86	0.38
237.59.1.59	0.19	7.14	43.70	0.81	1.62	44.91	0.84	0.01	0.74	0.31
239.47.1.42	1.27	19.09	44.80	4.49	2.32	23.06	2.99	0.97	0.60	0.40
239.47.1.49	0.77	19.46	42.90	3.15	2.71	26.00	2.99	1.37	0.27	0.37
239.47.1.55	0.60	20.37	43.33	3.95	1.46	26.57	2.30	1.09	0.03	0.29
240.60.1.3	0.43	6.49	38.02	1.07	2.48	48.31	1.78	0.76	0.32	0.33
240.83.1.32	0.60	8.71	46.01	1.58	2.68	34.69	2.70	1.26	1.12	0.63
240.44.1.6	0.08	4.36	35.49	0.54	1.20	52.93	2.09	1.51	1.51	0.29
240.66.1.8	0.31	7.46	39.10	0.78	1.56	48.53	0.73	0.53	0.72	0.28
240.56.1.13	0.20	8.52	39.34	0.89	2.37	46.07	1.02	0.92	0.14	0.51
240.56.1.12	0.10	7.85	30.68	0.37	1.73	55.98	1.54	0.99	0.31	0.42
245.525.1.8	0.48	9.12	47.42	1.94	3.09	32.12	3.52	1.64	0.15	0.52
245.525.1.9	0.16	7.61	39.22	0.50	2.61	44.69	2.14	1.78	0.72	0.57
245.130.5.9	0.46	17.76	48.72	8.25	1.55	21.38	0.64	0.49	0.18	0.56
245.500.4.42	0.27	8.97	43.37	0.99	2.24	41.98	0.71	0.78	0.23	0.46
246.61.1.33	0.15	7.73	37.91	0.39	3.92	45.63	2.36	1.26	0.02	0.63
246.61.1.34	0.41	10.18	32.70	0.98	1.06	47.94	1.33	0.76	4.34	0.30
246.61.36.16	0.38	9.79	43.30	1.09	1.83	38.05	2.75	1.20	1.10	0.50
246.78.51.2	0.27	9.40	45.19	0.79	2.33	36.79	1.11	1.03	2.47	0.60
2026.39.3.52	0.50	7.59	41.46	1.37	3.65	39.46	1.54	0.81	3.47	0.14
2026.39.3.53	0.19	6.67	43.06	0.79	1.96	42.59	2.25	0.87	1.35	0.27
2026.39.3.55	0.77	14.34	39.72	2.48	1.56	37.58	1.66	1.16	0.00	0.72

*Collected by wavelength dispersive spectroscopy, concentrations are the normalized average of 3-5 points per sample

The microstructure of the glaze paints was examined using backscattered electron (BSE) imaging. The thickness of the glaze paints ranges from 15μm to 50μm. The microstructures of the glaze paints with PbO concentrations lower than 30wt% show no evidence of a fused glass phase. The glaze paints with low PbO concentrations are non-homogeneous and generally porous, as seen in Figure 2a, with numerous quartz inclusions visible in the glaze paint layer. Glaze paints with PbO concentrations higher than 30wt% have a homogeneous glass phase and a layer of crystal growth at the interface between the glaze paint and the ceramic, as seen in Figure 2b. This interface is generally about 10μm thick, and consists of thin, elongated crystals. When a lead compound, such as galena (PbS), is ground to a powder and applied to the surface of a ceramic, the lead will react with quartz and clay from the ceramic body to form a lead oxide-alumina-silicate melt during firing, eventually forming a glaze through a process of dissolution and diffusion [15,16]. There will be diffusion of elements such as potassium, alumina and silica from the body to the glaze resulting in the growth of the crystalline phase seen at the interface [17]. Compositional analysis of this phase identified the crystals as lead-feldspars ($(K,Pb)AlSi_3O_8$) [15].

Also visible in numerous of the microstructures of the glaze paints are small circular, bright areas, about 2-4μm in diameter, as seen in Figure 2a and 2b. These areas have higher PbO concentrations than the surrounding glassy phase, and are likely remnants of the raw material used to make the glaze paint. An elemental x-ray map was collected of a glaze paint to confirm the identity of these bright areas. Figure 3 shows the results of the elemental x-ray map collected on a small area of a glaze paint containing a glass phase, a crystalline phase, and the bright circular areas. The sulfur (S) map and the lead (Pb) map (Figure 3a and 3b, respectively), show that the areas of highest lead concentration match with the areas of highest sulfur concentration, confirming the identity of the bright spots as un-fused galena (PbS). These results confirm that Pueblo I potters used galena ores as their raw material in the production of glaze paints. These results also suggest that Pueblo I potters were not roasting their ores to yield a lead oxide prior to application to the ceramic, as Pueblo IV potters were doing in the Rio Grande Valley [18,19].

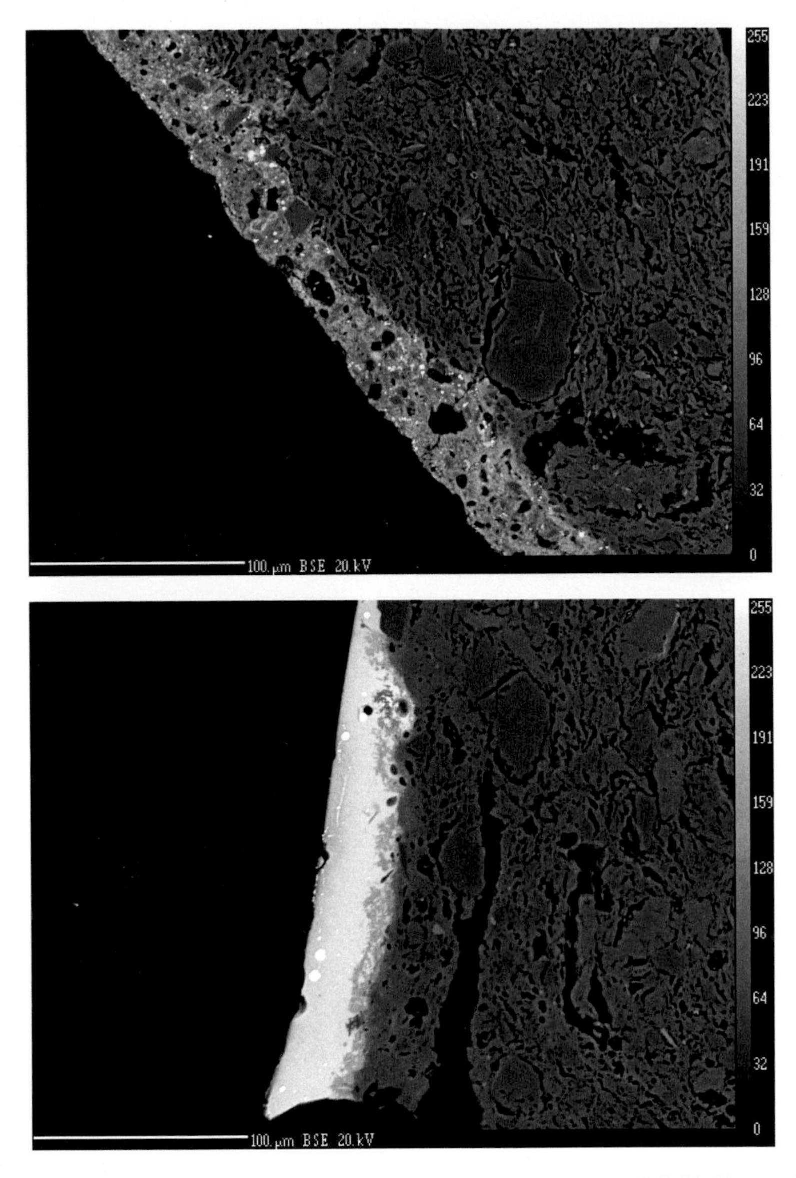

Figure 2 (a) Glaze paint with low PbO concentration (Sample ID 184.115.1.9) (b) Glaze paint with high PbO concentration (Sample ID 246.61.36.16)

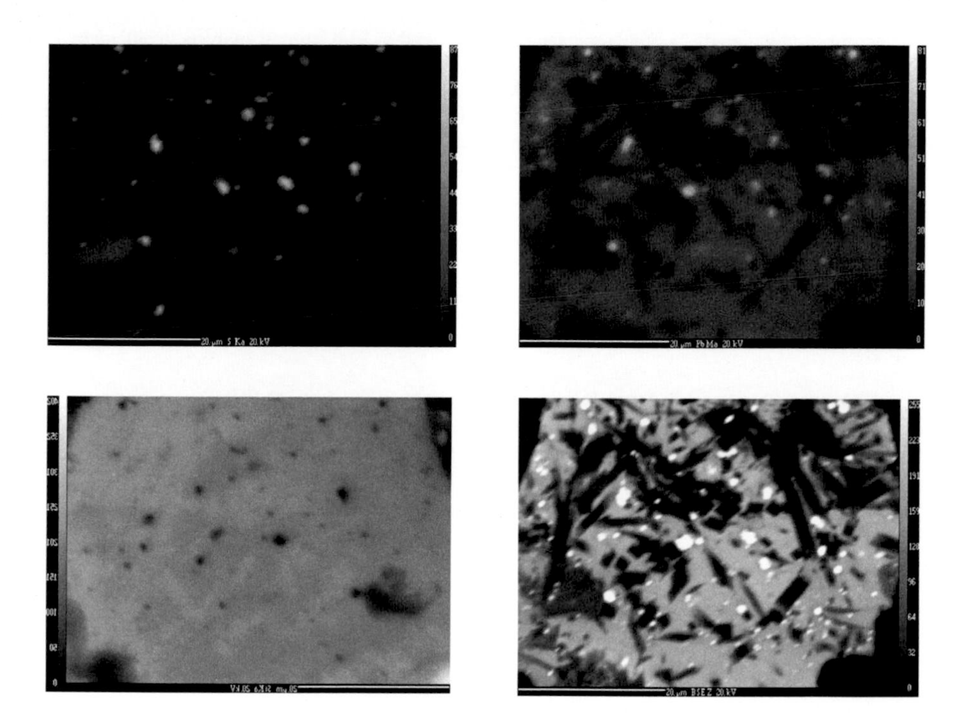

Figure 3 Electron microprobe element maps of glaze paint, clockwise: (a) sulfur (S) map (b) lead (Pb) map (c) silicon (Si) map (d)BSE image of area mapped, showing crystal growth and high-lead spots (Sample ID 245.525.1.8)

CONCLUSIONS

The results presented in this research suggest the existence of a patterned technological behavior in the production of glaze paints during the Pueblo I period, 750-825 CE. Although at this point it is not possible to determine with certainty whether potters were using distinct recipes, we can conclude that the production of glaze paints was intentional and not an accident resulting from experimentation with raw materials. Variation in visual appearance and composition suggests that while potters were aware of the effect of using a ground lead ore as a paint, they were not capable of controlling all of the variables necessary to produce a successful glaze. The compositional data for the set of Pueblo I glaze paints shows a different technological approach to glaze paint production than that seen during the Pueblo IV period, suggesting two instances of independent invention of glaze technology in the Southwest. Future work will include compositional analysis of a larger sample set, as well as lead isotope analysis to elucidate procurement strategies of raw materials.

Glaze paints emerged in the Upper San Juan during the Pueblo I period in a post-migration landscape that resulted in the aggregation of people with different culture histories into the first pueblo villages of the Southwest. The production of glaze paints mostly ceased after the rapid abandonment of sites in Ridges Basin and on Blue Mesa, and glaze paints were not produced again in the Southwest for nearly five hundred years. If glaze paints were used in this landscape as an integrative mechanism or as a way of signaling social identity, they failed, which explains the sudden disappearance of the technology.

ACKNOWLEDGMENTS

This research was funded by a Burgan Fund Grant from the Arizona State Museum and a Research Grant from the Arizona Archaeological and Historical Society. We would like to thank Kenneth Domanik (microprobe laboratory, Department of Lunar and Planetary Sciences, UofA), for his help with microprobe analysis. We would also like to thank the Anasazi Heritage Center (Cortez, CO) for providing us with the ceramic sherds used in this study.

REFERENCES

1. A.O. Shepard, in *Archaeological Studies in the La Plata District* (Carnegie Institution of Washington, Washington D.C, 1939), p. 249.
2. C.D. Wilson and E. Blinman, *Upper San Juan Region Pottery Typology* (Museum of New Mexico: Office of Archaeological Studies, Santa Fe, 1993), pp. 18-22.
3. J.A. Habicht-Mauche, in *The Social Life of Pots: Glaze Wares and Cultural Dynamics in the Southwest, AD 1250-1680*, edited by J.A. Habicht-Mauche, S.L. Eckert, and D.L. Huntley (The University of Arizona Press, Tucson, 2006), pp. 3–16.
4. L.S. Cordell and J.A. Habicht-Mauche, in *Potters and Communities of Practice: Glaze Paint and Polychrome Pottery in the American Southwest, AD 1250-1700*, edited by L.S. Cordell and J.A. Habicht-Mauche (The University of Arizona Press, Tucson, 2012), pp. 1–7.
5. J.R. Allison, *Animas-La Plata Project: Ceramic Studies* (SWCA Environmental Consultants, Phoenix, 2010).
6. R.H. Wilshusen and S.G. Ortman, Kiva **64**, 369 (1999).
7. J.M. Potter and J. Chuipka, Kiva **72**, 407 (2007).
8. J.M. Potter and T.D. Yoder, in *The Social Construction of Communities: Agency, Structure, and Identity in the Prehispanic Southwest*, edited by M.D. Varien and J.M. Potter (AltaMira Press, Lanham, MD, 2008), pp. 21–40.
9. J.M. Potter, in *Animas-La Plata Project: Final Synthetic Report*, edited by J.M. Potter (SWCA Environmental Consultants, Phoenix, 2010), pp. 3–18.
10. T.R. Fenn, B.J. Mills, and M. Hopkins, in *The Social Life of Pots: Glaze Wares and Cultural Dynamics in the Southwest, AD 1250-1680*, edited by J.A. Habicht-Mauche, S.L. Eckert, and D.L. Huntley (University of Arizona Press, Tucson, AZ, 2006), pp. 60–85.
11. D.L. Huntley, *Ancestral Zuni Glaze-Decorated Pottery: Viewing Pueblo IV Regional Organization Through Ceramic Production and Exchange* (University of Arizona Press,

Tucson, 2008).
12. K. Schleher, D.L. Huntley, and C.L. Herhahn, in *Potters and Communities of Practice: Glaze Paint and Polychrome Pottery in the American Southwest, AD 1250-1700*, edited by L.S. Cordell and J.A. Habicht-Mauche (The University of Arizona Press, Tucson, 2012), pp. 97–106.
13. N.F. Eisenhauer, V.H. Hensler, K.R. Adams, S.S. Murray, and E.M. Perry, in *Animas-La Plata Project: Ridges Basin Excavations: Eastern Basin Sites*, edited by T.D. Yoder and J.M. Potter (SWCA Environmental Consultants, Phoenix, 2010), pp. 203–236.
14. E. Blinman and C. Swink, in *The Prehistory and History of Ceramic Kilns*, edited by P.M. Rice (American Ceramic Society, 1997), pp. 85–102.
15. J. Molera, T. Pradell, N. Salvadó, and M. Vendrell-Saz, Journal of the American Ceramic Society **84**, 1120 (2001).
16. M.S. Walton and M.S. Tite, Archaeometry **52**, 733 (2010).
17. M.S. Tite, I. Freestone, R. Mason, J. Molera, M. Vendrell-Saz, and N. Wood, Archaeometry **40**, 241 (1998).
18. K. Schleher, The Role of Standardization in Specialization of Ceramic Production at San Marcos Pueblo, New Mexico, Ph.D. dissertation, Department of Anthropology, University of New Mexico, 2010.
19. E. Blinman, K. Schleher, T. Dickerson, C.L. Herhahn, and I. Gundiler, in *Potters and Communities of Practice: Glaze Paint and Polychrome Pottery in the American Southwest, AD 1250-1700*, edited by L.S. Cordell and J.A. Habicht-Mauche (University of Arizona Press, Tucson, 2012), pp. 107–116.

Mater. Res. Soc. Symp. Proc. Vol. 1656 © 2015 Materials Research Society
DOI: 10.1557/opl.2015.825

Analysis and Replication of Jianyang Tea Bowls from Song Dynasty China

James D. Morehead[1] and Pamela B. Vandiver[1]

[1] Department of Materials Science and Engineering Department, University of Arizona, Tucson, AZ 85750 USA

ABSTRACT

Black-glazed tea bowls from the Jian area of Fujian province, China, were analyzed to understand the physical basis of their visual appearance and the special glaze effects of nucleation, crystal growth, control of glaze flow, and hare's fur and spotted patterns that have frustrated modern and ancient factories that are unable to produce acceptable replicas. The black-glazed Jian bowls are divided into two distinct groups called "Hare's Fur" and "Oil Spot". Black glazes and bodies from the Jian kilns are rich in iron and calcium oxides, made from a plentiful local refractory dark red clay, and fired in hill-climbing dragon kilns. Twenty-six sherds were analyzed from the collection made by James Plumer at the kiln site in 1935 [1]. Analyses were conducted using optical microscopy, Xeroradiography, scanning electron microscopy (SEM-EDS) and electron microprobe analysis (WDS), and petrographic thin section analysis to reverse engineer some of the microstructure, composition and thermal history of Jian ware.

INTRODUCTION

The name Jianyang refers to the Jian county of Fujian province, China, where the tea bowls were produced at 28 or more workshop and kiln sites. These tea bowls are also referred to as" temmoku". The name derives from the fact that during the Song Dynasty (ca. 960-1279) Japanese Buddhist priests made pilgrimages to a temple on Mt. Temmoku (Tian mu shan) in the north-eastern part of the Zhejiang Province and brought back with them the Jianyang bowls to Japan. The kiln complex, Jian-yao, in Fujian is said to have originated in the early Song dynasty (960-1279 CE), and its production was active throughout the Song and the Yuan (1271-1368 CE) dynasties [2].

Two distinct groups of Jianyang tea bowls were investigated in this study: oil spot (OS), which is either a black glaze with bright white or golden crystals or a red under-fired glaze with black specks, and hare's fur (HF), which is a black glaze with streaks of white, gold, red, brown or orange-brown [3]. Another rare Jian type is called "yohen temmoku" and samples were not available for this study. The first technical study of this ceramic was conducted by Yamasaki in 1985 [2]; he reported crystals that appear to almost float in a liquid and crystalline pool of glaze that formed at the surface of the glaze.

EXPERIMENTAL METHODS

Twenty-Six bowl sherds were provided by Frederick Matson from the James Plumer collection made by Mr. Plumer who traveled to Fujian to search for the Jian kilns during the spring of 1935[1,3]. He found one marked "guan," or official, and established that the bowls were conveyed to the imperial court from Jian kilns, further confirming literary accounts of the high value of the bowls. Literary accounts describe their use for serving green tea.

The experiments detailed in this paper consist of a series of analyses with the objectives of further understanding the Jianyang tea bowl appearance and composition. The experiment began with a description of all the sherds based on observations under low magnification (10-35x) using a Leica EZ 4 HD microscope and photographically documenting the range of variability. Of these sherds, seven that represent the range of variability were chosen for further evaluation: 3 oil spot, 1 underfired matte-brown glazed sherd, and 3 hare's fur examples. Next, a fresh fracture of each sherd was prepared for imaging with a scanning electron microscope (Hitachi S-3400 Type II or S-4800) with ThermoNoran NSS-EDS elemental identification. Thin sections were prepared and studied with a Nikon polarizing microscope and Media Cybernetics Image Pro Plus software at 100-400x. For quantitative chemical analysis, chemical imaging and backscattered imaging, a Cameca SX 100 Electron Beam Microanalyser with wavelength dispersive x-ray spectroscopy (WD) was used with NIST geological standards. The goal was to establish the phases present in the glazes and bodies, their distribution and variation. About 300 WDS data points were analyzed and 6 areas measuring 150x120μ were elementally imaged for 6 elements, each taking about 3 hours. A Xerox Medical Systems Xeroradiograph 125 was used to image porosity and fabric texture that are indicative of methods of manufacture. A charged selenium plate sensor exposed to x-rays capacitive discharges, yielding radiographs with edge-enhancement. The replication experiment is explained near the end of the text.

Sample Inventory and Descriptions Based on Optical Microscopy:
Descriptions of the seven samples follow:

A. J4: Oil spot, golden crystals in black glaze; black body,
B. J5: Oil spot, reddish crystals in black glaze; gray body,
C. J6: Oil spot, red crystals in red surface, black interior glaze; black body and white crystalline interface at glaze-body interaction zone,
D. J8: Under-fired matte brown glaze on black body; brown, semi-vitreous glaze that has crawled on one surface; no crystal development in glaze or at the glaze-body interface, and partially reacted raw materials may be present, Refiring a fragment in 4 hours to 1200°C produced a shiny bubble-free black glaze.
E. J18: Hare's fur, reddish, well-developed crystals in black glaze; black body,
F. J20: Hare's fur, reddish to silvery crystals in black glaze; grayish black body,
G. J27: Hare's fur, reddish brown overdeveloped crystals on reddish brown glaze surface with black interior; black body; glaze is quite thin compared to others and the inner surface is somewhat weathered; the glaze looks dry and does not have as much contrast between the red crystals and the reddish-brownish glaze, thus, probably overfired or fired for too long a duration at peak temperatures; alternately, the glaze may have been applied too thinly.

Photographs of the exterior surface of each tea bowl are shown in Figure 1.

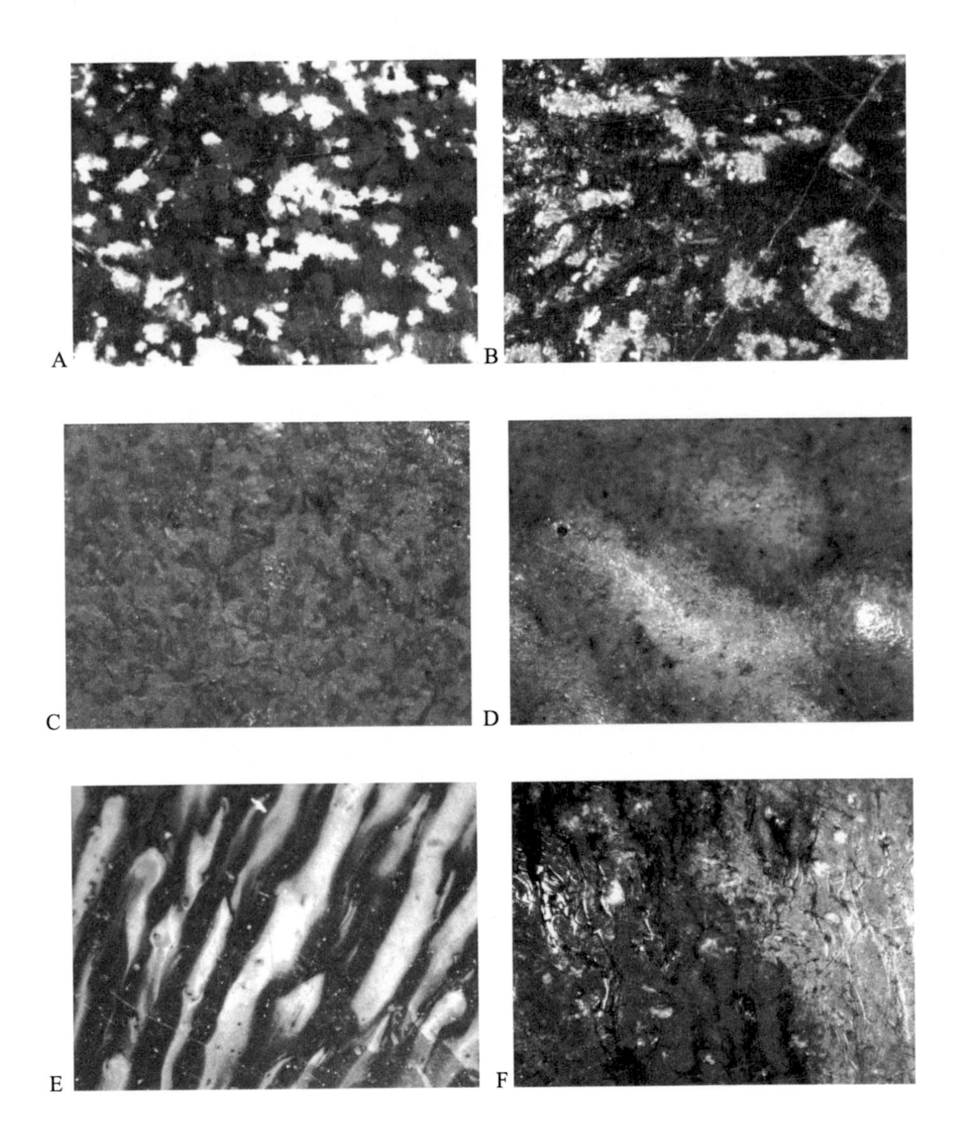

Figure 1: Optical Microscope Images of A-F, as Described in the Inventory Above

Figure 1: Optical Microscope Images, as Described in the Inventory, continued.

G

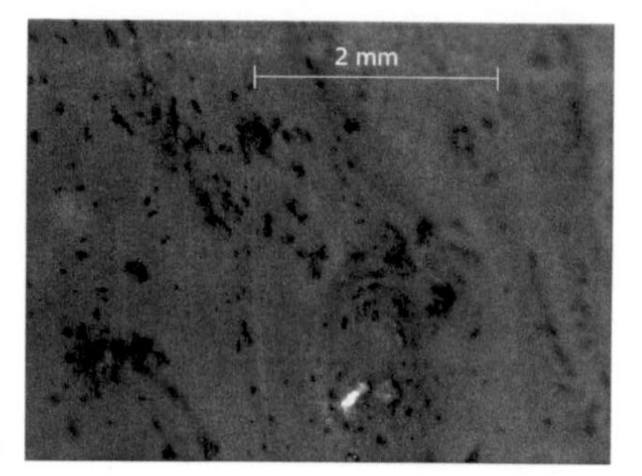

RESULTS AND DISCUSSION

Scanning Electron Microscope Images:

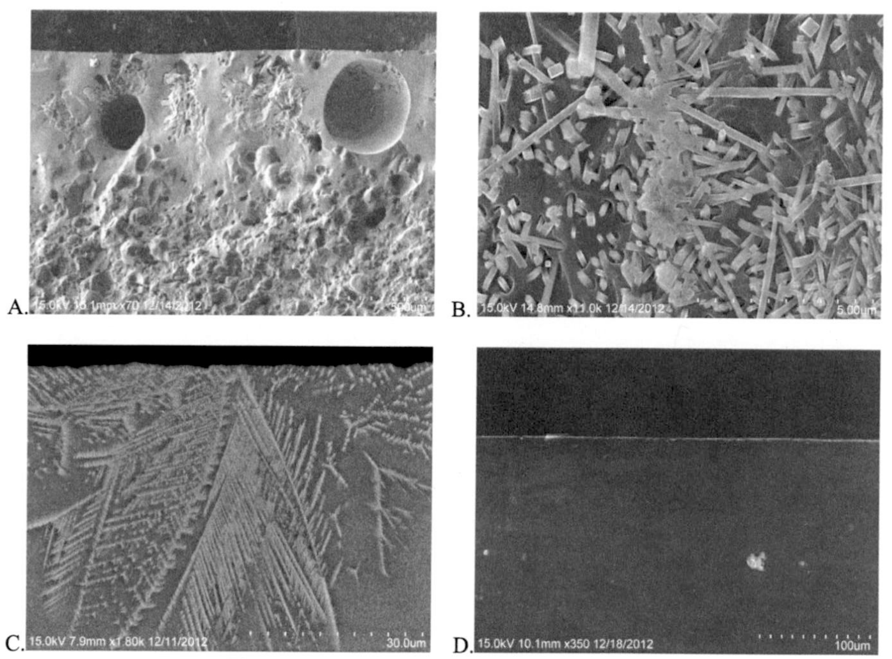

Figure 2: Scanning Electron Microscope Images

Scanning electron microscope images shown in Figure 2 are described below. A. Fractured cross section of J4 oil spot at 70x magnification showing glaze surface (top) and body (bottom) with clumps of crystals in the lower part of the glaze and some needlelike radiating crystals in the upper glaze. B. Body of the same J4 oil spot at 1000x showing the acicular crystal growth field of mullite or anorthite. C. Near surface cross section of the J6 oil spot glaze with red crystals in a black glaze matrix showing a dendritic crystals, possibly iron-containing, growing from the glaze surface into the interior of the glaze at 1800x magnification. D. A glaze only view of J18 hare's fur showing absence of crystal growth and only minor contrast in the upper left of glaze at 350x.

Thin Section Images

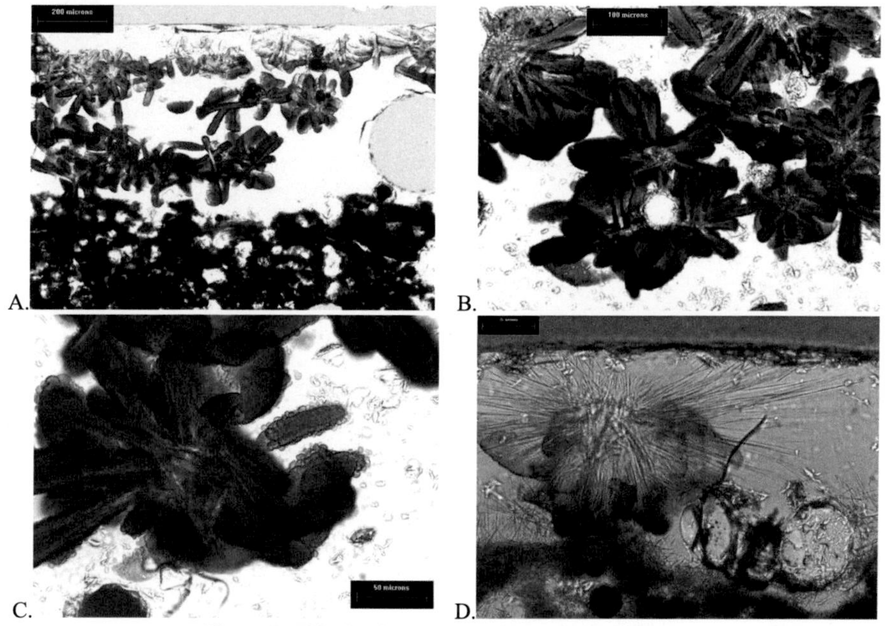

Figure 3: Thin Section Images of Sample J4 Oil Spot

Sample J4 with golden crystals in black glaze is imaged in thin section shown above in Fig. 3. A. Growths of clumped, radiating crystals appear in dark brown with a surrounding black zone, presumably rich in iron, shown at 100x. B. At 200x the crystal clusters are lighter at the center than the exterior, presumably indicating diffusion of iron outward is necessary for the central rod phase to crystallize. C. At 400x magnification the light colored interior crystal is observed as well as the presence of rough and zoned interfaces. D. Also at 400x, an anorthite crystal cluster formation is radiating outward and a dark crystal region is beginning to form around the cluster center; also noted were some small needle-like crystals growing at the glaze-body interface.

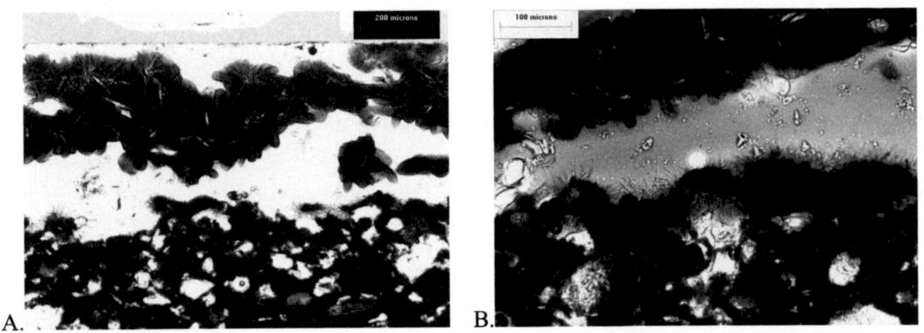

Figure 4: Thin Section Images of Sample J5 Oil Spot with Reddish Crystals in Black Glaze

Sample J5 is shown above in Figure 4. A. At 100x the dark brown to black crystals appear fully mature and slowly growing in the middle of the glaze layer. B. At 200x, fine needle-like crystals are growing from the body into the glaze.

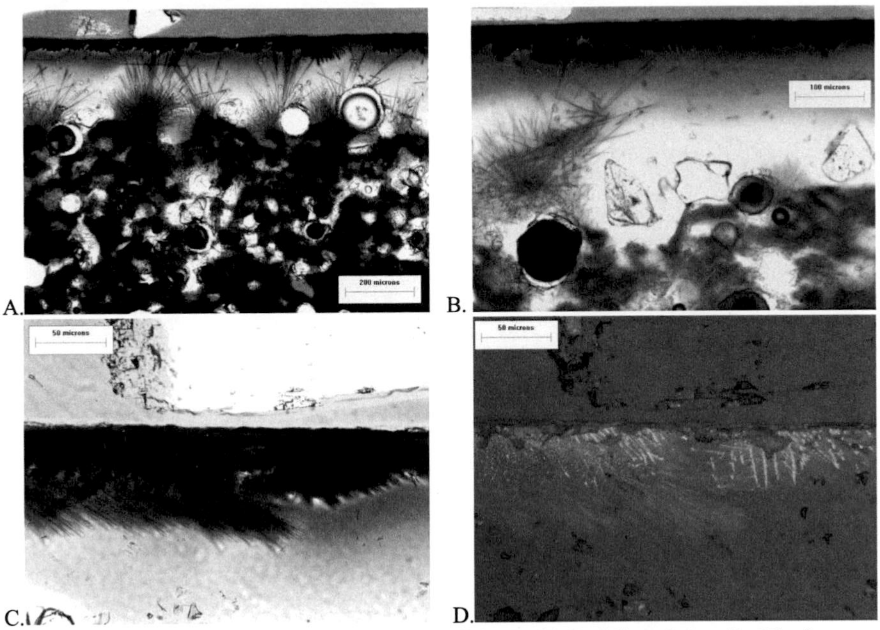

Figure 5: Thin Section Images of Sample J6 Oil Spot with Red Crystals in a Red Surface.

Sample J6 is shown above in Figure 4. A. At 100x, clusters of long needle-like anorthite crystals are growing rapidly from the body through the glaze toward the surface. The upper surface of the glaze is red. Some crystals are thicken as they grow toward the surface. B. Under

200x magnification a matte-like area of anorthite crystal formation is shown. C. At 400x this view shows the thick crystal structure at the surface of the glaze, as well as evidence of flow to the left. D. At 400x, the same image was taken as C but under reflected light, this image shows the dense regions of crystals growing from the surface inward. The red color may be hematite.

Figure 6. Thin section image of J8 UF

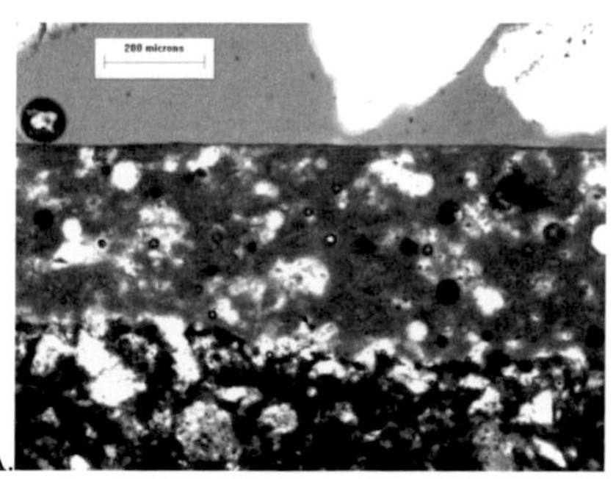

A.

Sample J8 UF, the underfire, mottled, matt sample, is shown above as A at 100x. The glaze has begun to melt but not homogenized, and no evidence of crystals growth is present. The black body indicates reduction early in the firing cycle.

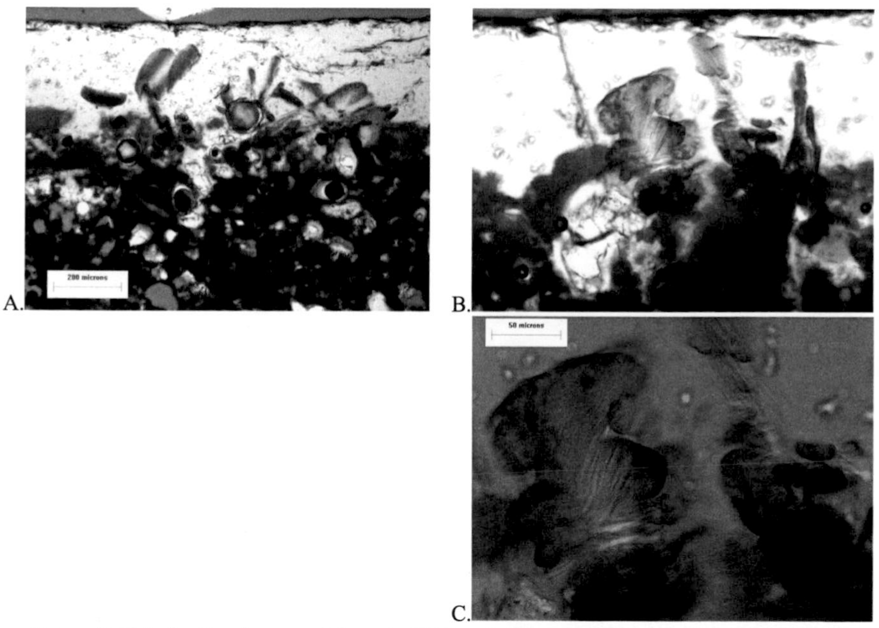

Figure 7: Thin Section Images of Sample J18 Hare's Fur with Red Crystals in Black Glaze

Sample J18 hare's fur shown in Fig. 7. A. The image at100x shows complex interaction of body and glaze without a distinct glaze-body interface. B. At 200x image shows the dark crystals growing outward from the body into the glaze. A weathered craze line is present toward the left. C. At 400x, dark zoned crystal clumps are present, but the crystals appear translucent.

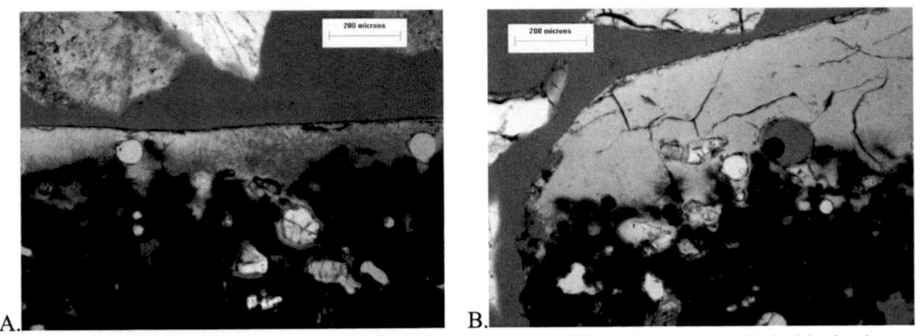

Figure 8: Thin Section Images of Sample J20 HF with reddish to silvery crystals in black glaze.

Hare's fur sample J20 is shown above. A. At 100x, an unusually uneven and curved glaze-body interface is documented, and thin acicular crystals grow outward from the body into the glaze. B. A 100x magnification view of the end of the sherd shows this rough glaze-body interface. Even smaller scale, acicular structures are shown growing outward from the body in both images.

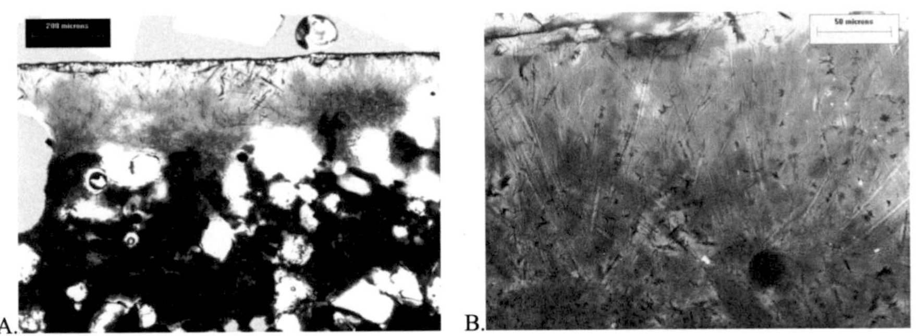

Figure 9: Thin Section Images of Sample J27 HF, probably overdeveloped.

Sample J27, the hare's fur red crystals on brown glaze, the example with overdeveloped crystal growth, is shown above. A. At 100x, the image shows quartz and pores in the body and high iron and extensive crystal growth in the glaze. Extensive acicular crystals identifiable anorthite are growing outward from the body into the glaze. B. At 400x, the cross section shows details of extensive anorthite crystal growth with a dark opaque phase precipitating at right angles on the better developed, thicker areas of anorthite. The dark phase is probably a crystal containing iron.

Microprobe Backscattered Images and Data

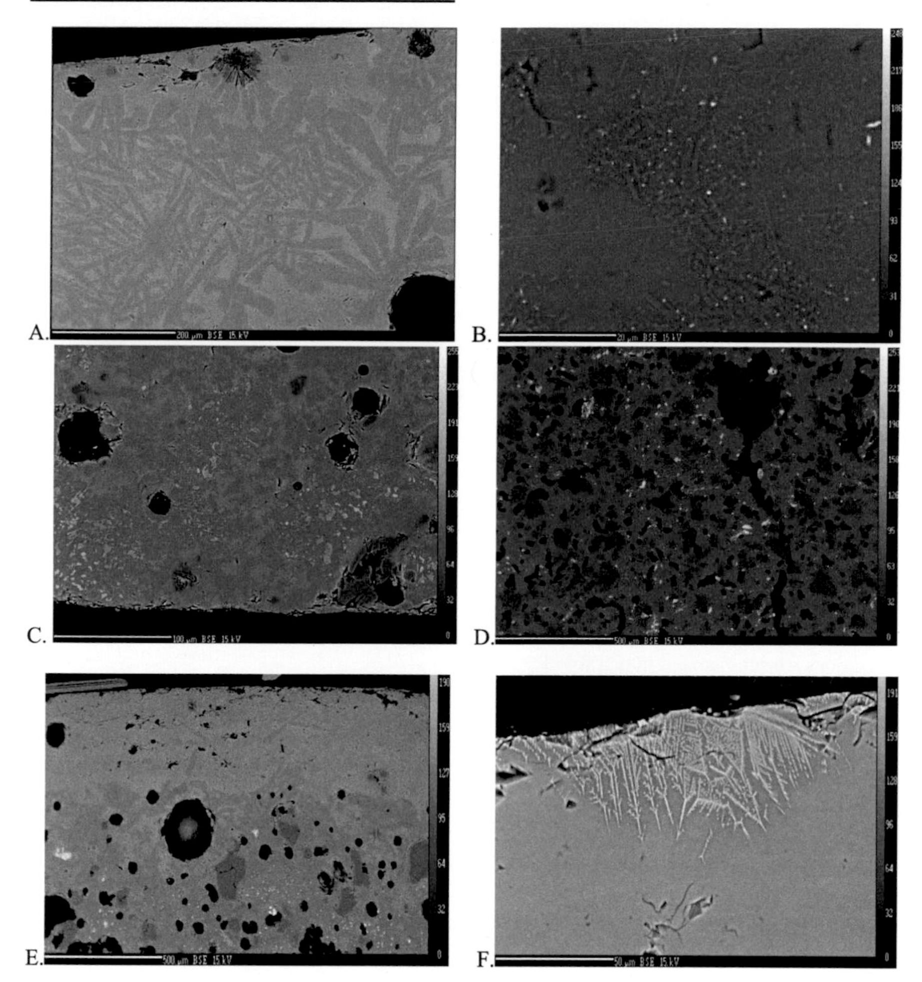

Figure 10: Back Scattered Images of Thin Section Samples

Figure 10 contains backscattered SEM images of the thin sections. A. The J4 OS glaze has dark crystal clusters of lower average atomic number that are growing independently in the glaze as radial clusters. Thus nucleation occurs in the bulk. The Cameca system has a BS contrast detection limit of average atomic number of 2. B. J5 OS image of the body. Anorthite appears to be growing in the body. Anorthite crystals were observed in all of the samples with the exception of Sample J8. C. The image is of the bottom side of Sample J8 bowl sherd with the glaze edge being at the bottom of the image. The glaze appears to have voids and appears as a non-homogeneous mixture of materials. D. Sample J20 body image. Voids in the material are present and also the various colors represent different materials which make up the body of the ceramic. E. Sample J18 image of the glaze and body. The glaze is at the top of the image and the body at the bottom of the image. No indications were present of the Hares Fur streaks. Several data points were taken by the microprobe which confirmed the composition of mullite. This image shows several individual mullite crystals. F. Sample J6 glaze surface. Dendritic crystal formation appears at the surface.

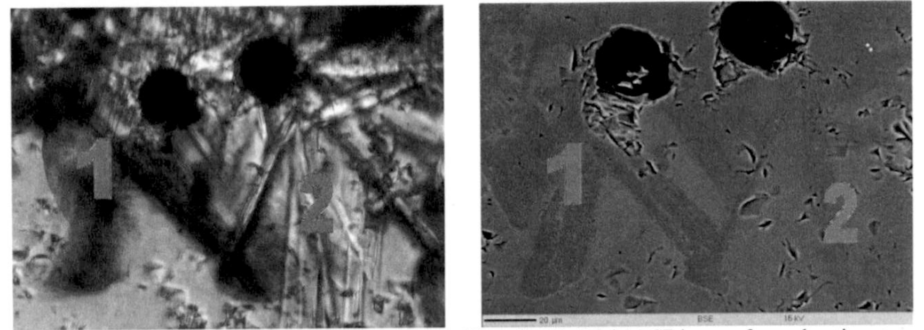

Figure 11: (Left) Optical Microscope image of Sample J18, (Right) BSE image of same location

Figure 11 shows two images that were taken of Sample J18 from the microprobe. The dark crystal (shown as #1) appears alongside the transparent needle structures of #2. Data point compositions are summarized in in Table 1.

1. Sample J18 Possible Feldspar

MgO	Al_2O_3	Na_2O	SiO_2	CaO	Fe_2O_3	MnO	K_2O	TiO_2	Cr_2O_3	P_2O_5	Cl	Total	
1.244	**19.671**	**0.170**	**61.570**	**8.679**	**4.569**	**0.500**	**2.021**	**0.462**	**0.010**	**1.418**	**0.001**	**100.315**	**Avg**
1.039	18.528	0.126	60.798	8.181	3.728	0.437	1.937	0.399	0.000	1.196	0.000	100.227	Min
1.454	21.094	0.192	62.291	9.490	5.307	0.565	2.103	0.551	0.022	1.683	0.004	100.419	Max
0.194	1.142	0.031	0.659	0.600	0.671	0.052	0.072	0.067	0.011	0.203	0.002	0.079	Std Dev

2. Sample J18 Anorthite Crystal

MgO	Al_2O_3	Na_2O	SiO_2	CaO	Fe_2O_3	MnO	K_2O	TiO_2	Cr_2O_3	P_2O_5	Cl	Total	
0.162	**33.605**	**0.121**	**47.248**	**17.898**	**0.951**	**0.137**	**0.277**	**0.035**	**0.006**	**0.148**	**0.003**	**100.590**	**Avg**
0.130	33.124	0.111	46.661	17.608	0.941	0.123	0.243	0.029	0.000	0.085	0.000	100.246	Min
0.199	34.017	0.137	47.765	18.061	0.969	0.148	0.313	0.042	0.013	0.182	0.005	101.127	Max
0.029	0.367	0.012	0.549	0.200	0.012	0.011	0.032	0.005	0.007	0.043	0.002	0.377	Std Dev

3. Sample J18 Glaze Composition

MgO	Al_2O_3	Na_2O	SiO_2	CaO	Fe_2O_3	MnO	K_2O	TiO_2	Cr_2O_3	P_2O_5	Cl	Total	
2.043	**18.564**	**0.073**	**60.184**	**5.802**	**7.089**	**0.802**	**2.855**	**0.474**	**0.009**	**1.584**	**0.001**	**99.482**	**Avg**
1.904	17.607	0.062	59.355	5.580	6.210	0.730	2.712	0.399	0.000	1.350	0.000	99.088	Min
2.131	19.071	0.088	62.636	6.137	7.672	0.860	3.029	0.507	0.029	1.835	0.004	99.965	Max
0.063	0.399	0.009	1.010	0.179	0.476	0.035	0.087	0.031	0.012	0.145	0.001	0.262	Std Dev

Table 1: Sample J18 Hare's Fur Compositions

The data show compositional differences between the dark crystal and the transparent needle structure. Reviewing possible compositional matches it appears that the needle structure is an anorthite crystal and the dark crystal is probably another feldspar crystal. For comparison, the glaze composition where no crystals occur is included in Table 1. The glaze (CaO + Al2O3+SiO2 = 84.5% and others are 9.2%) and possible feldspathic mineral CAO + Al2O3 + SiO2 = 90%; others are 12.6%) are shown to have a somewhat similar composition. Anorthite has a stoichiometric composition of 18% CaO, 33.6% Al2O3 and 47.2% SiO2, others 1.2%.

Dr. Heixiong Yang calculated a chemical formula for this composition using 8 oxygen atoms, as follows:

(CaO 0.41, K 0.12, Na 0.01) (Si 2.83, Al 0.96, Fe 0.11, P 0.05, Ti 0.01)O8

The framework is definitely microcline (as the Si/Al = 3:1), but the large cation part is close to anorthite with a considerable amount of vacancies. He assumed the analysis volume consisted only of the crystal of interest, although glass is most likely present between and below the crystals. Another possibility is that mica may be one of the precursor materials.

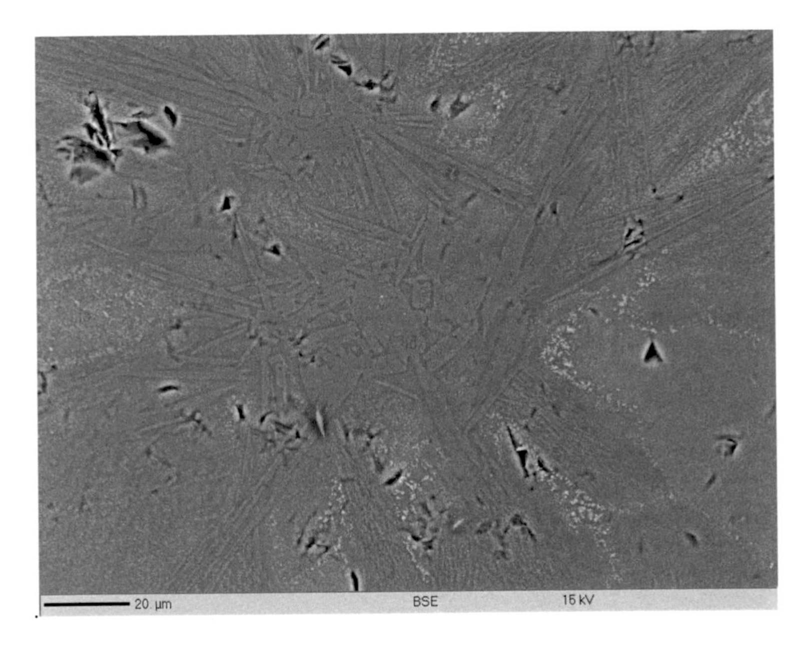

Figure 12: Sample J5 Oil Spot Crystal Cluster

Figure 12 shows a crystal structure with two types of crystals. Microprobe points were analyzed at the larger crystal sections and at the glaze in between the crystals. Data is provided in Table 2. Angular black pores in Fig. 12 are artifacts from thin section preparation and electron beam damage.

1. Sample J5 Anorthite Crystal

MgO	Al_2O_3	Na_2O	SiO_2	CaO	Fe_2O_3	MnO	K_2O	TiO_2	Cr_2O_3	P_2O_5	Cl	Total	
0.170	**32.035**	**0.160**	**47.379**	**17.305**	**1.991**	**0.143**	**0.409**	**0.051**	**0.012**	**0.230**	**0.001**	**99.886**	Avg
0.094	29.936	0.115	45.096	16.336	1.723	0.118	0.258	0.032	0.000	0.078	0.000	99.629	Min
0.243	34.147	0.198	49.472	18.207	2.298	0.169	0.561	0.071	0.027	0.366	0.005	100.244	Max
0.055	1.712	0.038	1.812	0.735	0.221	0.021	0.128	0.015	0.010	0.116	0.002	0.211	Std Dev

2. Sample J5 Glaze Composition Next to Anorthite Crystal

MgO	Al_2O_3	Na_2O	SiO_2	CaO	Fe_2O_3	MnO	K_2O	TiO_2	Cr_2O_3	P_2O_5	Cl	Total	
2.386	**13.815**	**0.092**	**63.488**	**4.271**	**6.896**	**0.943**	**4.102**	**0.668**	**0.009**	**2.291**	**0.001**	**98.962**	Avg
1.651	12.749	0.047	59.702	3.189	5.143	0.674	3.244	0.546	0.000	1.968	0.000	98.510	Min
3.308	15.450	0.157	66.775	5.909	10.975	1.306	4.574	0.775	0.016	2.622	0.006	99.320	Max
0.608	1.062	0.035	2.616	1.015	1.895	0.235	0.504	0.091	0.005	0.229	0.002	0.314	Std Dev

3. Sample J5 Glaze Composition

MgO	Al_2O_3	Na_2O	SiO_2	CaO	Fe_2O_3	MnO	K_2O	TiO_2	Cr_2O_3	P_2O_5	Cl	Total	
1.998	**18.257**	**0.094**	**60.824**	**5.215**	**5.808**	**0.818**	**4.026**	**0.525**	**0.006**	**1.262**	**0.004**	**98.838**	Avg
1.261	16.942	0.084	59.634	3.343	5.146	0.429	3.763	0.404	0.000	0.592	0.000	98.438	Min
3.315	18.768	0.101	63.833	6.489	7.443	1.244	4.283	0.885	0.018	1.559	0.011	99.268	Max
0.690	0.583	0.006	1.419	1.109	0.687	0.281	0.162	0.162	0.006	0.360	0.003	0.277	Std Dev

Table 2: Sample J5 Oil Spot, Red Crystals in Black Glaze, Data Points

Data point summaries are shown in Table 2. The crystals identified in #1 are anorthite. The glaze composition for sample J5 is shown in #3 for completeness. The data points next to the glaze appear as a similar but different composition compared to the glaze or the anorthite crystals.

Hares Fur Glaze Data

MgO	Al_2O_3	Na_2O	SiO_2	CaO	Fe_2O_3	MnO	K_2O	TiO_2	Cr_2O_3	P_2O_5	Cl	Total	
1.740	**18.420**	**0.088**	**60.554**	**5.775**	**6.922**	**0.578**	**2.864**	**0.659**	**0.009**	**1.169**	**0.001**	**98.781**	**Avg**
1.149	16.931	0.062	58.709	4.500	5.851	0.327	2.617	0.399	0.000	0.331	0.000	96.492	Min
2.131	19.827	0.162	64.601	6.429	7.672	0.860	3.921	0.936	0.029	1.835	0.013	99.965	Max
0.259	0.744	0.025	1.418	0.377	0.519	0.160	0.239	0.170	0.008	0.361	0.003	0.991	Std Dev

Oil Spot Glaze Data

MgO	Al_2O_3	Na_2O	SiO_2	CaO	Fe_2O_3	MnO	K_2O	TiO_2	Cr_2O_3	P_2O_5	Cl	Total	
1.684	**18.793**	**0.088**	**61.153**	**5.195**	**6.360**	**0.638**	**3.440**	**0.583**	**0.006**	**1.216**	**0.003**	**99.160**	**Avg**
0.930	16.386	0.046	57.920	3.033	3.979	0.318	2.475	0.403	0.000	0.198	0.000	98.438	Min
3.315	23.346	0.126	64.878	6.770	10.422	1.244	4.283	0.893	0.018	1.758	0.011	100.217	Max
0.506	1.322	0.017	1.671	0.948	1.173	0.224	0.531	0.150	0.006	0.410	0.003	0.416	Std Dev

Previous Microprobe Work

Average Glaze Data from All Samples

MgO	Al_2O_3	Na_2O	SiO_2	CaO	FeO	K_2O	TiO_2	P_2O_5	Total		Fe_2O_3	
1.478	**21.083**	**0.078**	**61.910**	**5.858**	**6.279**	**2.913**	**0.642**	**1.014**	**101.254**	**Avg**	**6.980**	**Avg**
0.750	17.680	0.000	56.780	1.970	4.060	1.910	0.350	0.270	99.870	Min	4.510	Min
2.160	31.970	0.130	66.450	12.220	8.260	3.600	1.020	1.720	103.100	Max	9.180	Max
0.365	3.169	0.026	1.842	2.207	0.951	0.386	0.164	0.365	0.837	Std Dev	1.060	Std Dev

Average Body Data from All Samples

MgO	Al_2O_3	Na_2O	SiO_2	CaO	FeO	K_2O	TiO_2	P_2O_5	Total		Fe_2O_3	
2.004	**22.007**	**0.371**	**60.468**	**5.162**	**6.919**	**2.665**	**0.990**	**0.672**	**101.256**	**Avg**	**7.380**	**Avg**
0.470	13.730	0.000	40.980	0.510	3.460	0.190	0.320	0.100	96.260	Min	4.700	Min
11.700	30.540	2.600	74.770	11.230	11.800	3.760	3.930	1.500	103.110	Max	10.390	Max
2.621	5.391	0.779	6.460	3.407	2.039	0.710	0.782	0.482	1.497	Std Dev	1.550	Std Dev

Table 3: Comprehensive Composition Data for Test Samples

The data in Table 3 is a comprehensive combination of data from this study of 7 sherds and data from a previous analysis of the same collection in 1988 by one of the authors [4]. The Smithsonian Institution 5-spectrometer probe of Eugene Jarosevich in NMNH was used for 650 analyses of 27 sherds, or about 22 analyses per sample, stepping across each polished section in 20 μincrements with each analysis at a 10μ defocused area and without benefit of SEM imaging. The results showed that the hare's fur and oil spot glazes had the same compositions, and the inference was drawn that the differences had to be processing and thermal history. Fifty-six analyses were acquired from the bodies or at the glaze-body interfaces and so may not be representative of the body interiors.

The hare's fur (J18, J20, and J27) and oil spot (J4, J5, and J6) glaze compositions of this study are a virtual match to the 1989 results. The average body data is very similar to the glaze data, and so may not be accurate. Another difference is that the current study chose Fe_2O_3 when the microprobe calculations were being performed. Since the previous study used FeO the calculation to Fe_2O_3 is provided and show that the iron oxide concentration is quite similar to the microprobe data obtained in this study.

Microprobe Elemental Maps: Elemental chemical imaging scans were performed on several thin section samples. Each scan shows the distribution of a single element in each sample. Elements scanned include Si, Ca, Al, Ti, K, Mg, Fe, and P. Figure 13 shows representative examples of maps obtained for oil spot J4 (above and inverted A through D: Ca, Fe, Al, K) and hare's fur J18 (below E through H: Si, Al, Fe, Ca):

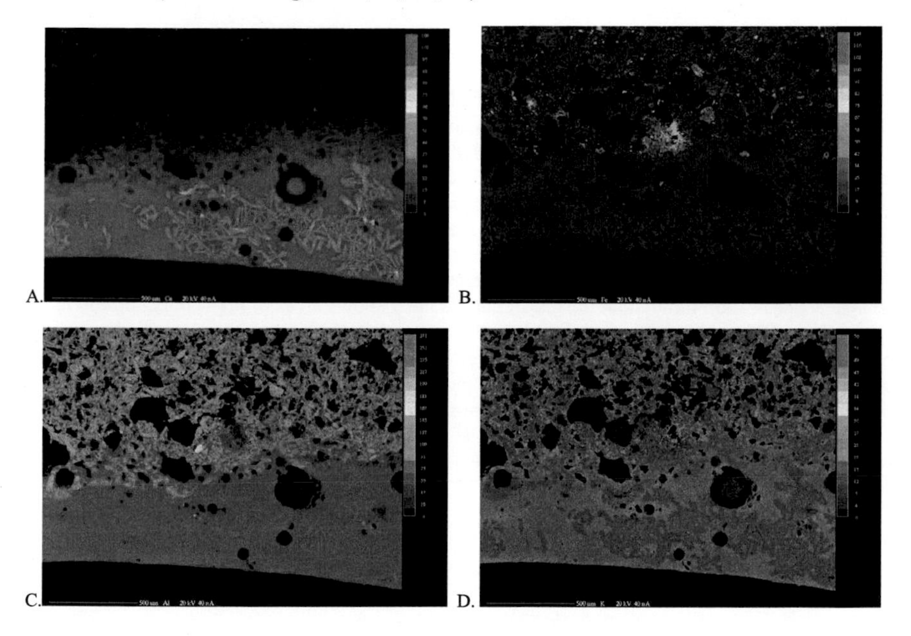

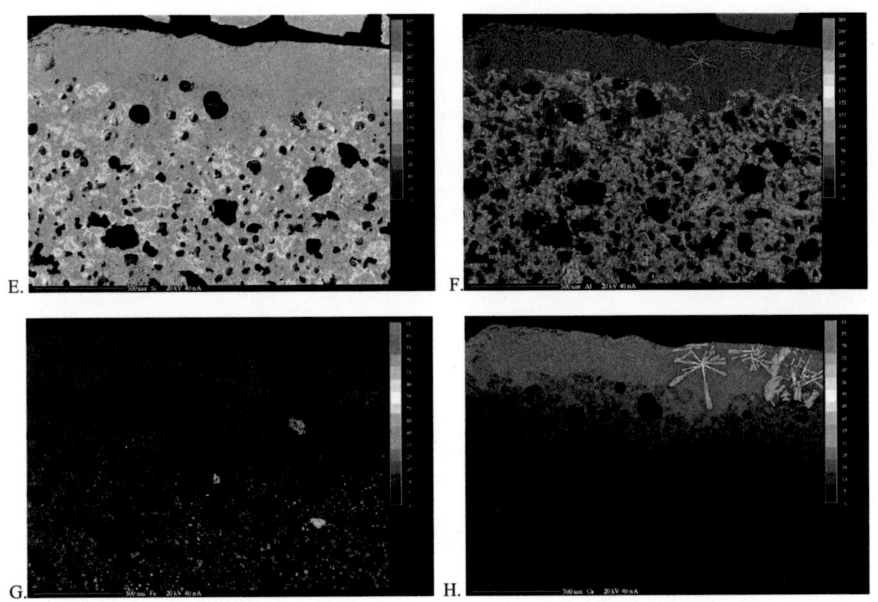

Figure 13: Elemental Maps of Thin Section Samples as Described Below

Figure 13 shows sample J4 oil spot interior of bowl: A. Map of calcium. Notice the high Ca concentrations through the crystals and specifically at the crystal cluster centers. B. Map of iron. High concentrations identified within the body and low concentrations in the crystal clusters near the glaze surface (near the bottom of the image). C. Map of aluminum. Nearly Consistent elemental concentration throughout the glaze with slight elevation in the crystal clusters, and somewhat higher concentrations occur in the body. D. Map of potassium. Mostly uniform elemental concentration through the glaze, slight depletion in the crystal clusters that matches the body concentration. Figure 13 also shows sample J18 hare's fur: E. Map of silicon. Consistent concentration throughout the glaze, but the body shows areas of high concentration in quartz particles. F. Map of aluminum. Consistent concentration throughout the glaze, with slightly higher concentrations within the glaze crystal clusters and the body. G. Map of iron. Mostly consistent low concentrations in the glaze and body, but iron-rich inclusions in the body have a bimodal distribution of inclusions. Most are small, but a minor number are large. The glaze crystal formations show a depletion of iron. H. Map of calcium. Highest concentrations shown in the crystal clusters and specifically at the center of the growths. Almost no Ca in the body.

Xeroradiograph Images and Forming Process: Xeroradiograph images were taken of the larger fragments as a way of determining similarity of construction techniques of the bowls. The radiographs (Figs. 14 and 15) were exposed at 45 kV and 5 mA for 1 minute using a Xeroradiograph 125 (Xerox Medical Systems).

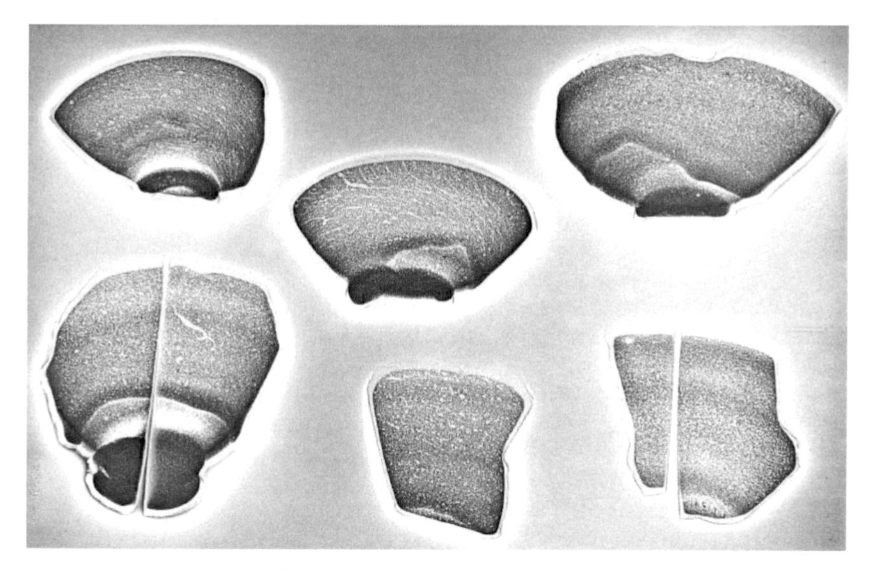

Figure 14 – Xeroradiograph Images of Jian Sherds

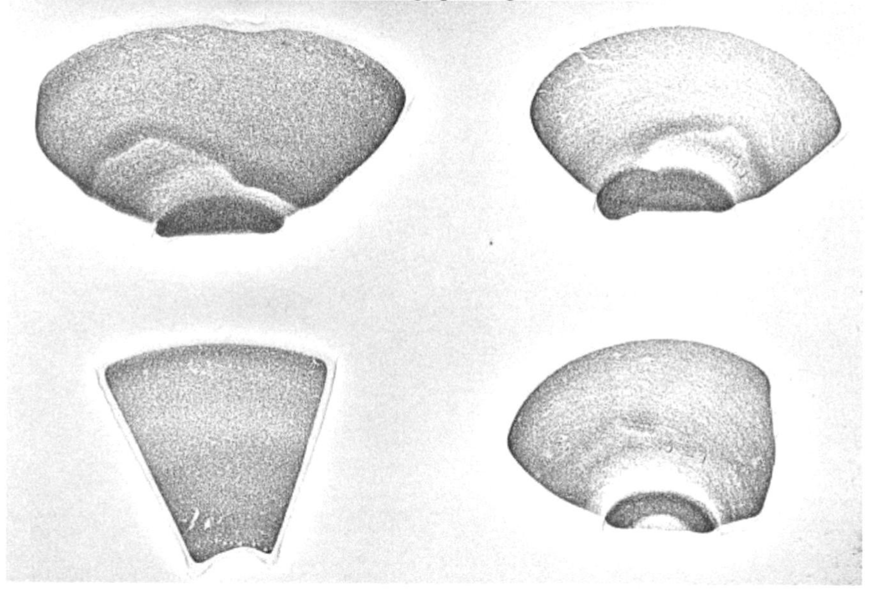

Figure 15. Xeroradiograph Images of Jian Sherds

The xeroradiography results show that the elongated pores in the body are an angle of 30-45 degrees to the horizontal base and rotate counterclockwise. This confirms the surface textural evidence that the bowls were thrown on a rapidly rotating wheel, as the diagonal pore alignment represents the sum of the applied forces. Small, round bubbles appear in the glazes, showing that the surface materials melted quite rapidly such that at least some gas bubbles did not make it through the viscous glaze before it became solid.

Phase Diagram Analysis: Three phase diagrams were consulted based on the composition of the glaze and presence of anorthite: the first, SiO_2-CaO-Al_2O_3; the second, SiO_2-FeO-Al_2O_3; and third, SiO_2-CaO-FeO [5]. Table 4 shows the compositional range used as boundaries. In each set of data the three oxide compositions were adjusted to equal 100%. It is assumed that the major iron oxide phase present is fully oxidized in the glass network or as Fe_2O_3 substituting in an alumina lattice. For the second phase diagram, the FeO concentration was calculated and appears in the Table 6 below.

Anorthite N=18	Al_2O_3	SiO_2	CaO
Minimum	31.3	44.7	17.1
Maximum	36.3	51.7	19.0

Table 4: Microprobe Composition Data from Anorthite Crystals, 18 Analyses

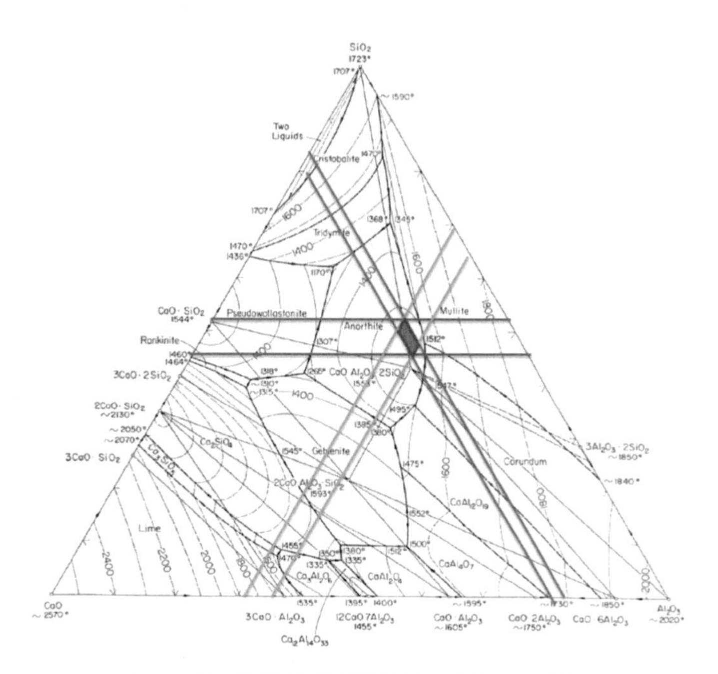

Figure 16 – SiO2-CaO-Al2O3 Phase Diagram [5].

The band shows the area where all 18 data points are contained. This band shows a temperature of approximately 1500°C but the first liquid at equilibrium forms at 1170°C. The phase diagram assumes no additional materials and equilibrium state. Refiring the Jian glazes produces rounding of a cube of glaze at a 1200°C firing temperature, but as the glaze cooled anorthite would have grown until reaching 1170°C, the lowest temperature for anorthite growth is 1170°C per the diagram. With the addition of fluxes, the glaze material analyzed is most likely anorthite that grew prior to the lowest temperature noted on the diagram. This diagram also has points in the mullite phase field, thus explaining finding mullite growths splaying from the anorthite crystals in some thin sections where calcium was deficient, as well as in the bodies.

Glaze N=58	Al_2O_3	SiO_2	CaO
Minimum	18.9	66.8	4.3
Maximum	25.5	75.0	10.8

Table 5. Microprobe Composition Data from Glaze, 58 Analyses

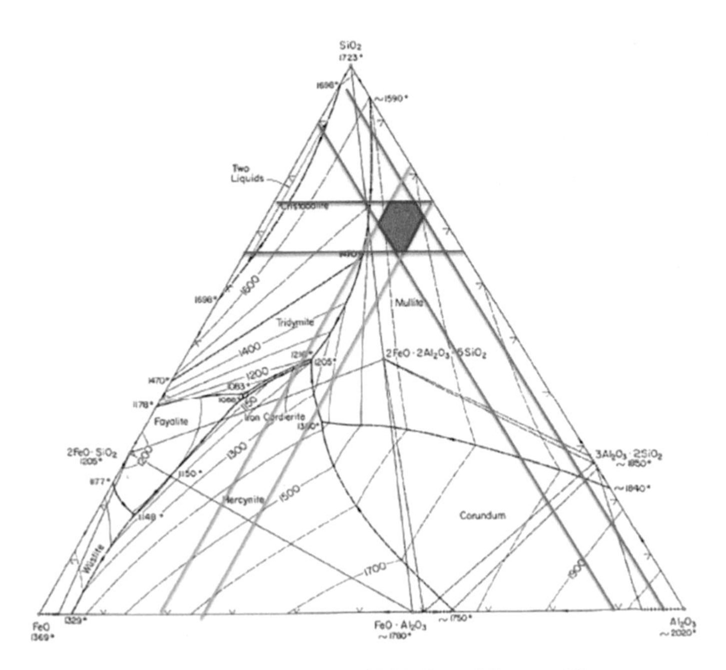

Figure 17. 1SiO2-FeO-Al2O3 Phase Diagram [5].

The dark band shows the area where all 58 analyses are contained. This area shows a temperature of approximately 1600-1700°C with eutectics in the SiO_2-$2FeO{\cdot}SiO_2$ and $2FeO{\cdot}2Al_2O_3$-$5SiO_2$ phase field of 1205°C and 1210°C. This phase diagram assumes no other materials and a long-term steady state. The Jian glazes likely saw around 1200°C firing temperature. It is of interest to note that when the other elements are removed, the formulation correction was nearly 10%, with approximately 7% from the CaO in the glaze. As CaO is a flux and K_2O, Na_2O and P2O5 are also present, they likely decreased the melt temperature significantly. The diagram also shows mullite may crystallize as one of the compatibility triangle components.

Glaze N=58	SiO_2	CaO	FeO
Minimum	78.6	4.2	5.8
Maximum	87.7	9.2	14.7

Table 6: Microprobe Composition Data from Glaze, 58 Analyses

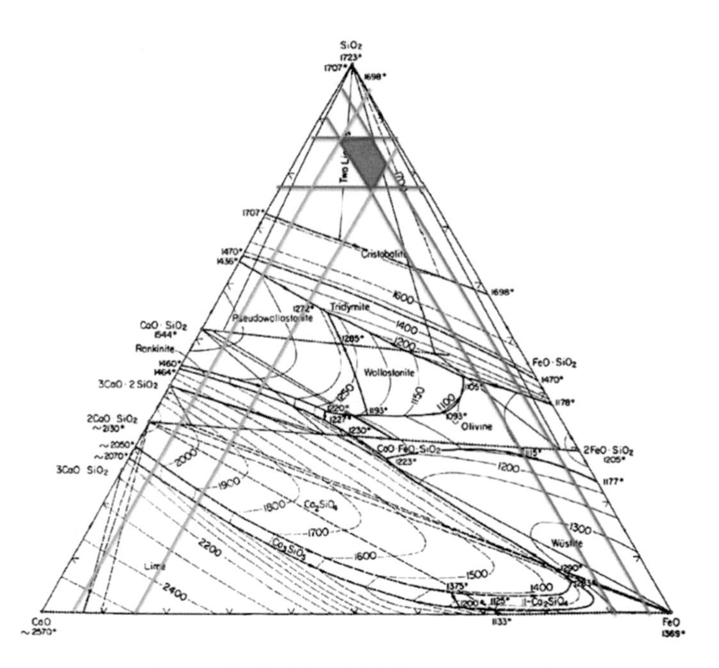

Figure 18. SiO2-CaO-FeO Phase Diagram [5]

For the third analysis, the SiO_2-CaO-FeO phase diagram was used. The summation of these three components equated to ~70 % of the glaze composition, indicating need of the quaternary. The compositional points land in the two-liquid region of the phase diagram with a temperature near 1700°C. Our composition is not at equilibrium, but metastable and kinetically active. The missing oxides, or some of them, likely acted to flux or reduce the temperature during glaze formation.

Replication Experiments: The replication experiment was undertaken to accomplish several goals: first, to provide further insight into the complex tea bowl production process, and second, to test multiple glazes of wide compositional range that are used as temmoku glazes by modern potters. Our texts involved varying temperature and atmosphere. Below is the list of glazes (Laguna Clay Co., Pomona, CA) and observations:

1. V-17 KCNST – Red and black speckle, like a teadust glaze
2. WC-525 Temmoku #1 – Black and shiny glossy
3. WC-526 Temmoku #2 – Black and shiny, with some brown crystals (Fig. 19)
4. WC-556 Iron Red – Black with red where thinner and around edges
5. V-2 Bright Brown-Black – Shiny bluish black glaze
6. Using the average analyzed oxide glaze composition, we calculated a rational, custom glaze formulation with the following materials: Albany slip (875.2 grams), Kingman potassium feldspar (57.6 grams), flint (42.7 grams) and red iron oxide (Fe2O3) (24.5 grams). This

formulation was satin-matte and not sufficiently glossy. This formulation should be shifted to less clay, more feldspar to lower the viscosity to produce more glossy, less viscous glazes.

Approximately 40 tea bowls of a dark brown cone-10 stoneware clay (Laguna Clay Co., WC-373, dark brown fired-color and composed of a medium dark coarse, refractory clay with 60 mesh sand temper additive) and a few white porcelain bowls were thrown that when completed showed the black glazes to be opaque. The tea bowls were bisque-fired (unlike those from Jianyang), then dipped in glaze. Six to seven of each glaze type were produced. A new Olympic, 4 cubic foot, natural-gas fueled, updraft test kiln was used for three firings: first in reduction, then oxidation, and third, in a reduction atmosphere. The kiln was arranged such that 5 kiln shelves held the tea bowls, and a stack and damper was built on top of the kiln to narrow the top and bottom temperature range.

The kiln had issues reaching the desired temperature (1250°C) during the first heavy reduction firing. A temperature of only 1000°C was achieved in 6 hours. The oxidation firing reached approximately 1300°C in 4 hours, and an oxygen probe verified the oxidizing atmosphere. A third firing in reduction reached 1250°C in 5 hours with oxygen probe verification. Glazed draw trials provided evidence of nucleation and crystal growth. Temperature was varied in 20 minute intervals to mimic a wood fired dragon kiln, and reduction was initiated prior to sintering of the glaze.

Figure 19. Brownish Yellow Crystalline Bowl Surface with Temmoku #2 Glaze

The bowls did not appear to accurately match the visual appearance of the original Jianyang tea bowls when fired in oxidation or reduction. Formulation #2 (Temmoku #1) in oxidation came the closest to matching the oil spot appearance of sample J6. The course crystal growth at the surface appears similar to the Sample J6. However, sample J6 is a deeper red in actual appearance.

CONCLUSIONS

Bowls in various sizes and shapes were used for serving ground green that looks brilliant emerald green when mixed rapidly into a frothy, steaming translucent liquid against a shiny black glazed bowl. This effect is enhanced by the presence of small golden yellowish or red crystals flickering in the glaze. The two types of ware most commonly produced were called oil spot and hare's fur. Oil spot glazes contain circular clusters of red or yellow iron-bearing crystals, and hare's fur glazes have elongated rivulets of vertical red or yellow crystals due to flow of the glazes during firing. The glazes on most bowls have flowed down the sides and have pooled into the center of the interior. A sharp change in the exterior curvature of the bowls near the base where trimming commenced usually stops the flow of the glaze and leaves a thickened meniscus, indicating the glaze was quite viscous when it flowed. Many other ancient kilns and some modern ones have tried to replicate these seeming rough ceramics that were prized by monks and used at the imperial court in tea contests and ceremonies. To view one of the largest, most beautiful of the Jianware bowls (Ht: 8.8 cm, W: 19.2), seek the online images of collection number F1909.369 at the Freer and Sackler Galleries of Art, Smithsonian Institution. The crystals grade from yellow to red with the height of the vessel. The scale of the production must have been amazing, as one of the author's was onsite at the Jian kiln #28 at a conference at the end of its excavation in fall 1995. Waster piles of ware, saggers and setters were 7 to 10 meters high and longer than 100 meters, and everywhere the earth was a rich iron-red, a plastic refractory clay weathered from volcanic rock.

The oil spot glazes (samples J4 and J5) contained a large but variable amount of dark crystals growing in the glaze. In the developed crystals the data show that these are anorthite clusters, with considerable calcium oxide at the center and with iron content increasing at the edges. It is most probable that the white or golden crystals shown in the optical images of Figure 1 are dense and well-formed anorthite crystals. The crystals that are a feldspar mineral, but not anorthite, could be providing an intermediate material used in the growth of the anorthite crystals or they could be an artifact of the raw materials used.

Oil spot glaze sample J6 appeared to have a crystal layer on the glaze exterior surface. This layer was not observed on any other of the oil spot glazes, and this is the least reduced of the samples, and has a dark gray body color. The glaze composition below this crystal layer was consistent with the other oil spot glazes and is included in the glaze formulations in Table 3.

The hare's fur glazes (samples J18, J20 and J27) also showed anorthite growths but these do not appear to form into dense, radial crystal clusters like the oil spot glazes. Some of the glaze surfaces and near-surface regions are enriched in iron, based on the color of the glass. Some secondary crystal growth is also present on some anorthite needles and rods. The glaze-body interfaces are often rough and undulating, probably intentionally so. No observations were made parallel to or at right angles to a "fur" streak, and this research is underway currently. Several small mullite crystals were identified but they do not appear to affect the visual appearance of these glazes. The glaze compositions of the hare's fur matches that of the oil spot glazes. This would indicate that the same material was used to produce both glaze types; however, the microstructures are different indicating probable processing and heat treating differences.

Microprobe data from the 1988 analyses shows the average glaze composition to be a close match to the current study. The microprobe data also shows the average body matrix composition to closely match the glaze data, but microstructural data indicate considerable quartz inclusions are present in the bodies, amounting to about 15vol%. We conclude that the body and the glaze raw ingredient sources were likely the same. Our access to combined and simultaneous elemental imaging, backscattered imaging, and point analysis has produced a better understanding of the microstructure and composition than was available to the earlier studies cited in the bibliography. A full treatment of this study is available in reference 2.

ACKNOWLEDGMENTS

We gratefully acknowledge the help and mentoring of University of Arizona colleagues: Prof. David Killick, Archaeometallurgy Lab and petrographic microscopy facility, Department of Anthropology; Dr. Kenneth Domanik, Microprobe Lab, Lunar and Planetary Sciences; Steven Hernandez, University Spectroscopy and Imaging Facility (SEM-EDS), and Dr. Hexiong Yang, Department of Geology. We appreciate their comments, help and advice. We are grateful to Frederick Matson who encouraged and provided samples in 1987 in the early stages of this study. We wish he were here to review it.

REFERENCES

1. Plumer, J., *Temmoku: A Study of the Ware of Chien,* Marunouchi, Chiyoda-ku, Tokyo: Idemitsu Art Gallery (1972).
2. James Dustin Morehead, *Reverse Engineering the Physical Chemistry of Black Oil Spot and Hare's Fur Glaze from Fujian, China in the Song Dynasty (ca. 10-13th Centuries CE),* M.S. Thesis, University of Arizona, April 2014.
3. Yamasaki, Kazuo, "Scientific Studies on Special Temmoku Bowls, Yohen and Oil Spot," *Proceedings of the International Conference on Ancient Chinese Pottery and Porcelain*, Shanghai, China: Science Press, 1982.
4. Vandiver, P.B., "Compositions of Hare's Fur and Oil Spot Glazes from Jianyang Tea Bowls of the Song Dynasty," *International Symposium on Ancient Chinese Pottery and Porcelain*, Shanghai, China: Science Press, 1989, presented as poster and attached as addenda, and P.B. Vandiver, "Ancient Glazes," *Scientific American,* April 1989.
5. Ernest M. Levin and C.R. Robbins, *Phase Diagrams for Ceramists: Equilibrium Diagrams of Oxide Systems*, volume 1, Columbus, Ohio: American Ceramic Society, 1964.

Mater. Res. Soc. Symp. Proc. Vol. 1656 © 2015 Materials Research Society
DOI: 10.1557/opl.2015.838

The Technological Development of Decorated Corinthian Pottery, 8th to 6th Centuries BCE

Jay A. Stephens[1], Pamela B. Vandiver[2], Stephen A. Hernandez[2] and David Killick[1]

[1]Department of Anthropology, University of Arizona, Tucson, AZ 85721 USA
[2]Department of Materials Science and Engineering, University of Arizona, Tucson, AZ 86721 USA

ABSTRACT

Polychrome slipped and decorated pottery from Corinth, Greece, developed over two centuries from monochrome, dark brown slips and washes on a calcareous yellow clay body to a wide range of decorative techniques. Once significant experimentation with color variability began, five colors, each with various levels of gloss, were produced. Some slip colors involve multiple-step processing to control glass content and degree of sintering; the control of particle size to produce variable roughness and a matte or semi-matt or glossy appearance. Considerable evidence supports nearly continuous development and engineering of the ceramic slips, although no data support the improvement in composition or processing of the ceramic bodies. For instance, significant macro-porosity consistently is present in the bodies. We present the results of study of 27 sherds with 59 examples of Corinthian polychrome paint layers, measuring 5 to 35 microns in thickness, that were collected by Marie Farnsworth in the late 1950s and 1960s from Greek archaeological sites. Black, red, white, wine red (or purple) and overlying, matte banded slips and paints were studied by optical microscopy, petrographic and scanning-electron microscopy with semi-quantative energy dispersive x-ray analysis, as well as wavelength-dispersive electron microprobe (EPMA) elemental mapping and analysis.

INTRODUCTION

From the 8th to 6th centuries BCE, Corinth was at the center of a social and technological revolution. With the revival of major trade networks during the Middle and Late Geometric (800 – 720), Corinthian people began importing materials from the Eastern Mediterranean that helped shape the culminating Protocorinthian and Corinthian styles, motifs, and technological features [1, 2]. As a port city on the isthmus between the Peloponnese and mainland Greece, Corinth was home to a vibrant network of traders and importers, and became the major area for ceramic production at the close of the Geometric period. This, in turn, allowed Corinthian pottery workshops to draw from the styles of the eastern areas, especially from the imported black and red on buff ceramics. Much of this influence came from eastern metalworking and is mostly seen in the styles and motifs that became popular during the Protocorinthian period (720-625 BCE) [3]. This entanglement of Corinthian and Eastern Mediterranean people and goods promoted the evolution of Black Figure style, beginning in the 7th century BCE in middle and late Protocorinthian ceramic products. The use and further development of an imported polychrome style became one of the defining artistic features and illustrates technological mastery of Corinthian workshops. Predominantly featuring a miniature animal and figurative style, Corinthian potters were one of the first regions to develop a polychrome palette of black, white, red, wine red and purple, beginning in the 7th century BCE, simultaneously with the black figure motifs [1, 4, 5]. Glossy black, originating from a fine particle clay slip, was the predominant slip color on the polychrome vessels with white, red, wine red and purple used in conjunction with sgraffiato incisions to add detail and imply three-dimensionality to their pottery. Corinthian potters developed new colors, purple and white, added a range of textures, matte, semi-matte, satin-matte and glossy, and were able to layer slips over one another, controlling firing to maintain their stability and clarity. The use of this style continued through the 6th century in the Corinthian phase (625-535 BCE), but died out shortly after, as their Athenian neighbors took control of the markets and trade [1, 6]. For the Protocorinthian and Corinthian periods, however, the pottery of Corinth was popular, widely sought and traded across the Mediterranean and into southwest Asia.

In searching for details regarding the production of Protocorinthian and Corinthian vessels, little information is available. Most literature has been focused on Athenian ceramics. Many potters, painters and workshops have been identified, and they are the more famous of the ancient Greek potters. Marie Farnsworth identified the colorants as based on iron pigments [7].The clays used by Corinthian potters have been studied by Farnsworth using neutron activation and x–ray diffraction [8, 9]. She reported that the clay most commonly found in Corinth and the surrounding plain is a calcium-rich illite, occasionally with some montmorillonite. The montmorillonite component of this clay is believed to be exaggerated, as her tested samples were levigated to include only particles less than 2 microns [8]. However, she was puzzled by the high calcium carbonate content that is inconsistent with the high earthenware firing temperatures of 800-1000°C that she anticipated having been used [10]. For a study of technological change in Corinthian amphoras, Vandiver [11] surveyed and analyzed similar clay resources, under the direction of Charles Williams who had also helped Farnsworth, and found calcium ions intercalated in the illitic clay, rather than present as inclusions of calcium carbonate that would have caused spalling of the body. In 2003 Ian Whitbred added further to these conclusions [12].

PROBLEM STATEMENT AND METHODOLOGY

The polychrome style marks a major technological development that had significant impact on future styles of Greek pottery. This study utilized a collection of ancient Greek ceramics assembled by Marie Farnsworth that consist of 151 labeled bags of site-specific sherds. Each bag contains between 1 and 12 sherds, thus allowing for the sampling and testing of pigment and slip types on a large scale. Corinthian style sherds were catalogued and sorted to separate and organize them by period and technological variability. Twenty-seven were chosen for characterization.

Our focus is to understand the ceramic technology and production sequences and methods by materials analysis of Corinthian sherds. The slips ranged in thickness from 5 to 35 microns, and analytical techniques had to be optimized to produce results. In the end, several techniques were found to produce complementary results that answered microstructural and compositional questions, and aided in developing an understanding of the range of variability of the decorative treatments.

An inventory of the 27 polychrome slipped and painted sherds is presented in Table 1 with approximate dates. Fifty-nine different colors and textures on these sherds were analyzed using optical microscopy (Leica EZ4 HD 8-35x with integral digital camera) and scanning electron microscopy with energy-dispersive semi-quantitative x-ray analysis (Hitachi S-4800 Type II with Tracon EDS). Other techniques (electron microscopy, optical petrography, and elemental mapping) were applied to a subset of representative samples for further understanding of the technology and methodology that produced these ceramics.

Sample	Site/Date	Black	Red	White	Wine Red/Purple	Banded
MF4	Corinth/700 BCE	X		X	X	
MF7(1)	Corinth/625 – 600 BCE	X			X	
MF7(2)	Corinth/625 – 600 BCE	X			X	
MF7(5)	Corinth/625 – 600 BCE	X			X	
MF17	Corinth/8th C. BCE	X		X		
MF24	Aegina/ Mid 6th C. BCE	X		X		
MF27(1)	Corinth/6th C. BCE		X			X
MF27(2)	Corinth/6th C. BCE		X			X
MF28	Corinth/6th C. BCE	X	X			
MF30(1)	Corinth/6th C. BCE	X	X			
MF30(2)	Corinth/6th C. BCE	X	X			
MF34	Corinth/6th C. BCE		X	X		
MF100(1)	Aegina/ Mid 6th C. BCE			X		
MF100(2)	Aegina/ Mid 6th C. BCE			X		
MF101(1)	Aegina/7th C. BCE	X		X		
MF101(2)	Aegina/7th C. BCE	X		X		
MF104	Aegina/7th C. BCE	X	X	X		
MF124(1)	Corinth/7th C. BCE	X	X	X		
MF124(2)	Corinth/7th C. BCE	X		X	X	
MF124(3)	Corinth/7th C. BCE		X			
MF124(4)	Corinth/7th C. BCE	X		X	X	
MF124(5)	Corinth/7th C. BCE	X				X

MF124(6)	Corinth/7th C. BCE	X		X		
MF149	Corinth/7th C. BCE	X	X	X		
MF151(1)	Aegina/7th C. BCE	X	X	X		
MF151(2)	Aegina/7th C. BCE	X		X	X	
MF151(3)	Aegina/7th C. BCE	X	X			
Sheerd Analysis Total		**20**	**11**	**15**	**7**	**3**

Table 1: Inventory of 27 Corinthian Polychrome Slipped and painted sherds, 8-6th century BCE incorporate the variability present in the assemblage, including optical petrography using thin sections (Olympus B-51 with Optronics software and camera, 100-1000x), and wavelength-dispersive electron probe microanalysis with elemental mapping (Cameca SX100 using a 1 and a 5 micron beam at 15 KV for 12 elements identified by SEM-EDS). Using a combined suite of techniques (OM, SEM-EDS, Petrography and WDS) allowed characterization each color and texture at various levels of scale and from many viewpoints. Xeroradiography (Xerox Medical Systems Xeroradiograph 125) of the sherds was employed to examine the question of the quality of clay preparation for these fine wares, and differential thermal analysis was used to examine the homogeneity of the various Corinthian clays in the assemblage and with quite consistent results.

RESULTS

Decorated polychrome surfaces were imaged with an optical microscope at 8-35x magnification to identify and describe variability, included the uses of the paint layers and their respective colors and textures. The polychrome paints were often applied on top of a base of yellow to red slip which would turn black after firing in reducing conditions (Fig. 1), or very occasionally a base of oxidized red (Fig. 2), and usually had a painted layer of red, white and/or wine red applied over the slip; as for instance the yellowish white applied over the black and shown in the upper left of Fig. 1. Many of the vessels also have incised lines to define the different areas of color or to outline figures, as shown in Figure 1. However, many sherds displayed unique characteristics that also demonstrate the skill of Corinthian potters.

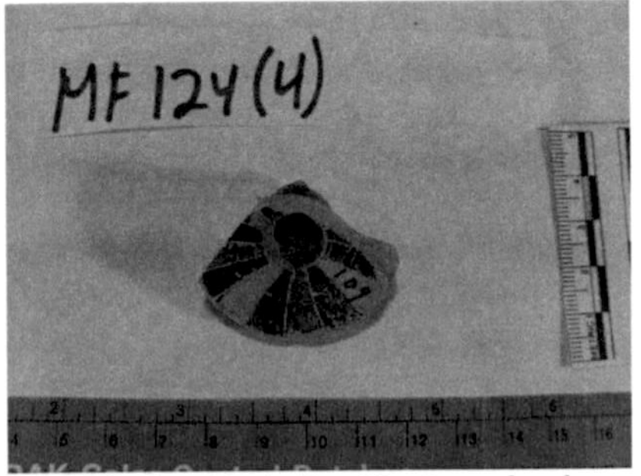

Figure 1: Base black slip with wine red and white added color, and incised decoration, MF124(4).

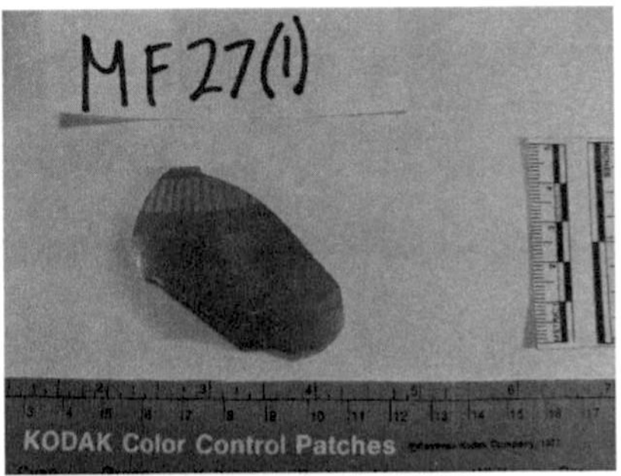

Figure 2: Red slipped sherd, MF27(1), illustrating the use of a base red sherd and alternating matte – gloss banding

Besides color, another source of variability are the banded patterns that have a different reflectance and often texture, but occur within a field of the same or very similar color, as shown in specular light in a black area (Fig. 3). Handling and turning the original vessel would have highlighted this the interplay of matte and reflective surfaces within a single color area; the vessel would have variously sparkled or displayed complex colored surfaces to the viewer. The banded patterns are present as glossy and matte bands of the same color, and raise questions about the production steps that went into making these ceramics. Two different forms of this banding were documented; glossy and matte red banding (Fig. 2), and glossy and matte black banding (Fig. 3).

Optical microscopy allowed deconstruction of the sequences of the production processes. In the Protocorinthian samples displaying a polychrome style, a yellow to red slip that turned black in reduction was applied first, followed by the white, wine red, purple and red slips. Fig. 1 shows specks of black seen below the white slip layer (bottom left). These colored areas were then incised to outline and emphasize the presence of the underlying yellow body. This initial investigation provided an excellent basis for SEM-EDS, Electron Beam Microprobe, and petrographic thin section analysis. Samples were taken from each of the major colorant classes (black, red, wine red, and white) as well as the major areas of variability (matte and glossy treatments) in order to deconstruct the technological processes practiced to produce these ceramics. Specific attention was paid to how these colorants interacted with each other in the ceramic as a way to evaluate the measure of control during application, drying and firing. Levigated clay washes and burnishing were occasionally identified on the clay body surfaces prior to application of polychrome slips and paints. This helped prepare and consolidate the vessel for later application of the polychrome decoration. Further data on the individual slip classes was then collected.

Black slip results: The black slip is the most common of the observed colors and is present on a wide variety of shapes during the Protocorinthian and Corinthian periods. It is typically applied as a matte or muted slip, but is also present as a high gloss slip that was burnished. Some instances of the black slip with red spotting were found. Cyril Smith proposed such black to red variation in Athenian pottery is caused by the thinness of the slip in the red areas [13]. Suggested explanations for these red spots are based on Cyril Smith's hypothesis, as well as Aloupi-Siotis. The first is that the Corinthian potters were levigating their clay but did not produce homogeneous particles for the clay slip. The second explanation is that these pottery did not apply an even coat of colored slip to the black decorated part of the pot before it was fired. Lastly, Aloupi-Siotis proposed that the cause of this spotting is due to the temperature of the firing; in this regard, the temperature is too low to allow the black to totally vitrify under reducing conditions, and some areas of it reoxidize [14]. The optical variations in the black slip presented a starting question for SEM-EDS analysis.

Figure 3: Black with Specular Reflection where Fine and Glossy and Matte Surface where Coarse particled, OM 20x, MF124(5)

Figure 4: Sintered Surface of Glossy Black Iron-Based Slip, SEM 10,000x, MF151(2)

The results of SEM-EDS analysis showed that the samples had a highly sintered iron-based slip (Fig. 4) with many fluxing ions, including Na, Mg, Cl, K and Ca (Fig. 5). Ti was also present, and along with S and C, and is known to modify the iron chromophore, but it was not present in sufficient quantity. The fluxing cations would have promoted sintering in the slip, as they lowered the temperature required for reaction to occur. Furthermore, particles are of single to sub-micron in size, which is known to be a prerequisite for the production of a reduced black slip [14], and demonstrates the high-degree of control in levigation that Corinthian potters practiced to produce these slip layers. Thus, slip thickness not variable particle size was the cause of the variegation.

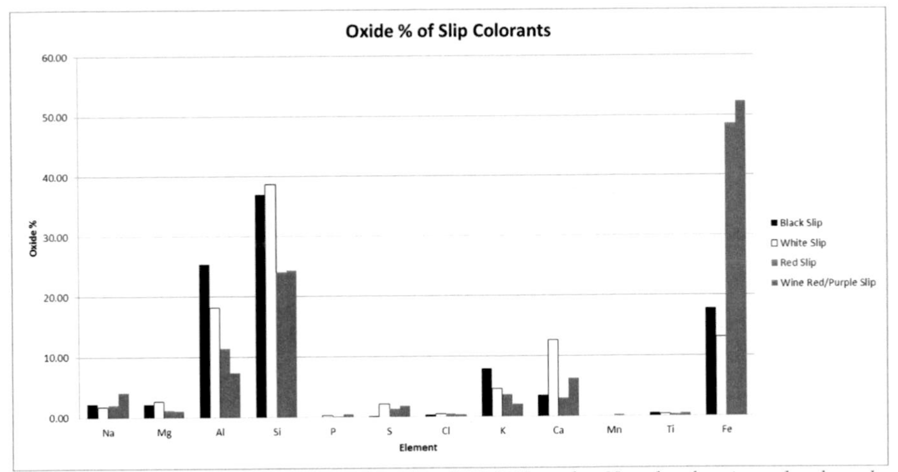

Figure 5: Oxide percent of Slip Colorants based on SEM-EDS results. Note that the wine red and purple paint layers have been merged, based on compositional, but not microstructural, similarity.

Petrographic analysis of the black-slipped thin sections showed the black slip of 10 microns was typically applied over the ceramic body, as shown in Fig. 6. Iron is the primary colorant in the coating, and formed as a result of reducing atmospheric conditions in the kiln when it was fired. These iron particles have reduced to magnetite particles and result in a black color. Under reflected light in the petrographic microscope, the magnetite particles appear as small dots of reflected light (Fig. 6), and under SEM-EDS they are present as micron and sub-micron cube-shaped particles embedded in and on the surface of the layer.

Added Red slip: Like the black slips, the added red slips are also composed of an iron-rich clay. However, the added red slip has a much higher percentage of iron oxide, (45.05%) than is present in the black slip (10.11%) (Fig. 5), and is much less sintered (Fig. 7). Fig. 5 shows that lower concentrations of the fluxing ions, Na, Mg, S, Cl, K and Ca, are present in the added red slip EDS spectra than in the black ones. Furthermore, the particles are still single to sub-micron in size, similar to the black, but they are more aggregated and the surface is somewhat rougher. Backscattered imaging was used to illustrate the prominence and clustering of iron oxide particles across the surface of the red slip (Fig. 8).

The SEM-EDS data for the red slip are reflected in the petrographic findings. When analyzed under transmitted light, the red slip layer is seen as bright red and opaque (Fig. 9). This color corresponds to highly oxidized iron, as hematite. However, we will see that this layer differs from the added wine red paint, as this layer does not shine bright white under reflected light in thin section. This indicates that the red is a high-iron-rich clay (compare Fig. 9 to Fig. 16) instead of an iron oxide-based paint. Furthermore, in cross section, we see that the added red slip layer, with an average thickness of 5 – 10 microns, typically sits on top of the black slip layer. With this in mind, the ceramic was either fired in multiple stages or in a single firing with multiple phases that promoted the formation of these colors.

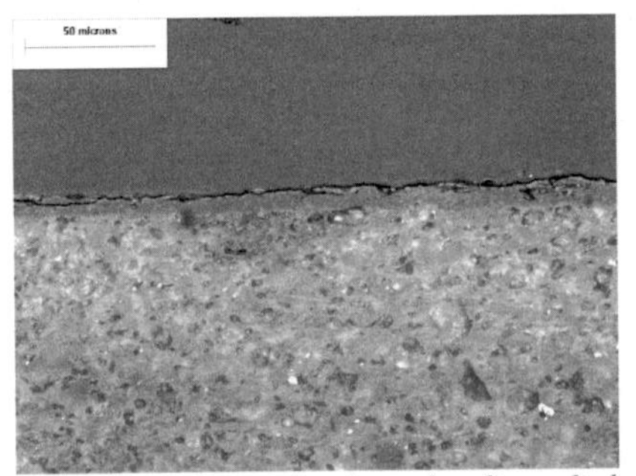

Figure 6: Black iron slip above clay slip applied on coarser body, PM 50x, MF28.

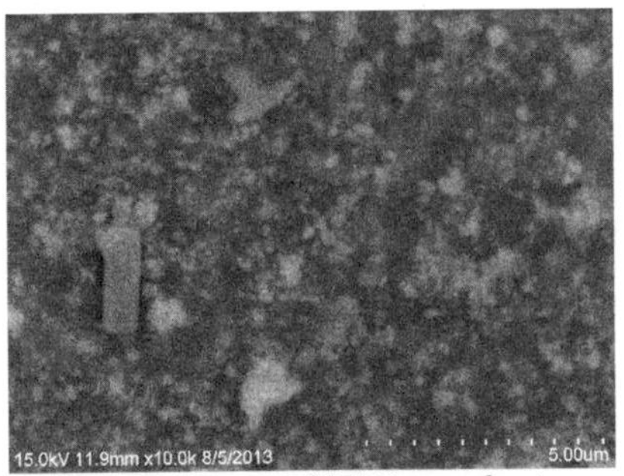

Figure 8: Backscattered SEM image showing red iron oxide aggregates, SEM 10,000x, MF27(2)

Figure 7: Surface of red slip with submicron, sintered particles, SEM 4,500x, MF149

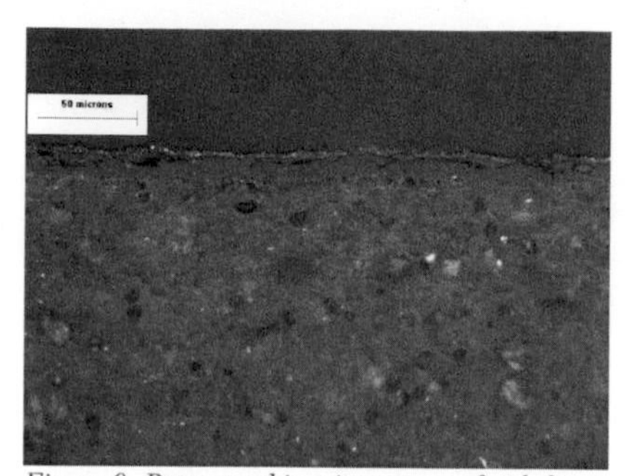

Figure 9: Petrographic microscopy of red slip with opaque red layer of ~5 microns, PM 50x, MF151(1)

White slip: The white slip in Figure 10 shows considerable sintering and fine micron and sub-micron particle size. The white slip areas have higher amounts of alumina and silica with minor amounts of Ca (refer to Fig. 5), leading to a preliminary conclusion that an illitic or kaolinite clay was used, perhaps one with a considerable admixture of calcium oxide (~20%) as an addition or as an impurity. The iron seen in the compositional spectra is too high to produce white color and may be a result of the surface analysis sampling volume of 3-5 microns that penetrated into the black iron-rich under layer, seen with optical microscopy. However, the slip in cross section is about 20 to 50 microns thick (Fig. 11).

Much like the SEM-EDS data, the white slip had very few impurities when viewed under the petrographic microscope. Though some of the sample was lost during sample preparation, enough remained to reconstruct its application. This white slip was applied over the base black slip (Fig. 11), and confirms the initial results obtained by optical microscopy, even though such a layering process seems counter-intuitive.

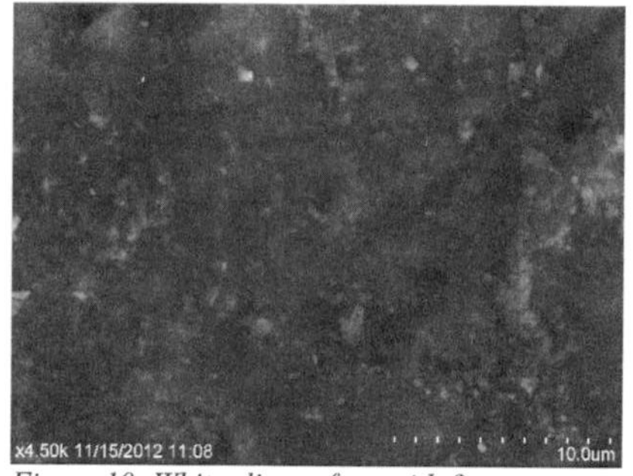

Figure 10: White slip surface with fine particulate structure aggregated in ridges, SEM 4500x, MF124(6).

Figure 11: Cross section showing thick white slip applied over black slip onto body, PM 50x, MF124(6).

Wine Red and Purple Paint Washes: The added wine red and purple paint washes represent common colorants on Protocorinthian and Corinthian vessels. In analyzing these sources with SEM-EDS, sintered aggregates and larger globular areas were seen across the surface of the layer (Fig. 12), in addition to partly sintered areas where iron particles of sub-micron size (Fig. 13) had agglomerated. Furthermore, several dendritic crystals were discovered around the highly sintered globular areas (Fig. 14). Although dendrite formations were not found on all wine red samples, they all primarily were composed of high concentrations of iron, as well as some Ca, K and Mg with minor Na and Cl (Fig. 5). Backscattered EDS was used to illustrate the prominent iron particle aggregates (Fig. 15) that are somewhat similar, but smaller, to those in the better sintered red slip (Fig. 8).

When viewing the wine red under a petrographic microscope, the suspected iron-based pigment is confirmed. Under reflected light, the paint layer shines bright white, as the iron oxide crystal facets reflect the light and confirms the hypothesis of iron oxide rather than iron-bearing clay (Fig. 16). Once analyzed under cross polarized light, this layer turns a deep red to orange color (Fig. 17). This color is an indicator of a hematite-based iron oxide, and although further testing is needed to confirm this result, it presents a basis to predict the range of firing atmospheres to which this ceramic was subjected. When analyzing the wine red paint in the content of the cross section, this pigment is seen to be applied, with an average thickness of 5 – 10 microns, on top of the black slip layer. However, this technique could not alone distinguish them. Analysis of these samples using elemental mapping with the electron microprobe will help clarify which of these two options this preparatory layer is.

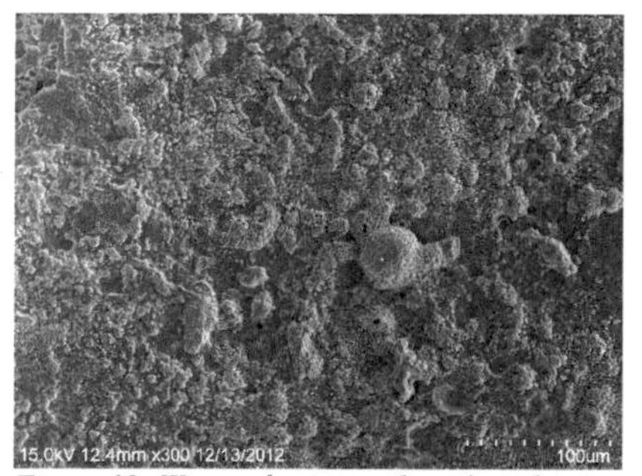

Figure 12: Wine red paint wash surface showing 5-50 micron globular sintered agglomerates, SEM 300x , MF4.

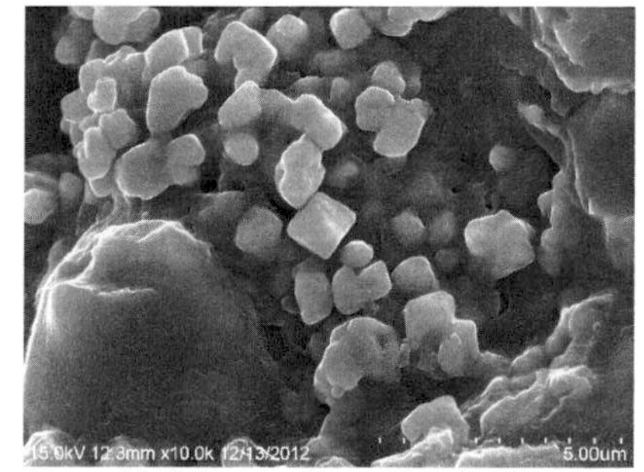

Figure 13: Submicron iron-rich cubical and rounded particles, and quartz (lower left), sintered clay to the right, SEM 10,000x, MF4.

Figure 14: Iron-dendrites in wine red paint wash, SEM 13,000x, MF124(4).

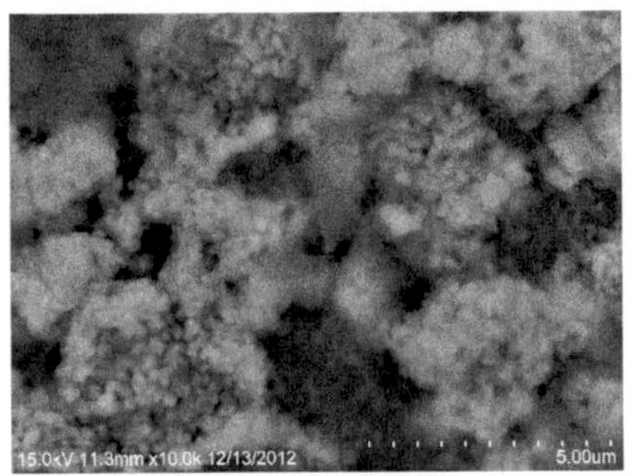

Figure 15: Backscattered image of agglomerated iron-containing particles that have partially sintered, SEM 10,000x, MF7(1).

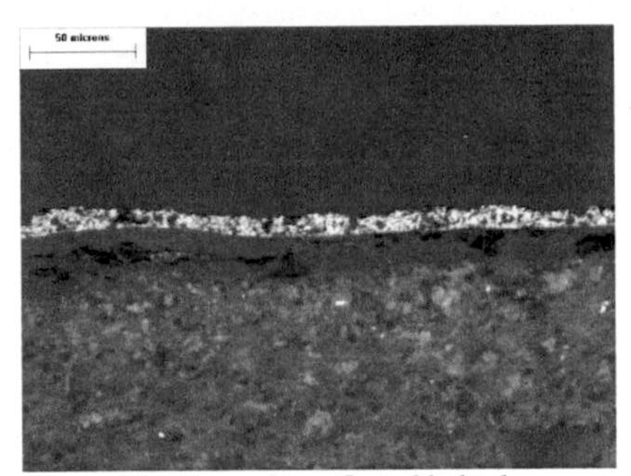

Figure 16: Wine red in reflected light shows iron-rich particles light scatter, over a black slip, PM 50x, MF151(2).

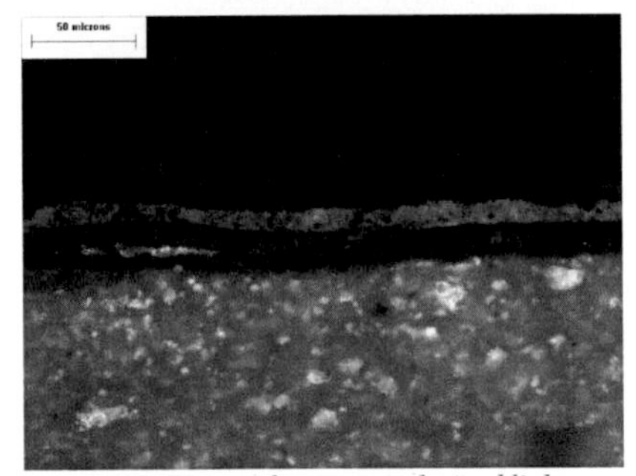

Figure 17: Wine red in cross polarized light shows red iron oxide concentration, PM 50x, MF151(2).

High and low gloss black banding: The variability of high and low gloss black banded slips was tested with SEM-EDS. The black high-gloss bands contained a high concentration of sub-micron particles of average size of about 0.5 microns with many particles being high in iron, as shown in the backscattered image in Figure 18. In low gloss and matte banded slips, micron-sized particles were only a minor constituent. Only minimal compositional differences between the matte and high gloss bands were obtained (Fig. 19). The high gloss band has slightly more alumina, silica and potassia, indicating the presence of a small amount more clay in the high gloss areas.

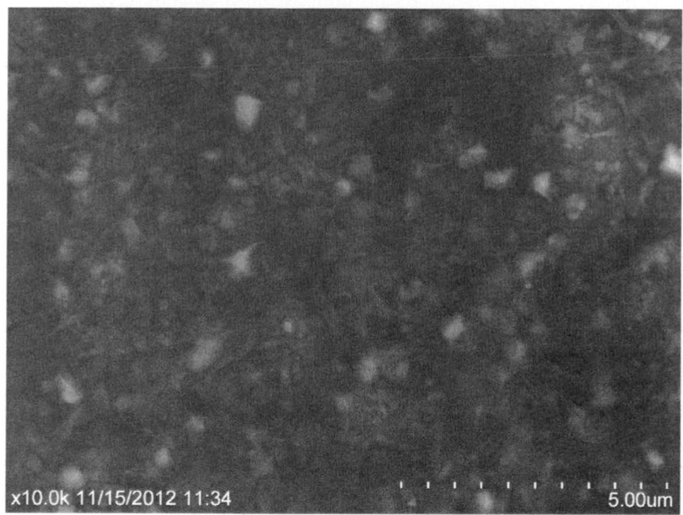

Figure 18: High gloss black slip banding with mostly half micron particles well dispersed and well sintered, SEM 10,000x backscattered, MF124(5)

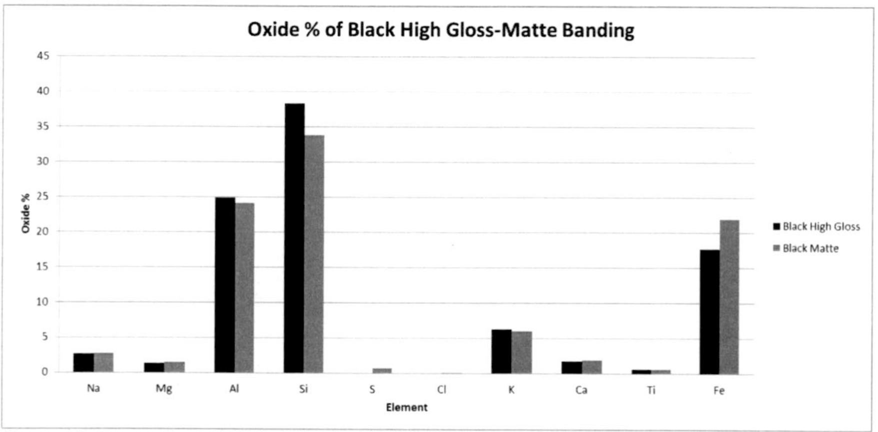

Figure 19: Comparison of EDS results for black high gloss and black matte banding.

Red high gloss and matte banding: The red high gloss and matte banding comparison displayed similar results to the black high gloss and matte banding, as a large amount of sintering and sub-micron particles are found in the high gloss band (Fig. 20). Furthermore, this band has similar higher concentrations of silica, alumina, and potassia, indicating increased clay relative to iron oxides (Fig. 21). Possibly the soda concentration was also slightly greater. The red matt band displayed less sintering, but a similar particle size. Calcia and iron oxide also increased.

Due to the thinness of the banded layers, petrographic results for the banded samples were inconclusive.

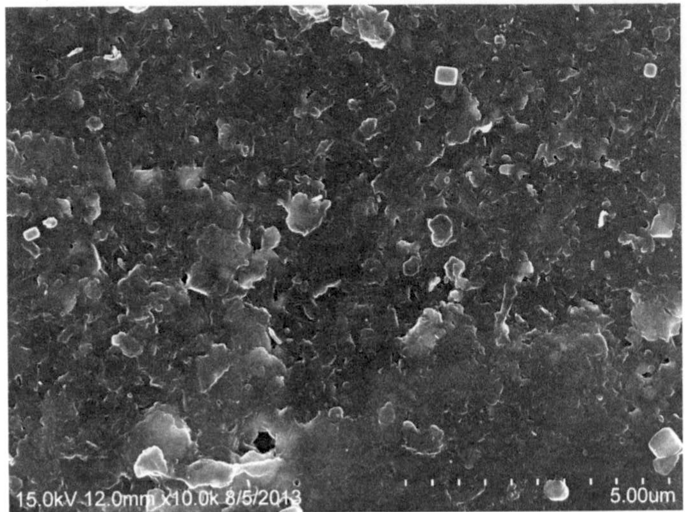

Figure 20: High Gloss Red Banding shows fine submicron particle size that is well-sintered with particles aligned parallel to the surface, SEM 10,000x, MF27(1).

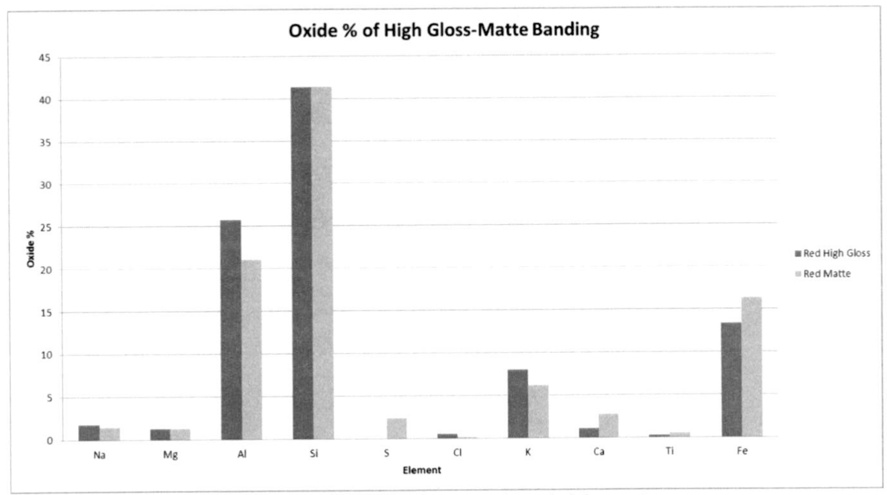

Figure 21: EDS results comparing high gloss and matte red banded decoration.

Data from Electron Microbe Analysis: The eleven thin sections (MF numbers 7, 24, 27(1), 28, 30(1), 34, 100, 124(5), 124(6), 151(1), 151(2)) were analyzed for 12 elements (Na, Mg, Al, Si, P, S, Cl, K, Ca, Ti, Mn and Fe). A 1-micron beam was used for point analyses and a 5-micron beam diameter was used for feature analysis. The added red and white slips showed considerable porosity, and the totals were lower than desirable, as shown in Table 2 that compares the normalized semi-quantitative EDS data with the standardized and calibrated WDS results. The WDS totals are a direct relation to the porosity of the slip. A third row that normalizes the WDS data provides a comparison of the two techniques, showing that general trends hold between the two techniques, although the numbers vary considerably. Because these slips and paints are very thin, close to detection levels, precision is illusive. Based on the microprobe data, a plot was drawn to show the compositional differences among the various colors of slip.

Average Oxide %

	Slip	MgO	Al_2O_3	Na_2O	SiO_2	CaO	FeO	MnO	K_2O	TiO_2	Cl	P_2O_5	SO_3	Total
WDS	Black	1.64	28.81	1.35	43.07	1.97	10.11	0.05	7.03	0.71	0.09	0.12	0.02	94.95
	Red	0.89	7.69	0.47	16.78	1.26	45.05	0.09	2.08	0.32	0.44	0.13	0.04	75.21
	Wine Red	0.47	7.92	1.05	20.00	0.54	54.14	0.05	2.79	0.34	0.13	0.15	0.03	87.60
	White	1.74	6.90	0.33	21.91	19.65	4.49	0.10	0.85	0.17	0.22	0.74	9.67	66.77
	Purple	0.49	7.66	0.46	17.86	0.33	58.81	0.08	3.52	0.27	0.02	0.32	0.01	89.82
WDS Std. Dev.	Black	0.53	4.54	0.84	6.17	2.44	1.98	0.03	3.36	0.42	0.12	0.07	0.03	6.40
	Red	0.61	3.56	0.34	8.45	1.13	16.07	0.04	1.38	0.26	0.28	0.17	0.03	18.58
	Wine Red	0.30	4.08	1.04	8.97	0.92	12.84	0.03	1.71	0.40	0.13	0.16	0.03	5.16
	White	1.11	4.80	0.29	16.50	8.35	2.31	0.05	0.47	0.09	0.21	0.29	13.21	24.50
	Purple	0.14	1.29	0.21	5.79	0.14	8.60	0.01	1.61	0.16	0.02	0.05	0.01	2.74
Normalized	Black	1.73	30.39	1.42	45.43	2.07	10.67	0.06	7.41	0.74	0.10	0.12	0.02	100.00
	Red	1.18	10.24	0.63	22.35	1.68	60.00	0.12	2.77	0.43	0.59	0.17	0.05	100.04
	Wine Red	0.54	9.04	1.20	22.83	0.62	61.80	0.05	3.19	0.39	0.14	0.17	0.03	99.92
	White	2.60	10.34	0.50	32.82	29.44	6.72	0.14	1.28	0.25	0.33	1.11	14.48	100.00
	Purple	0.54	8.53	0.51	19.88	0.37	65.47	0.09	3.91	0.31	0.02	0.36	0.01	99.88
EDS	Black	2.11	25.35	2.22	36.91	3.47	17.75	N/A	7.91	0.44	0.26	0.00	0.08	96.50
	Red	1.19	11.38	2.07	24.08	2.99	48.59	0.20	3.63	0.28	0.47	0.07	1.39	96.34
	Wine Red	1.13	7.36	4.09	24.35	6.26	52.30	N/A	2.05	0.45	0.37	0.43	1.78	100.57
	White	2.67	18.17	1.72	38.67	12.60	13.08	N/A	4.53	0.35	0.25	0.41	2.10	94.55
	Purple	N/A	N/A	N/A	N/A	N/A	N/A	N/A	N/A	N/A	N/A	N/A	N/A	N/A

Table 2: A Comparison of WDS and EDS analytical results. 49 points were analyzed for the black slip with WDS analysis and 14 with EDS analysis. 16 points were analyzed for the red slip with WDS analysis and 13 using EDS analysis. 16 points were analyzed for the wine red paint wash with WDS analysis and 17 using EDS analysis. 5 points were analyzed for the white slip with WDS analysis and 16 using EDS analysis. 6 points were analyzed for the purple paint wash with WDS analysis.

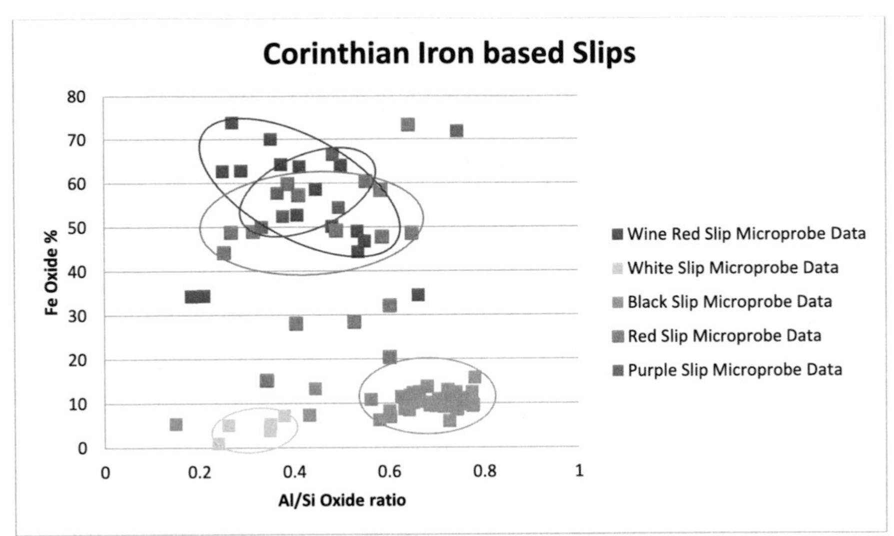

Figure 22: Plot of Al/Si oxide percent versus Fe oxide percent illustrates the differences in the polychrome decorative treatments. The white and black slips are different compositions from the red, wine red and purple slips and paint washes.

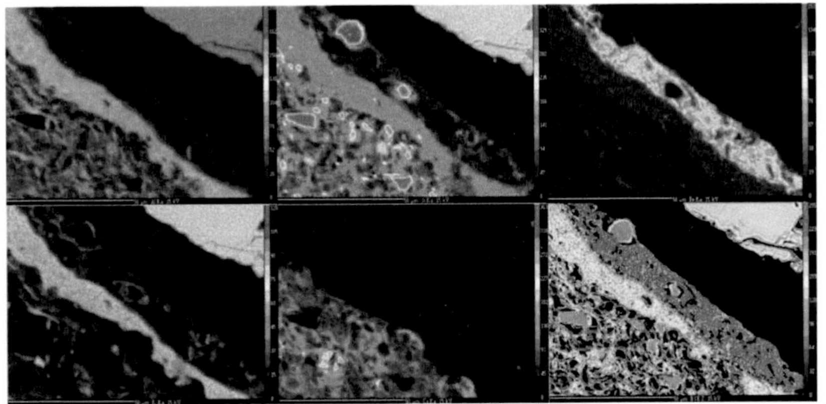

Figure 23: Sample MF30. Top layer, from left to right: Al, Si, Fe. Bottom layer, from left to right: K, Ca, Total compositional results. Elemental mapping results illustrate the difference between red and black Corinthian slips. The high iron red slip sits on the top layer of sample MF30, with the low iron black slip layer directly underneath it. The black slip has a much higher ratio of Al/Si.

The production of Athenian blank and red figure relied on the fact the black and red depictions were made of similar base components [15] although the black slip was made from a different source, it displayed similar chemical properties. The Corinthian's were producing an entirely separate added red paint that was much higher in iron (Fig. 22, 23). The added red paint is shown to have a different clay binder (Fig. 22) and be lower in flux content (Fig. 23). Petrographic analysis help to further illustrate the distinction between these slips, as the red slip turns bright red under reflected light, whereas the black slip is predominantly black or grayish with nanoparticles of magnetite distributed throughout the layer (Fig. 6).

During Electron Microscopy, an anomalous feature was observed on the upper left corner of sample MF24 and glowed bright white, indicating a metallic species. Upon further investigation using broad spectrum EDS and elemental mapping, the feature was determined to be composed of high levels of lead, tin, and antimony, along with a myriad of fluxes. These results are unprecedented in Corinthian, yet alone Greek, ceramics as no surface feature of this composition has been documented. Based on its unusual metallic nature, and presence on only one corner of the sample, the feature is not believed to be an aspect of the slip. However, its composition suggests it could have been related to a metallurgical function, such as solder.

CONCLUSIONS

Following the evolution of Greek civilization during the Geometric period (900 – 700 BCE), pottery production mirrors the rapidly evolving cultural climate and becomes a major field for experimentation and innovation. The Corinthian potters of the Protocorinthian and Corinthian period were at the forefront of this movement, producing complex polychrome slipped and painted wares for consumption and export around the Mediterranean. Optical microscopy, SEM-EDS, Electron microscopy, and optical petrography were used to better understand the technology and methodology needed for their production.

Black, red, wine red, purple, and white were the central colors of these Corinthian styles and were seen to be added in a systematic fashion. By studying the cross sections of Protocorinthian and Corinthian pottery, the chain operatoire of the surface treatments could be reconstructed. Beginning with a base wash of Corinthian clay, the surface of the Corinthian pot was smoothed to prepare it for the addition of polychrome slips. A base slip of black was then added, and took the form of either a decorative element, such as a figure or rosette, or a uniform surface coating. Red, wine red, purple, and white colors were then added to the surface of this base black slip as accents, and gave these Corinthian styles their character.

Iron was shown to be the key element in the production and character of these Corinthian slips and pigments. The produced black slip (10.11%) relied on a relatively low iron content in comparison to the added colors of red (45.05%), wine red (54.14%), and purple (58.8%) slips and paint washes. The species of iron also proved key in discriminating the materials used in the production of polychrome slips, as the black slip relied on Magnetite nanocrystals while the added accent colors used hematite-rich clays and manufactured paint washes to produce their distinctive colors. As a result, the black slip was concluded to be a product of an iron-containing slip fired under reducing conditions, while the added red was an iron-rich-clay species, different from that of the black, fired in oxidizing conditions. Wine red and purple paint washes required much more complex production methods to produce their distinctive color, given their iron-oxide nature. Thus, a possible method for the manufacture of this paint wash includes the treatment of smelted iron with acid and grinding the resulting metal species until fine, submicron particles were achieved. This pigment would then be mixed with a clay binder and appropriate fluxes to facilitate its use as a paint wash. Though the submicron particles would be difficult to obtain from a treated iron oxide, they would replicate the results observed through optical petrography; an orange layer under transmitted light and bright white layer under reflected. The white slip was shown to be an illitic or kaolinitic clay with calcium impurities.

Marie Farnsworth called into question the ability to form a black slip from the calcium rich Corinthian clays that she analyzed [8], however it is believed that this would have indeed been possible. W. D. Kingery presented conclusive evidence on the required characteristics of the Athenian black gloss, these being high iron oxide and potassium oxide with low calcium oxide, as well as simple techniques, levigating, that can be employed to greatly lower the calcia concentration in the selected clay [16]. Kingery's results followed the model established by Binns and Frazer, who were two of the first to question the production of the Attic black gloss [17], as well as Bimson, who discussed the need for a high potassium oxide and iron oxide illitic clay with low concentrations in calcium oxide [18]. While his samples were Athenian, these key characteristics are mirrored in this study of the Corinthian black gloss (Table 2). Aloupi-Siotis has also conducted extensive studies on the formation of a black gloss, and concluded that a separate illitic clay from the ceramic body was used in the formation of the clay slip [14] In her study on the production of Corinthian amphora, Vandiver was able to achieve chemical results, in both the type A amphora and sampled Lignite Quarry clay, that were similar to Kingery's required criteria to form an appropriate black gloss [11]. Thus, it is believed that these Corinthian potters would have been able to form an appropriate black slip using Corinthian clays. In the next phase of this study, Corinthian calcareous clays will be used in an attempt to replicate the Corinthian black slip and illustrate the ability of these potters to make use of local materials.

The evaluation of the diverse materials and production methods that were involved in the creation of Corinthian ceramics calls into question the firing conditions and techniques used. While the black layer required a reducing atmosphere to be produced, the added red, wine red, and purple layers all appear to require oxidizing atmospheres, as all three shown orange under transmitted light; indicative of an oxidized hematite iron species. At this point, it is unknown if a single firing with multiple phases or multiple firings with specific phases was employed. However, we can conclude that the production of Corinthian ceramics was a complicated multistep process that required a great deal of control and skill.

Perhaps one of the greatest accomplishments of these Corinthian potters was their meticulous control of particle size in their quest to achieve specific textures and results. As observed in the study of Corinthian polychrome slips, all five major slip and paint wash categories relied on careful control of their of their materials to obtain particles of single to submicron size. This conclusion would be solidified if the proposed method of producing Corinthian wine red/purple paint washes could be proven and replicated, as it grounded in the meticulous control of ceramic material to produce a desired outcome. Perhaps the greatest example of Corinthian material control is seen in the multi-textured Corinthian pottery. With both red and black matte – gloss banded examples, the production of the higher gloss slip is believed to be a result of greater control over the iron and clay particle sizes used. Future experimental archaeological and elemental mapping studies will help to confirm these results.

The results of this study display the impressive levels of innovation, experimentation, and dedication of Corinthian potters. However, their use of such complex procedures and extreme control over their material begs the question, why commit to such an intensive methodology for the production of these polychrome vessels? We, as scholars, understand that ceramics serve a function, be that decorative, utilitarian, or otherwise, however we must also consider the energy investment that went into producing such objects. The effort displayed by these craftsmen to produce specific results, on such a large scale, warrants an investigation into the possible motive behind their manufacture. One explanation for the production of these energy inefficient ceramics centers on their visual similarity to metal objects. In this way, Corinthian ceramics acted as skeumorphs for metal objects. M. Vickers proposed such a viewpoint of Athenian pottery, noting that they not only were less functionally sound than their metal counterparts but would have also been much more affordable as a result [19, 20]. In this respect, pottery is viewed as an affordable substitute for durable metal vessels, of which little survive. Furthermore, Vicker's identifies the use of "purple" and black slips on a ceramic vessel as imitating copper accents, such as those used on bronze statuary, and corroded silver [19, 20]. Vicker's astute conclusions about polychrome Athenian pottery and their relation to contemporaneous metal vessels lends credence to the assertion that these Corinthian wares acted in a similar manner. At a time when metal imports from the Near East [3] and craftsmen experimentation were at an all-time high, it is logical to think that the mass production of a complex ceramic ware served a social significance. Thus, it is proposed that complex manufacturing processes, as is reconstructed in this study, can be used as an indicator of greater social meaning.

In the next phase of the project, we will endeavor to replicate this data through experimental replication. The use of experimental archaeology will confirm and/or refute the theoretical conclusions put forth by this study, and represents a major leap forward in the quest to understand the decision making process of these past people.

ACKNOWLEDGMENTS

We recognize Dr. Marie Farnsworth who assembled the research collection in the 1950's and 60's for the scientific analysis of Greek pottery. We thank Prof. Mary Voyatzis for providing insight into the dating of the sherds, the stylistic trends of Greek pottery and recent archaeological finds pertinent to this study. Finally, we gratefully acknowledge the help of Dr. Kenneth Domanik, Lunar and Planetary Sciences, Electron Microprobe Lab.

REFERENCES

1. Cook, Robert M., *Greek Painted Pottery*, London: Methuen, 1972, p. 42-56.
2. Mee, C., *Greek Archaeology: A Thematic Approach*, Chichester, U.K.: Wiley-Blackwell, 2011.
3. Rasmussen, Thomas, "Corinth and the Orientalising Phenomenon" *Looking at Greek Vases*, Ed. Tom Rasmussen and Nigel Jonathan Spivey, Cambridge: Cambridge University Press, 1991, p. 58-67.
4. Robertson, Martin, *A History of Greek Art*, London: Cambridge University Press, 1975, p. 25-26, and M. Robertson, *The Art of Vase Painting in Classical Antiquity,* Cambridge; New York: Cambridge University Press, 1992.

5. Schaus, Gerald, "The Beginning of Greek Polychrome Painting," *Journal of Hellenic Studies*, cviii, 1988, p. 107 – 117.
6. Boardman, John, "The Orientalizing Style" in J. Boardman, *Early Greek Vase Painting: 11th-6th Centuries B.C.: A Handbook*, New York: Thames and Hudson, 1998, p. 86-87.
7. Farnsworth, Marie, "Coloring Agents for Greek Glazes," *American Journal of Archaeology*, 67(4), 1963, p. 389-396.
8. Farnsworth, M. "Corinthian Pottery: Technical Studies," *American Jour. of Archaeology*, 74(1), 1970, p. 9-20.
9. Farnsworth, M., "Greek Pottery: A Mineralogical Study," *Am. Jour. of Arch*., 68(3), 1964, p. 221-228.
10. Williams, Charles, personal communication, July, 1989.
11. Vandiver, P.B. and C.G. Koehler, "Structure, Processing, Properties, and Style of Corinthian Transport Amphoras," in *Technology and Style, vol. ll, Ceramics and Civilization,* ed. W.D. Kingery and E. Lense, Columbus, OH: American Ceramic Society, 1986, p. 173 - 217.
12. Whitbred, Ian K., "Clays of Corinth: The Study of a Basic Resource for Ceramic Production," *Corinth,* 20, 2003, p. 1-13.
13. Smith, Cyril S., *Seventy Two Objects Illustrating the Nature of Discovery,* MIT Press, Cambridge, MA, and the Smithsonian Institution Press, Washington, D.C., 1980.
14. Aloupi-Siotis, E., "Recovery and Revival of Attic Vase Decoration Techniques. What can it offer to archaeological research?" in Papers on *The Colors of Clay*, The J.P. Getty Publications, 2009
15. Maniatis, Y., E. Aloupi, and A.D. Stalios, "New Evidence for the Nature of the Attic Black Gloss," *Archaeometry* 35(1), 1993, p. 23-34.
16. Kingery, David, "Attic Pottery Gloss Technology," *Archeomaterials* 5(1), 1991, p. 47-54
17. Binns, Charles F., and A. D. Fraser. "The Genesis of the Greek Black Glaze." *American Journal of Archaeology*. 33(1), 1929, p. 1-9.
18. Bimson, Mavis. "The Technique of Greek Black and Terra Sigillata Red." *The Antiquaries Journal* 36(3-4), 1956, p. 200-204.
19. Vickers, M., "Silver, Copper and Ceramics in Ancient Athens." In: M.J. Vickers (ed.): *Pots & Pans: A Colloquium on Precious Metals and Ceramics in the Muslim, Chinese and Graeco-Roman Worlds*, Oxford; Oxford: Oxford University Press for the Board of the Faculty of Oriental Studies University of Oxford, 1985, p. 137-151.
20. Vickers, M., "The Cultural Context of Ancient Greek Ceramics: An Essay in Skeuomorphism." In: P.E. McGovern, M.D. Notis & W. Kingery (eds.): *Cross-Craft and Cross-Cultural Interactions in Ceramics*, Westerville, OH: American Ceramic Society, 1989, p. 45-63.
21. Jay A. Stephens, "Technical Analysis and Replication of Corinthian Polychrome Slips, 8th to 6th Centuries BCE", senior thesis, Department of Anthropology, University of Arizona, May 2014, 83 pp.

Mater. Res. Soc. Symp. Proc. Vol. 1656 © 2015 Materials Research Society
DOI: 10.1557/opl.2015.841

Ceramics at the Emergence of the Silk Road: A Case of Village Potters from Southeastern Kazakhstan during the Late Iron Age

MaryFran Heinsch[1], Pamela B. Vandiver[2], Kyra Lyublyanovics[3], Alice M. Choyke[3], Chandra Reedy[4], Perry Tourtellotte[5], and Claudia Chang[5]

[1]University of Chicago, Department of Anthropology, Chicago, IL 60637
[2]University of Arizona, Department of Materials Science and Engineering, Tuscon, AZ 85721
[3]Central European University, Department of Medieval Studies, Budapest, Hungary
[4]University of Delaware, Laboratory for Analysis of Cultural Materials, Newark, DE
[5]Sweet Briar College, Department of Anthropology, Sweet Briar, VA 24595

ABSTRACT

Between the fourth century B.C. and second century A.D., changes in climate, culture and commerce converged to extend networks of influence and intensify social stratification in communities situated along the Silk Road. The horse-riding nomads and agro-pastoralists of what is now Southeastern Kazakhstan were important actors in the unfolding of these events. The settlements and kurgan burials of the Saka and Wusun could be found dotting the alluvial fans north of the Tien Shan Mountains just a short time before Alexander the Great founded outposts in the Ferghana Valley and Chinese emissaries formalized relations with their periphery. In other words, the appearance of Iron Age Saka-Wusun sites anticipated the formation of the Silk Road's northern branch and subsequently helped mediate long-distance relationships connecting East and West. Historical accounts appear to confirm the presence of the Saka and Wusun in this role, but there is much that remains unknown regarding relationships both within and across their communities. Typological variability in their material culture has fed speculation concerning their position within trade networks, but there has been very little in the way of materials analysis to test the validity of these assumptions.

The ceramics recovered at Tuzusai near Almaty provide an excellent opportunity for examination of the impacts and implications of extended regional contacts throughout the region. Although no Persian or Chinese ceramic imports were identified, an extensive vocabulary of pot forms was locally produced. However, the pottery, particularly pitchers, drinking cups and bowls, and, especially with bright red surface decoration, is found in elaborate burial kurgans. The pottery is coarse, perhaps better called a "rock body" than a clay body, as very little clay is present. The frequency of sherds from the excavation (over 1000) and from surface survey is very low (e.g. 3 surface sherds for one-half days effort) compared with excavations in Southwest Asia or China. Rims are unusually worn. Thus, we suggest pottery was precious and high status, but difficult to make. A local survey of clay resources produced meager results. Tests showed that the finest sediments had perhaps 3% clay-sized particles. Among the adobe houses at Tuzusai is evidence of courtyard work areas for pottery production with fired remains of a possible firing pit or kiln and bone potting tools. Other courtyards were areas for dairying and spinning and some copper alloy and iron metal working. Our aim was to establish the life history, production sequences, status and uses of the pottery. Given our current understanding of local production resources and the technical difficulty associated with the production of thin walled forms using these materials, we suggest that these ceramics were high-status goods, many used in feasting activities, and valued not solely for their function in feasting activities, but for the labor and skill required to produce them. Study of the ceramics, clay sources, production methods, and decoration suggests greater social permeability of Saka-Wusun communities than was previously proposed and allows us to understand the formative dynamics of village along the Silk Road.

INTRODUCTION

The conventional view of socio-cultural developments in the Iron Age of Southeastern Kazakhstan (8th century B.C.E. to 5th century C.E.) has been dominated by archaeological discoveries from burial kurgans supplemented by ancient historical documents. These attest to the presence of diverse and highly mobile communities throughout the region. Although a number of permanent settlements attributed to semi-nomadic Iron Age groups are widely known elsewhere [1, 2], the prevailing emphasis on mobility and nobility produced only modest enthusiasm for additional exploration of settlements along the margins of the steppe and northern foothills of the Tien Shan Mountains. Beginning in 1994, Claudia Chang and Perry Tourtellotte in collaboration with Feodor Grigoriev initiated the Kazakh American Archaeological Expedition (KAAE) produced some of the first systematic field surveys of settlement sites in this region [3]. A substantial portion of their work has focused on an area 20 km to the east of Almaty in the Talgar alluvial fan. The 525 km^2 area of the fan is dotted with over 700 kurgan burials, and, to date, at least 70 Iron Age settlements have been identified [4].

Recently, a handful of these settlements have been excavated, although the site of Tuzusai has so far received the most attention (Fig. 1). Archaeological and magnetometry surveys of the site indicate Tuzusai was a large site, perhaps as many as 11 hectares in extent [5].

Figure 1. The location of Tuzusai in relation to alluvial fans north of the Tien Shan Mountains

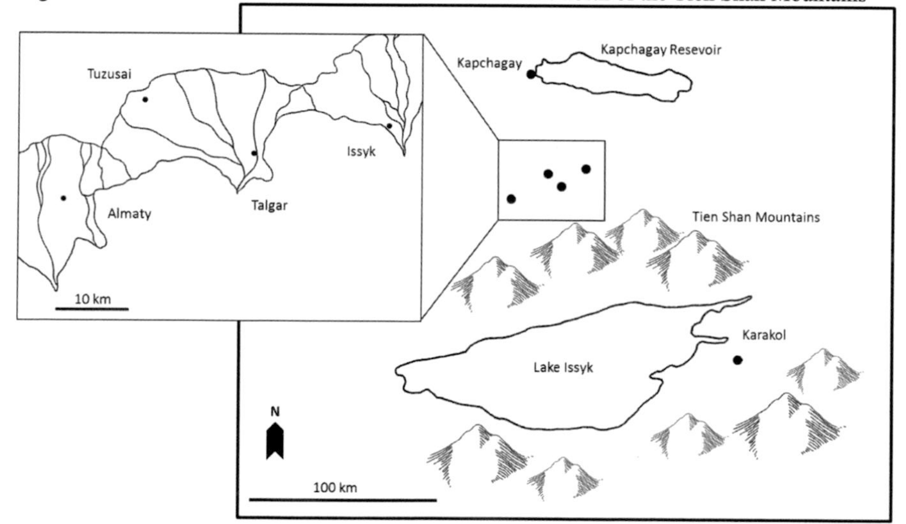

Well-excavated and carefully dated with a series of tree-ring and calibrated radiocarbon dates [6, 7], the initial occupation of this site spanned roughly four centuries from 400 B.C.E. to 1 C.E. These dates indicate that the settlement at Tuzusai was established after the Achaemenid annexation of Sogdiana (550 B.C.E.), and not quite a century before Alexander the Great established outposts in the Ferghana Valley (329 B.C.E). The opening of the Silk Road through the efforts of Han emissaries occurred as the site fell into abandonment, with the northern steppe route of the Silk Road tracing a path not far from the site. The community at Tuzusai was therefore witness to a series of events in which steppe empires of Central Asia coalesced and contacts between East and West intensified.

Because prior understanding of the Iron Age in this region derives mainly from kurgan finds, the material culture of Tuzusai provides an important opportunity to learn about the role of agrarian activities within regional subsistence patterns while also examining local production traditions at a pivotal time in the formation of long-distance economic and political relationships. Coordinated efforts to analyze material assemblages at Tuzusai are on-going and describe an agro-pastoralist community relying on irrigated cereal cultivation (including millet, rice and wheat) and mixed sheep, goat and cattle herding [8,9]. Excavations have so far yielded evidence of mudbrick structures and semi-subterranean domestic architecture. These were associated with hearths and activity areas producing paleobotanical and faunal remains, iron slag, spindle whorls and pottery [10, 11]. The ceramic finds were diverse, although the spatial density of individual sherds is considerably less than most Chinese or Southwest Asian sites of the same period.

These ceramics are not yet well understood. Owing to limitations in the pottery available for study from domestic contexts as well as difficulties in dating funerary finds, few attempts have been made to produce a periodization for Iron Age pottery in Southeastern Kazakhstan. This is the first technical study of the ceramics. While there has been some preliminary documentation of morphology and paste texture [12], this has yet to be evaluated in the context of available clay resources and evidence for production techniques. Because ceramic production traditions are shaped through the intersection of regional influences, local preferences and material constraints, describing these traditions for the Tuzusai ceramic assemblage is an important next step in elaborating our understanding of the position of this community within emerging networks of interaction.

PLAN OF RESEARCH AND METHODS

A catalog was made to document variability, including information on morphology, decoration, finish, and texture. Profile drawings, caliper measurements and Munsell designations were supplemented in the field with the use of both an Aven USB powered magnifier and Leica EZ4 microscope with internal camera to document surface wear, brushstrokes, surface indications of assembly technique and variations in both size and density of aplastic inclusions. Representative sherds were later Xeroradiographed Xerox Medical Systems Xeroradiobraph 125) to reconstruct sequences of manufacture, especially assembly and shaping. Relative Mohs hardness and open porosity estimates were also provided for these sherds. A selection of sherds was examined with scanning electron microscopy with energy dispersive spectroscopy (SEM/EDS, Hitachi S-340N Type II with ThermoNoran NSS-EDS) to estimate firing temperature and bulk chemical composition of clays. Information on mineralogy of inclusions, clays and porosity is provided by petrographic analysis. In addition to an examination of the ceramics themselves, micro-wear analysis of several bone tools from Tuzusai was conducted to match textures and inclusion sizes to show how they were used to form, burnish and scrape the pottery. Inclusions from the potters are embedded in the bone. Infrared spectroscopy (Bruker FTIR) was used to attempt identification of organic residues on white to tan interior pot residues, and blackened exterior ones, as well as on a red slip, but without results.

A preliminary survey was also made of available ceramic raw materials within three kilometers of the excavation at Tuzusai. Samples of potential mineral tempers and argillaceous or fine-grained soils that may be similar to materials used in ceramic production at Tuzusai were collected. These materials were evaluated in the field with an HCl test to judge relative calcium content and with Stokes' Law sedimentation tests to estimate particle size distributions with results that only 3% of the finest sediments consisted of clay. These poor quality raw materials were used to attempt a reconstruction of the bowls and cups that occur at high frequency within the assemblage. The replicate pottery made on-site using local resources proved how woefully inadequate those resources are. The environment is very active being at the foot of glaciated mountains that are moving in a northerly direction making extensive sedimentary weathering impossible. The overwhelming majority of forms appear to have been molded using assembly strategies that responded to shortcomings in available raw materials. A survey of a wider area to search for clay resources along the mountains and further north in the desert was planned, as well as visits to the two pottery workshops and one brick yard in Kapchaygay (Fig. 1).

The pottery corpus of over 1000 excavated pottery sherds and pot profiles, studied in the field by optical microscopy, were compared to contemporaneous museum collections, for instance from the nearby mound burial of the so-called Golden Warrior at Issyk. The intent was to establish the range of uses for the ceramics based on whole vessels from burials that presumably were the best that were produced. No Persian or Chinese imported vessels or sherds were found in these museum collections.

INVENTORY OF FINDS AND FIELD ANALYSIS

Ceramics Classification: A total of 992 accessioned ceramics with point provenience information and 649 level bags consisting of small, non-diagnostic fragments were excavated from two adjacent 8 meter by 8 meter trenches in the 2012 and 2013 field seasons. Most sherds were found thickly encrusted with caliche which often obstructed observation of surface features. Despite this, a wide variety of pottery shapes and finishes could be identified. Of the 902 accessioned finds, at least 141 distinct rim profiles and 22 different handles were recorded with drawings and photos. Two different base shapes were also identified and photographed. Repair holes, usually 0.5 cm in diameter, were observed on at least 7% of the collection. The ratio of unique elements relative to the total number of sherds speaks to the morphological diversity and idiosyncrasy of the finds. Some aesthetic parallels can be observed between Tuzusai pottery and that recovered from Iron Age burials within both the Talgar and neighboring Issyk alluvial fans. The Tuzusai assemblage is greater than that in the available mortuary assemblages. Table 1 shows the wares and their characteristics.

The Fine Orange Ware category describes a range of smaller vessels, generally cups and small bowls without handles (Fig. 2). These well oxidized ceramics are lightly burnished and profiles typically reveal a fine grained fabric with only a few small pores. Fine Orange pottery

Category	Percent of Accessioned Inventory	Wall Thickness	Color	Morphology
High Frequency Ceramics				
Fine Orange Wares	9.8% (n = 87)	≤ .7 cm	Light red to red. - 10R 4/8 10R 6/8 2.5YR 5/8 2.5YR 7/6	- Small cups and bowls no larger than 13 cm in diameter and 9 cm deep. - No handles. Round bases only.
Red Washed Wares	84.9% (n = 753)	.7 cm ≥ 4 cm	Reddish yellow, light red and red. – 10R 6/6 2.5YR 6/6 5YR 6/6 7.5YR 8/6	- Extensive range of forms including mugs, bowls, pitchers, beakers, cooking pots (garshok), small and large storage vessels (homcha). - Wide array of handles. Both rounded and flat bases.
Low Frequency Ceramics				
Corrugated Ware	2.3% (n = 20)	.6 cm ≥ 1.2 cm	Weak red to reddish brown 10R 5/3, 4/3 2.5YR 5/4, 4/3	- Globular bodies with or without a tall neck. - Coil handle joined to exterior near neck. Rounded bases.
Reduced Red Ware	0.1% (n = 1)		Gray and black with a thick bright red wash or slip	- Only one example. - Possibly a small cup no larger than 8 cm in diameter.
Punctate Ware	1.5% (n = 13)	.3 cm ≥ .6 cm	Light red 2.5YR 6/6	- All fragments belong to one vessel. - Small, nearly complete jar with round bottom, roughly 9cm in diameter.
Transitional – Red Washed	1.5% (n = 13)		Pink 7.5YR 8/4, 7/4	- All fragments belong to one vessel. - Globular shape with small lugs and perhaps asymmetrical opening.
Transitional – Gray Ware	2.3% (n =20)		Gray and Black	- All fragments belong to one vessel. - Globular shape with lug handles.

Table 1. Preliminary categories of diverse ceramics in the Tuzusai assemblage.

may have sharply carinated profiles, or they may have either raised parallel bands or narrow circumferential incised lines on their exteriors. Those displaying carinated rims or shoulders are found more frequently in the earliest levels of the site. These have the appearance of metal skeurmorphs. More common are Fine Orange wares with either raised bands or incised lines. Unlike carinated forms, these have only a thinly tapered lip with no distinct rim. Variations with incised lines are more fragmentary and less well represented in the assemblage. Some incised bowls have traces of a red wash carefully applied within the incised grooves that circle both the lip of the vessel and the base.

This same red wash is found in greater abundance on a variety of thick-walled forms (0.7-0.4 cm), here classified as Red Washed Wares (Fig. 3). This category is the largest and most abundant at Tuzusai. It includes some smaller objects such as mugs and bowls, but more frequently describes larger vessels in a wide variety of shapes and sizes including pitchers, beakers, cooking pots (garshok), and storage jars (homcha). The storage vessels can be quite large. At least one homcha excavated at Tuzusai stands nearly a meter tall with walls just over 4 cm thick near the base. While round bases can still be observed for Red Washed mugs and bowls, some forms appear with uneven, flattened bases. Shallow circumferential ridges on interiors and "wipe-marks" on the exterior are widely observed on these wares. All Red Washed wares are oxidized with significant amounts of surface spalling and in many cases, smoke clouds or soot. Large portions of the exterior and the lips of rims are often heavily eroded, exposing a coarse and sandy fabric that sometimes contains very large lime grains.

Where the surface has not eroded or spalled, as many as three different finishes can be seen. The use of a false slip or re-wetted exterior predominates. Less commonly, a thin cream colored slip may have been applied. This treatment appears reserved for a small number of bowls or basins and has not been observed on beakers and storage jars. Red wash was then applied to the dry surface, frequently restricted to accents at the lips of rims and the margins of handles. This red wash was thinner than most slips and in some instances, the red wash is applied in a design or pattern. A cross-hatched design was applied with dry strokes on the body of a deep basin. The largest homcha was covered in an overlapping scroll pattern using a heavily loaded brush or rag (3.5-7 cm) that produced long drips. In addition, there are a few examples of pieces that are thickly and completely covered with red wash. These include a pitcher or tankard shape with a strap handle and a simple bowl with inverted rim.

Ceramics tentatively designated as Corrugated Ware are markedly different in appearance from either Fine Orange Wares or Red Washed Wares and comprise a small minority of the entire assemblage (2.3%) (Fig. 4). Wide, regular, circumferential ridges cover all or part of the exterior. Fabrics are coarse with fairly uniform sized inclusions and appear low-oxidized and unburnished. Vessels are either decorated or coated with a thick white slip. One example has a globular body with a restricted neck and angular everted rim. The neck of this pot is corrugated and covered in white slip while the body is decorated with white slip painted in diagonal lines to form triangles.

There are only isolated examples of the remaining five wares listed in Table 1. The only accessioned piece of Reduced Red ware is very thin-walled and dark gray in color. This carinated rim fragment displays a thick and smoothly applied red wash that is flaking off a well-burnished exterior (Fig. 5a). One small, but nearly complete Punctate Ware jar is light red in color with very thin, grainy walls and uneven walls (Fig. 5b). An open punctate decoration diagonally and imprecisely radiates from the rim. Traces of red wash were visible under a microscope at x20 suggesting that it may have been either decorated or covered in red wash on its exterior. Both Transitional Red Washed and Transitional Gray wares are so named because, while they are similar in many respects to the Red Washed ceramics that predominate at Tuzusai, they display features often associated with regional Early Medieval ceramics (5th century C.E. – 8th Century C.E.). The one example of Transitional Red Washed ware displays a single row of closed punctates near the shoulder, above which is found a single row of stamped impressions (Fig. 5c). A similar cross-in-field or wheel stamp can be found on medieval period ceramics at other nearby sites including Talgar and Charyn [13]. There are ample traces of soot which cloud the exterior below the shoulder and partially obscure faint indications of red wash that haphazardly cover the entire body. The only example of Transitional Gray ware was a nearly complete, heavily reduced, lightly burnished pot with a globular shape, narrow rim, and a series of 5 small evenly spaced nodes or lugs at the shoulder (Fig. 5d). This vessel was found in situ near the Transitional Red Washed ware in relatively late deposits at Tuzusai, and is one of only two clearly reduction-fired ceramics in the Tuzusai assemblage.

In addition to these potsherds, a significant number of ceramic spindle whorls were included in the site's ceramic inventory. A representative collection of 32 of these were accessioned. The spindle whorls were undoubtedly formed from discarded Red Washed body sherds, with the edges of the pot fragments abraded to give them their rounded shape. Each varied in size from 5.2 mm to 15.8 mm in thickness and between 25.2 mm and 57.6 mm in diameter. A few were incomplete, that is, they did not have a hole drilled in their center. Because most of the ceramics were heavily covered with caliche, the exposed surfaces and edges of the spindle whorls were useful in helping to describe the range of paste textures and fired color in much of the assemblage. Together with the available evidence from the potsherds, as many as six variants of paste textures could be identified while in the field (Table 2 and Fig. 6).

A representative sample of sherds was selected on the basis of these paste texture categories, the preliminary ware categories outlined above, as well as sherd thickness and shape. The sample was weighted to reflect the greatest range of technical and stylistic variation possible, and included at least one representative sherd of each of the low frequency ware categories. This strategy yielded a collection of 120 potsherds, a little more than 13% of the accessioned inventory. These are being subjected to more intensive analysis including additional microscopy, petrography, xeroradiography and SEM-EDS. In addition to these sherds, a selection of bone tools and locally available raw materials were also collected for analysis.

Figure 2. Fine Orange Ware

Figure 3. Red Washed Ware

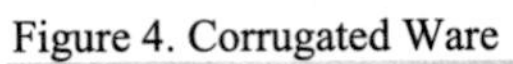

Figure 4. Corrugated Ware

Figure 5. Other low frequency ceramics

Figure 6. Paste texture categories

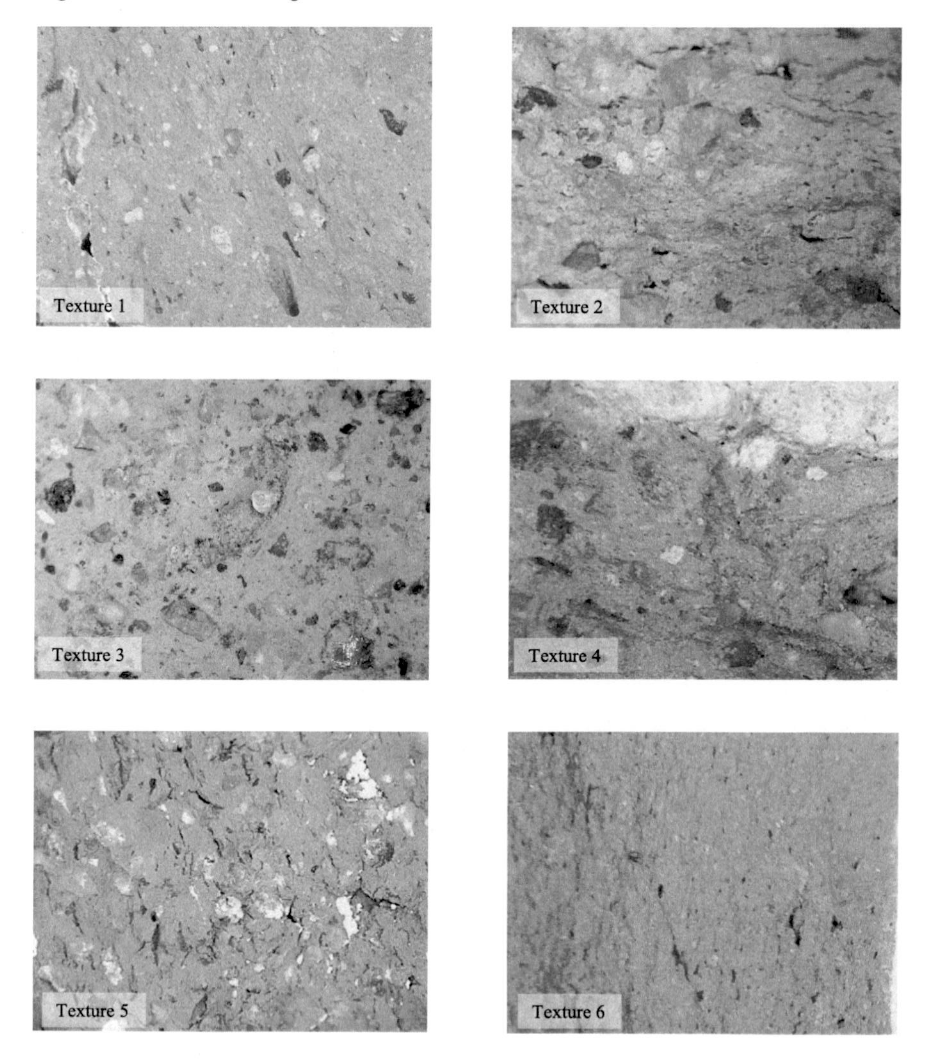

Table 2. Description of texture categories

Texture Category	Color	Description
1	Reddish Yellow and Light Red	Straw impressions and a wide variety of small grains including quartz, mica, dark black inclusions, white particles (lime). Low density of larger quartz inclusions
2	Reddish Yellow	High density of angular inclusions in a wider range of grain sizes, including quartz, granite and a red-brown mineral not yet identified. Little or no mica present.
3	Reddish Yellow, Light Red and Red	Similar to category 2 but with a wider variety of mineral inclusions containing mica, cherts, and olivine or serpentine.
4	Light Red and Red	Similar to category 1, but slightly more variety in mineral inclusions and the occasional appearance of very large grains of lime (in one instance as wide as 1.7 cm.).
5	Light Red	Similar in density of inclusions to categories 2 and 3 but with high concentrations of mica and quartz.
6	Red	Very fine and silty matrix. Very few, small inclusions.

Bone Tools: A small number of sheep bones from the Tuzusai faunal assemblage were identified as possible tools for ceramic manufacture. These were remarkable for their rounded and abraded surfaces. Ten potential bone tools were accessioned and included in the sample for analysis (Fig. 7).

Figure 7: Possible bone tools for ceramic manufacture.

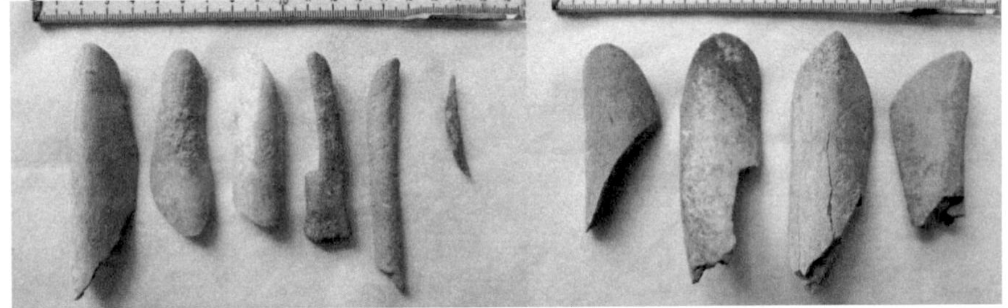

Argillaceous Soils and Sand: A total of 23 samples of potential ceramic raw materials were included for analysis. These samples were collected during brief prospecting surveys during both 2012 and 2013 field seasons within 3 kilometers of Tuzusai. Fine soils were sampled from multiple locations along the Big Almaty Canal, the nearby site of Tsegonka, as well as from the site of Tuzusai itself. Fine soils were collected from multiple points below the layer of the topsoil from eroded embankments alongside both active and dry streambeds as well as from the sterile soil layer beneath the oldest excavated cultural horizon at Tuzusai itself. Samples of mud brick from excavated architectural features were also included as were hardened reddish orange lumps adhering to iron slag. These latter samples were associated with a fired structure that may have served as a small kiln or metal-working site and will be analyzed to estimate the firing temperatures for this as well as for the pottery. Three of the 23 samples were of alluvial sands near active streams that bore similarities to inclusions already noted in the ceramics.

Available Resources: The Talgar fan, in which lies the site of Tuzusai, is notable for its fertile chernozem soils, though not particularly for its clay resources. The northern fringe of the fan is characterized by thick layers of loess loam overlying heterogeneous gravels within a matrix of silt and sand [14]. This loess loam is formed through the alluvial redeposition of loess deposits much further upstream. During hot and dry summer months silt particles from the margins of the fan and desert are blown southward by seasonal cyclonic winds. This aeolian material is trapped by the mountains and is quickly redeposited at altitudes between 1000 and 1300 above sea level [15]. The rivers supplying the Talgar, Almaty, and Issyk fans are each characterized by large sediment charge rates, particularly during periods of glacial melting in the spring and summer [16]. At these times, montane loess may become turbated with other fine particles including eroded granite sediments, mica and perhaps two-layer clays before being deposited in the fan.

The samples of potential ceramic raw material extracted from these sediments were not expected to be particularly rich in clay suitable for pottery-making. Any glacial clays that may have washed downstream along with the loess are known to be refractory, stripped of fluxes, salts and fine particulates. In addition, clay deposits with good sintering qualities have been said to be nearly non-existent in the greater part of Kazakhstan, with many of the available clays sintering near temperatures of 1300°C [17].

Of the 23 samples of potential ceramics raw materials collected, 15 argillaceous and silty soil specimens, including the architectural mud brick fragments, were examined for clay content. None of these appeared to be suitable for pottery production. The samples were tested in the field with 1 N HCl solution for lime with positive results, but pH paper indicated the samples were nearly neutral, contraindicating the presence of clays. The salt content of these samples was poor, amounting to a minor amount of 1 to 5% calcium oxide in various hydration states. Ten centimeter shrinkage bars indicated very little linear drying shrinkage in these materials (1-2%), yet coil tests indicated low plasticity and dried samples were extremely friable, breaking apart when picked up. A Stokes law sedimentation test indicated that these samples contained between 20 and 50% fine sands (ISO criteria $0.063 \geq 0.2$ mm in size) and a wide range of silt sized particles (ISO $0.02 \geq 0.0063$ mm), but very little or no clay sized particles. The clay should have formed a colloid suspension that required nearly a week to settle. Only the specimens from Tsegonka, 1 km to the east of Tuzusai produced as much as 2% clay by volume. The water for the other specimens was essentially clear within 48 hours. Given the possibility that calcium bentonite may have been present in these samples, this test was repeated with the addition of a sodium carbonate (soda ash) solution. The results of the sedimentation test in this case were in every way the same, confirming that clay was not present in the form of flocked particles of a bentonite.

While proper clays could not be located within the vicinity of Tuzusai, alluvial sands did appear to closely match the range of observable aplastic inclusions in the ceramics themselves. Sands from Tsegonka and the Big Almaty Canal proximal to Tuzusai contained similar mineralogical heterogeneity and range of grain size (Fig. 9).

PRELIMINARY THIN-SECTION PETROGRAPHY

A small selection of 13 sherds was examined by thin-section petrography. These were mainly Red Washed wares, including three transitional sherds, along with single examples of Fine Orange and Punctate wares. All of the samples examined are very similar petrographically, with very weathered granitic inclusions, although there are differences in the amount of inclusions and in the presence or absence of some accessory minerals. Amount of inclusions sand-sized or larger (≤ .063 mm) ranges from 10 – 30%, with maximum size ranging from 1 to just over 2 mm in length. Regardless of size or percentage of abundance, the shape of these larger inclusions remains very similar, ranging from an aspect ratio of 1.8 – 2.1 (relatively elongated) and roundness of 1.6 – 2.3 (rounded rather than angular). The size and shape of macropores varies more than do the granitic inclusions. These can range in maximum size from 1- 5 mm in length, and are much more elongated than the mineral inclusions, with aspect ratio ranging from 2.2 – 3.9. In some samples these pores are rounded (roundness of 2.1) but can range up to 3.7 (angular).

The ubiquitous granitic inclusions consist primarily of quartz plus feldspars (more plagioclase than alkali feldspars), with total feldspars being greater than total quartz (and hence likely to have derived from the breakdown of granodiorite). Feldspars are heavily weathered (with surfaces altered to clay and/or sericite), and sometimes have a perthitic texture. There is a mix of some rock fragments and many separated polymineralic components. Other common minerals include hematite, biotite, and hornblende, with muscovite, orthopyroxenes, clinopyroxenes, calcite, polycrystalline quartz, microcrystalline quartz, magnetite, sphene, leucoxene, epidote, chlorite, zircon, and charred organic material also seen.

With a dearth of good clay resources in the vicinity of the site, and many of the ceramic samples found to be low in clay content, it is possible that plasticity was improved by the presence in the hematite-rich matrix of abundant fine micas (tiny shreds of fine sericite, with some fine biotite) and sometimes very small, rounded chunks of glauconite. Examples with the appearance of a wash-type coating seem in thin section to have a layer that is simply bleached out rather than being an actual wash or slip; though lighter in color these layers contain the same large-grained granitic inclusions as found on the interior of the sherds.

FORMING PROCESSES AND SEQUENCES

The Clay Body: Prior to the thin section analysis that isolated glauconite and sericite as sources of plasticity, we attempted in the field to make some bowls by a variety of methods using the separated silt constituent with about 25% by volume of sand-size particles, after some experimentation, as the best clay body we could produce that resembled the paste textures observed in the Tuzusai ceramic assemblage (Fig. 10).

Regardless of similarities in texture however, it quickly became clear that this blend of materials would be insufficient to produce the wide range of vessel forms observed at Tuzusai. Efforts to reconstruct five vessel shapes including a mug, two bowls, and two small garshok, revealed that even the clay body was not plastic and instead was thixotropic, such that the pots slumped and the clay body would not stay in place to allow additions. Reconstructed vessels cracked at contact points when touched or shifted. Similar cracks, and repairs to such cracks by smearing clay on the interior, can be seen throughout much of the Tuzusai assembly, but especially on thick-walled Red Washed wares.

Figure 9. Sampled alluvial sands

Figure 10. Left: Tuzusai sherds (1 and 3 from left), Our replica sherds from broken bowl (2 and 4)

Vessel Assembly and Shaping: While the dearth of good clay resources in and around Tuzusai raises suspicions that pottery could have been transported to the site from elsewhere, this remains only a remote possibility. Not only do ethnographic parallels suggest that this was unlikely, but the finding of bone fragments with ceramic microwear at Tuzusai recommends that in fact, pottery was being made at this site. Microscopy of the bone implements revealed abraded surfaces with fine troughs or ridges consistent in size with the range of sand sized particles prevalent in most of the observed paste textures. (Fig. 11). All ten of the bones yielded similar results.

The use of these bones as modelling or shaping ribs is also supported by similarities in the size of ridges on the scraped interior surfaces seen on nearly all of the Tuzusai pottery, particularly Fine Orange, Red-Washed, and Corrugated Wares (Fig. 12). A similar bone implement of problematic provenience was used to achieve a similar pattern of scraping and shaping marks by using the rounded edge on each of the five attempted pottery reconstructions. The flat surfaces were used to smooth the exterior surface. The most successful bowls were made inverted over a male mold, with the exterior tooled, then when dry enough to handle the interior was smoothed and scraped. Strip forming worked for larger pots.

Evidence for the shape and size of individual clay elements was examined in the field with a grazing light, loop and probe. Xeroradiography was later carried out on a sub-sample of sherds to compare their internal structure with surface and profile observations and suggest a range of assembly strategies. Flattened coils or strips were among the most frequently identified clay elements from the surface and profile features (Fig. 13a). These were particularly apparent on a large fragment of corrugated ware which showed thin circumferential linear pores. Wide overlaps between these strips are visible in step fractures, particularly in larger, thicker vessels (Fig. 13b). Coils or strips are also seen near rims and these tightly overlap in thicknesses of 2 to 3 layers. Additional coils are observed in profile which served to thicken and reinforce the wall at joints between strips.

Both round and flattened bases can be seen on vessels with this type of strip assembly (Fig. 14). In both kinds of base, clay strips forming the base of the walls abut a flattened slab at roughly a 45° angle. Based on information about the quality of the clay body and the raw

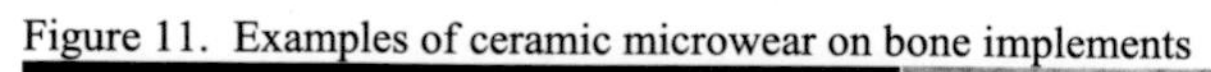

Figure 11. Examples of ceramic microwear on bone implements

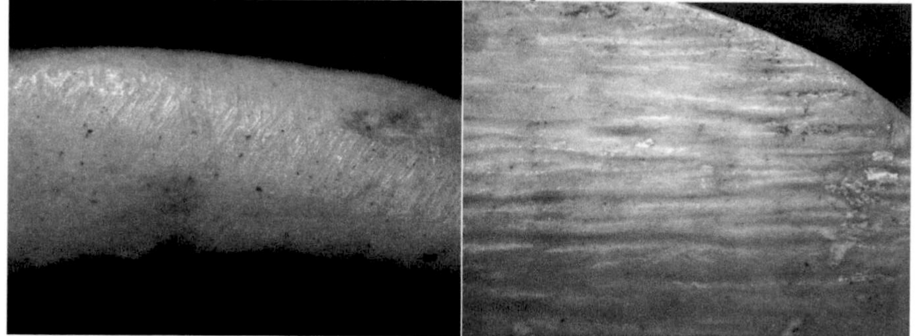

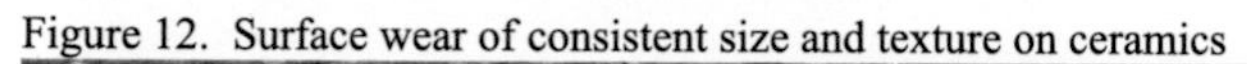

Figure 12. Surface wear of consistent size and texture on ceramics

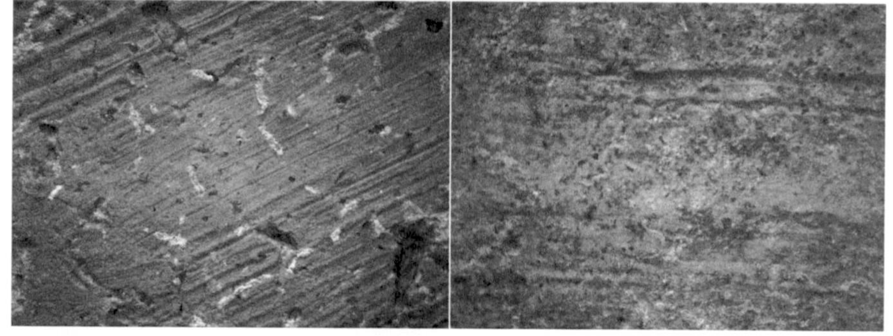

Figure 13. Linear pores at the margins of flattened coil joins and step fractures

materials, molds, both male and female shapes, were used as a support during assembly. Differences in the ways these molds were used likely correspond to differences in base shape. A number of vessels had profile thicknesses that were continuous and curvilinear with the thickest part of the wall at maximum diameter. This feature is especially consistent with the use of an male mold in the construction of a round-bottom pot. Limited evidence for textile impressions (Fig. 15) supports the use of molds, since these impressions were found on the interior of the vessel and may indicate the use of fabric to separate the pot from the mold.

Figure 14. Comparison of rounded and flattened bases

Figure 15. Textile impressions on the interior of vessels, especially bowls

Xeroradiographs of this pottery also argue for the use of molds, even if only for the lack of positive evidence for an alternate assembly technique (Fig. 16a, b, c, and d). At first look, the porosity is distributed throughout these sherds and is not directional as would be expected of coiling, strips, or slab joints. The pores themselves are not only formed around aplastic mineral inclusions but also result from some fine fibrous vegetal material that included other small rounded organic material, similar in size to millet and sorghum seeds. Additional pores are associated with fine cracks at the margins of clay strips or coils. Internal vertical cracking that likely resulted from handling of leather-hard pots, and short fine cracks that are consistent with surface friction during the removal of partially dried pottery from a mold also contributed to the porosity observed in the xeroradiographs. Evidence of strips to build up these vessels is most clearly shown near the rims, although banding patterns in the thickness and density of the sherds also attests to this (Fig. 16a and c). That these strips were joined against a mold is suggested in the high frequency of small rounded pores observed on the interior where insufficient scraping occurred.

Figure 16. Xeroradiographs

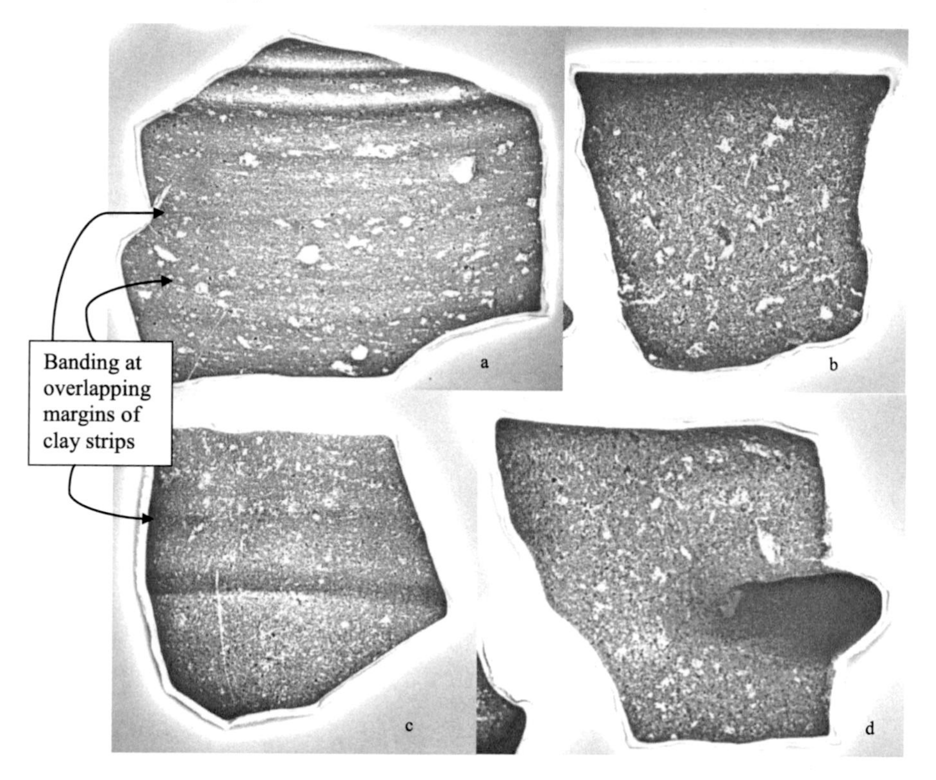

Evidence for other forms of wall assembly was lacking; for example, no variations in thickness or faceting as from the use of paddle and anvil or diagonal elongation of porosity as found in throwing on a wheel. There were no radiating patterns of fine fracture lines characteristic of paddling that occurred when we undertook this process. Likewise, no elongated pores were found with a 30 to 45 degree incline from horizontal which would have indicated a wheel thrown pot. There were also no repeated circular or near-circular indentations or areas of thinning. Only one other assembly technique was represented in these data. The Punctate Ware in our sample provides indications of small disks or lumps of clay pinched together. The rounded base and exterior walls of this vessel are fairly smooth, but the interior walls reveal numerous finger impressions (Fig. 17a). Overlapping pieces of clay are likely in the radiograph in Fig. 17b. This mode of assembly was not observed for any of the other ceramics at Tuzusai.

Figure 17. Pinched joins in two different pieces of Punctate Ware

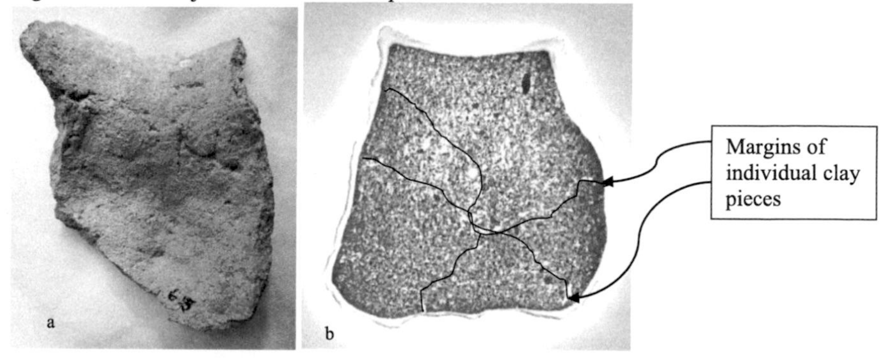

The Tuzusai potters became flamboyant in their design of handles and winged or horned attachments. Whereas wall assembly and shaping strategies were homogeneous, more diversity and design variation occurred in the crafting of handles. These may be characterized separately as knobs, lugs, and coils. Knobs are here meant to describe the small projections typically found at the shoulder of globular shaped vessels. Some are more pyramidal while others are more rounded. None are very large, less than 2cm deep (Fig. 18). Small elongated pores between knobs and the vessel wall suggest that the knob was added as a small lump of clay and then shaped and smoothed into the wall surface. They are not very strong and often contain porosity at the joint.

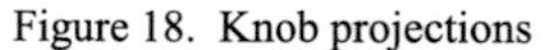
Figure 18. Knob projections

The lug handles have a wider range of shapes and sizes. Flat round disks with a very small center opening are observed, as are elongated narrow ledges (Fig. 19 a and b). There are thick lobes that extend at a wide range of angles, some of which are vaguely zoomorphic in character (Fig. 20). The largest lug handles are more than 10 cm deep and look like horns (Fig. 20b). Many of these lug handles were bonded flush to the wall surface, the joints healed with additional clay and slipped over in much the same way that the knob projections may have been added. Other lug handles, however, were inserted into an incision in the wall surface (Fig. 21). The clay elements comprising the lug are much thicker at the interior edge, yielding the impression that they may have been inserted as tapered coils or cones from inside the pot and then given their shape as they were worked into the vessel wall (Fig. 21b). This tabbed lug uses a metal construction technique to handle attachments.

Figure 19. Disc and ledge lug handles

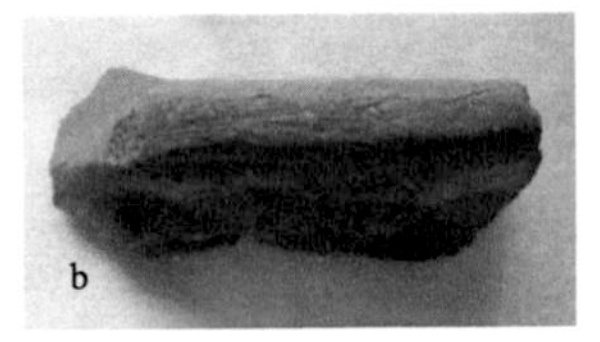

Figure 20. Lobe and horn or wing lug handles that were inserted through the wall as in metal construction and joining of a handle to a vessel.

b

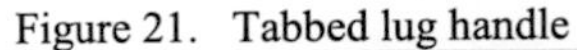

Figure 21. Tabbed lug handle

The coil handles also include a wide range of shapes and methods of assembly. These handles may accommodate only a finger's width or the entire palm and are attached at varying angles between 45 degrees and perpendicular to the vessel wall (figure 22). Forming the coils for the largest of these handles may have required near constant attention to maintain their shape since the silt-sand pastes were highly susceptible to slumping and cracking. Butt joins between handle and wall surface can be observed, particularly on one example of corrugated ware where the seam between handle and neck was not over-slipped and smoothed into the wall (figure 23). Vertically attached coil handles are ovoid in profile, attached to both the wall surface and the lip of the rim and well joined to the vessel wall with no obvious separation between the wall and handle (figure 24). In addition to these, several examples of horizontal coil handles were attached to the vessel either by inserting them through an incision in the wall or using a smaller coil inserted into the wall as an anchor for attaching the handle (figure 25). Like the tabbed lug handles, this riveted type of attachment is again reminiscent of metal crafting techniques.

Figure 22. Coil handles of varying sizes

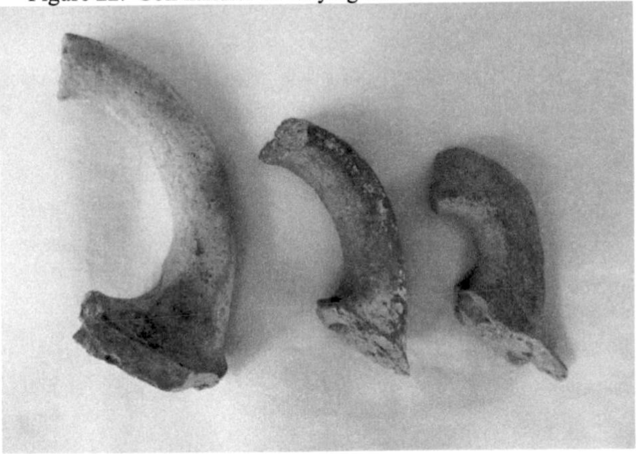

Figure 23. Butt join for coil handle through the wall Figure 24. Vertical handles

Figure 25a. Horizontal riveted coil handle, similar to that used for metal work; b, pot reassembled

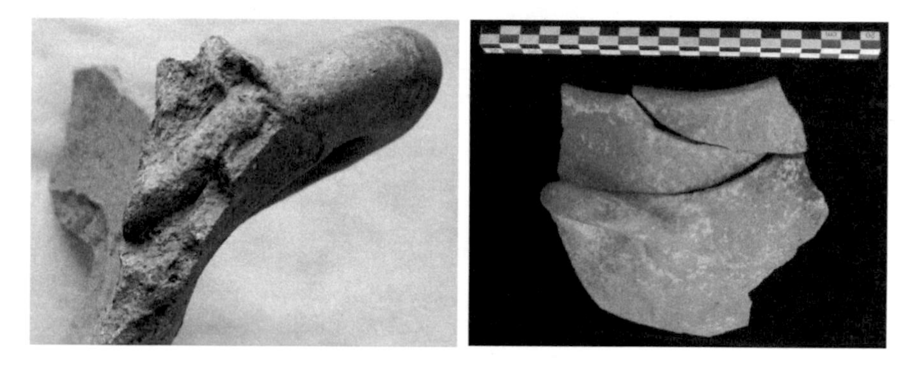

Figure 26. Unique two-fingered handle shape (top to the right)

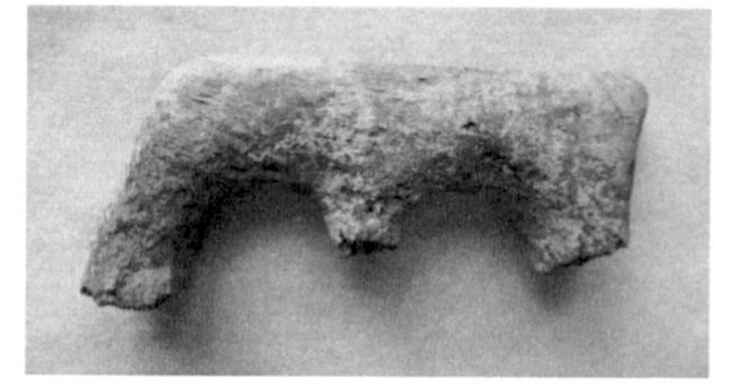

Surface Treatments: Many surface treatments are observed in the Tuzusai ceramic assemblage. These include the use of burnishing/shaping, false slip, cream slip, red wash, incisions, stamps and punctate impressions. Fine Orange, Reduced Red and Transitional Gray wares all bear characteristic marks of burnished exteriors. This finish was produced with a narrow, rounded tipped implement that shaped the exteriors of these pots as it also aligned clay or silt particles (Fig. 27a). A similar surface texture was achieved using a bone tool as a rib to smooth and shape the leather hard surface of one of the reconstructed vessels (Fig. 27b).

Figure 27. Burnishing and shaping indentations

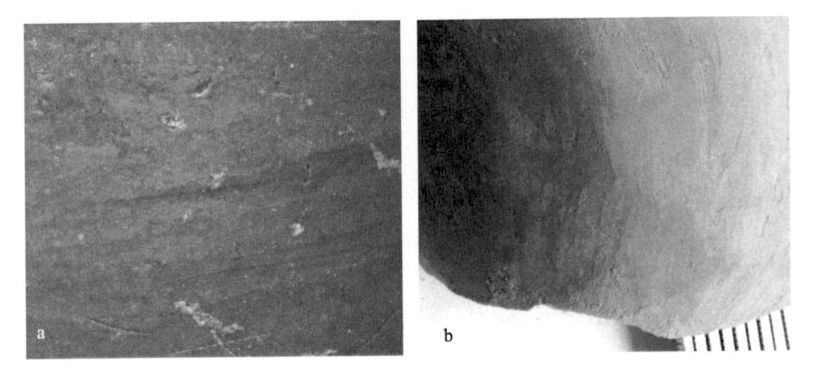

None of the Red Washed Ware displayed any burnishing or shaping of this kind. Instead, a false slip is seen on a large majority of these ceramics. Whether this effect was created simply by rewetting the surface or by applying a slip made of the same materials as the silt matrix, the use of a cloth or rag appears to be probable. The fine, multidirectional slip trails that result are favorably comparable between artifacts (Fig. 28a) and reconstructed vessel (Fig. 28b).

Figure 28. False slip with multidirectional drip lines on artifact and on reconstruction

The use of a cloth in the application of cream slips is even more likely. These slips have been identified on both Corrugated Ware and select basins and small bowls of Red Washed Ware. In one example, impressions from fine fibers or hairs were seen embedded in the surface (Fig. 29).

Figure 29. Fine gauge fiber impressions in cream slip

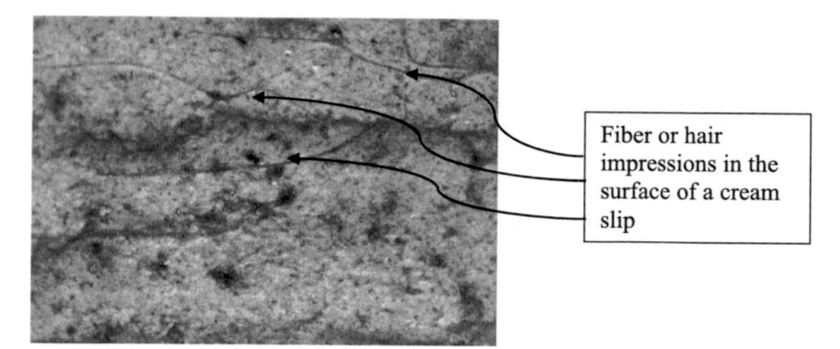

Red washes could also have been applied with a cloth in many cases. This may be particularly true for pieces that are thickly covered in red wash, a Red Washed Ware tankard handle for instance (Fig. 30a), or the piece of Reduced Red Ware (Fig. 30b). Results were consistent with a reconstructed pot in which a red wash was made using the finest sifted fraction of the available sand and silt materials and daubed onto the exterior with a cloth rag (Fig. 31). In rare instances, however, there is some evidence to support use of a brush. The margins of individual application strokes consistent with the use of a brush are visible under a microscope at x25 (Fig. 32).

Figure 30. Daubed applications of red wash

The red washes likely include pigment quality hematite. These were applied to a leather hard or wet silt body resulting in the expansion of the surface and the formation of fine cracks. The 200 to 500 micron thick red paint found on the Reduced Red Ware was also applied prefire. It does not fluoresce in short- and long-wave infrared illumination, a test that indicates the presence of some organic compounds. Furthermore, although the red paint on this specimen was flaking off at the margins, it survived gentle boiling in water and subsequent soaking during the

Figure 31. Reconstructed red wash with a daubed application

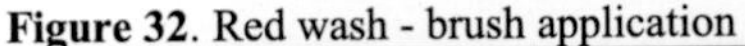

Figure 32. Red wash - brush application

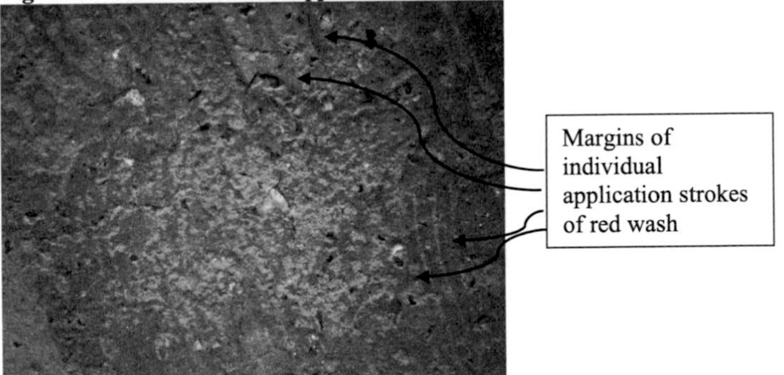

ASTM test for open porosity. In almost every example, the original pattern for both Reduced Red and Red Washed Wares cannot be determined in sufficient detail due to past wear.

Decoration in the form of incised or modelled patterns was rarely observed, and only a few late period potsherds that demonstrate their use. Two vessels had a punctate design, but even in this, the punctates were technically different from one another. The Transitional Red Wash Ware had a single row of closed punctates created with a solid object such as a bone or stick (Fig. 33a), whereas the Punctate Ware was clearly decorated with a hollow reed or other tube to produce an erratic constellation of open punctates (Fig. 33b). Circumferential incised lines were typically restricted to a small number of Fine Orange Ware bowls (Fig. 34); however, one rare example of coarse Red Washed Ware has a row of teardrop shaped incised motifs across the shoulder of the vessel (Fig. 35). Unlike the incised Fine Orange Ware, the incisions in the teardrop motif were not filled with red wash. With stamped designs, a wheel or cross-in-field pattern was filled first with cream slip and then the relief portion of the pattern was painted with red wash (Fig. 36).

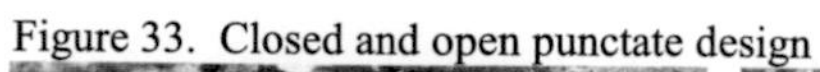

Figure 33. Closed and open punctate design

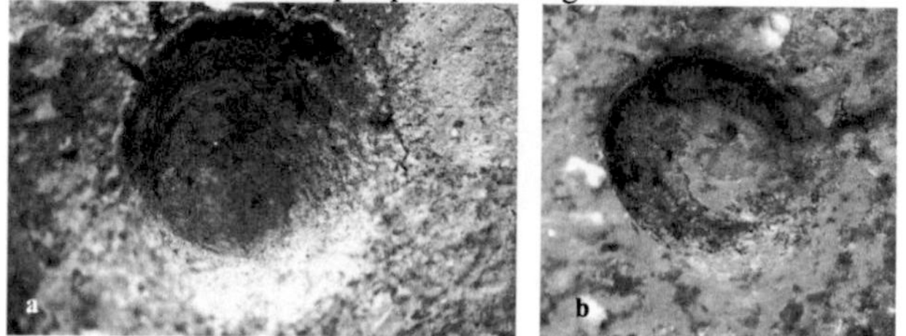

Figure 34. Incised lines

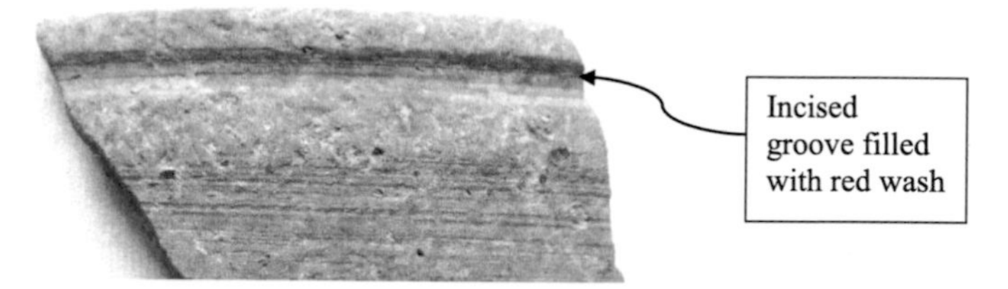

Figure 35. Teardrop incised lines

Figure 36. Stamped designs

Firing Conditions: Firing conditions were estimated from measurements of open porosity, hardness, oxidized fired color of locally available raw materials, and refiring of sherd fragments. SEM analysis of refired samples is forthcoming. The temperature range is 900 to 1000°C.

The spalling around mica and lime particles frequently observed for Tuzusai pottery, offers the first indication that firing temperatures exceeded 800°C. The high degree of porosity indicated in xeroradiographs of the Tuzusai pottery is consistent with estimates of open porosity using a simple ASTM test. This involved gently boiling the sherds for 2 hours and then allowing them to remain in water for 24 hours. The weight of each sherd was recorded before and after this process, taking note of the percentage increase in weight. A representative subsample of 27 potsherds was tested. Some potsherds were more weathered than others, admitting more water, and thus yielding an artificially high value. With only a few exceptions open porosity was in the range typical of most earthenware (Fig. 37). The mean open porosity for all the Tuzusai potsherds tested was 12.5%, while the range of open porosity varied between different categories of wares (Table 3). A sample of reconstructed pottery was also tested for open porosity after a sequence of trial firings. A replicate pot fired to 950°C was a light red color similar to the bodies of Red Washed and Fine Orange Wares and was more porous than 22 of the 27 potsherds in the subsample. Because lower percentages of open porosity usually correspond to higher firing temperatures, the higher porosity of the reconstructed pottery compared to the Tuzusai potsherds suggests firing temperatures were likely to meet or exceed 950°C.

Despite having porosity consistent with earthenware, nearly all of the pottery in the Tuzusai assemblage made a clinking sound when struck unless significant internal cracking or corrosion was present. This clinky quality of pottery usually corresponds to higher firing temperatures or longer soaking times producing a harder ceramic. The Mohs hardness for many earthenwares ranges between 2.0 and 3.0, while the Tuzsai subsample ranged between 2.5 and 4.0 (Table 3). The majority of the subsample, 23 of 27 in all, could be assigned a Mohs score of 3 or more (Figure 38), but the mean hardness of the sherds varied with category of wares (Table 3).

Reconstructed pots ranged in hardness with firing temperature. Each sample was fired in increments of 50° C between 800°C and 1200°C, with a soaking time of 12 hours. A replicate pot fired to 900°C, while light red in color, was still somewhat chalky and soft, thudding when struck. At 950°C, the replicate pot was more clinky, maintained a light red color and rated a 3.0 on the Mohs scale. At 1050°C the body color become brown in color, changing from light red to

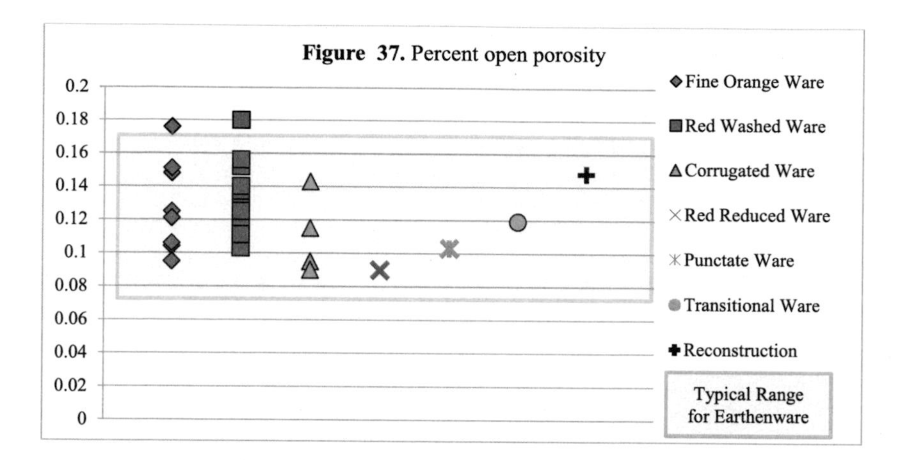

Table 3. Comparison of open porosity and hardness

Ware Category	Open Porosity		Mohs Hardness	
	Range	Mean	Range	Mean
Fine Orange	9.5% - 17.6%	12.8%	2.5 – 4.0	2.9
Red Washed	10.3% - 18.0%	13.3%	3.0 – 4.0	3.1
Corrugated	9.0% - 14.3%	11.1%	3.0 – 4.0	3.5
Reduced Red	9.0%		3.0	
Punctate	10.3%		3.0	
Transitional	11.9%		3.0	
Reconstruction	14.8%		3.0	

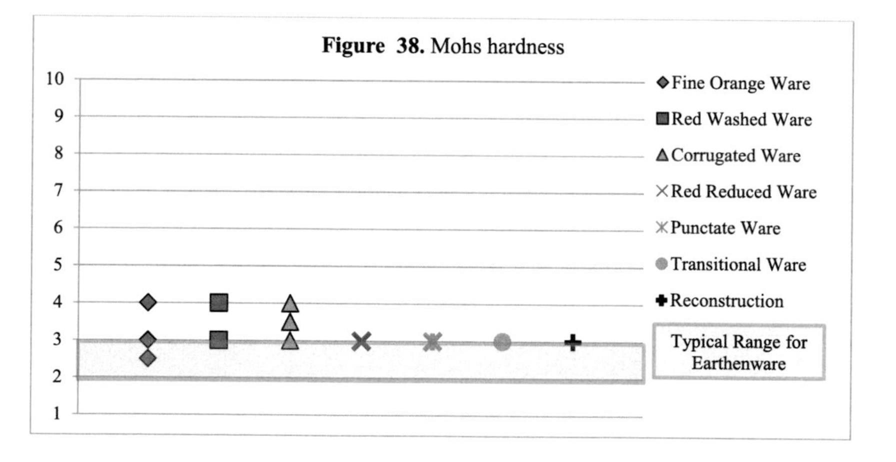

weak red, and the hardness increases from a 3.0 to a 4.0. These samples are consistent with the weak red color, lower porosity, and higher mean Mohs scores found for Corrugated Ware potsherds. The combined results of these preliminary descriptive analyses suggest that the Tuzusai pottery was fired either in the upper range for earthenwares, or else at a somewhat lower temperature but for a relatively long time. The Corrugated Ware and Reduced Red Ware were fired at higher temperatures than the Fine Orange or Red Washed wares. One suggestion has been that the Fine Orange may be an import from the Syr Daria region [24], but similarity paste forming, texture and inclusions, and shapes of the Fine Orange with the Red Washed ware that dominate the assemblage mean that importation is not likely, but more study is required.

One of the fired amorphous lumps from the pottery-making courtyard was determined to have been fired to 1000oC, the same range as the pottery and unlike the adobe used to construct the homes that disaggregates when put in water. This is one lump of almost 26 lbs. of similar pieces that were excavated from only the one pottery-making courtyard.

In Summary: Nothing about the manufacture of appearance of these pots is standardized. The amount of variability of manufacturing is excessive even for a group of village potters working in courtyards near their homes. Even though the raw materials are certainly not optimal, the potters were able to overcome raw material constraints with inventive solutions. Pottery making is not an established tradition, but quite experimental and even fun and risky. Many of the handles imitate horn or wings, and some use solutions from metalworking, such as rivets and through-wall joints. The firing is quite high for earthenware and consistent throughout the corpus. Lesser firing temperatures for this clay body lead to pots that break easily as the clay body is quite and little glass is formed to hold the sand and rocks in the body together. The firing temperature range is consistent, and the fired pots represent otters with experienced practices.

CONCLUSIONS

Between the fourth century B.C. and second century A.D., changes in climate, culture and commerce converged to extend networks of influence and intensify social stratification in communities situated along the Silk Road. The horse-riding nomads and agro-pastoralists of what is now Southeastern Kazakhstan were important actors in the unfolding of these events. The settlements and kurgan burials of the Saka and Wusun could be found dotting the alluvial fans north of the Tien Shan Mountains just a short time before Alexander the Great founded outposts in the Ferghana Valley and Chinese emissaries formalized relations with their periphery. In other words, the appearance of Iron Age Saka-Wusun sites anticipated the formation of the Silk Road's northern branch and subsequently helped mediate long-distance relationships connecting East and West. Historical accounts appear to confirm the presence of the Saka and Wusun in this role, but there is much that remains unknown regarding relationships both within and across their communities. Typolgical variability in their material culture has fed speculation concerning their position within trade networks, but there has been very little in the way of materials analysis to test these assumptions.

Difficult for pottery production at this site is the scarcity of adequate clay resources. Reconstruction efforts using a variety of materials both collected at the site and nearby produced pots with some practice (Fig. 39). The very small percentage (about 3%) of clay-sized particles in local soils makes it unlikely that elutriation techniques would have yielded sufficient clay.

Preparation of the ceramic paste must have required careful selection and collection of fine sediments. The clay fraction was identified by petrographic microscopy as glauconite and micaceous sericite with feldspar and other inclusions more numerous than quartz. Better clays are neither available in alpine meadows or lakes, nor in the red lakebed sediments in the nearby foothills. Most lowland areas with active water accumulation contain calcium salts that bleach the pottery surfaces to a white color during firing, and these are common in medieval pottery, but not at Tudusai. Near Kapchaygay lake are some sediments currently used for bricks and tourist pottery, but these deposits are quite distant, over 100 km. and beyond the range expected to be collected and transported by such village potters who wished to make pottery but who lived outside centers with a pottery tradition.

Figure 39. Reconstruction and molded bowl fragment (left) compared with similar sherd

The sequence of manufacture and the details of manufacture make investigating each pot a new experience. A wide variety of handforming methods were used, including slab building, moulding and coiling of rims. Handles and decorative surfaces are special, both in their diversity of form and the many ways in which joints were made. Obvious is the inventiveness and variation in strategies to produce pottery. The uses of round and flat bases relates to differences in the use of molds during construction and support during drying. In contrast, other Iron Age ceramics found in Southeastern Kazakhstan outside the Semirechye region, beyond the Talgar fan have predominantly flat bases [22]. Given the technical problems associated with the raw materials in this area, the difficulties in forming a pot from a flat base are considerable. The sophisticated use of molds and supports in pottery production suggests forming of metal in molds (repousse and raising), and some pots may be metal skeuomorphs. The thin walls and sharply carinated profiles of many Fine Orange Ware cups and bowls, emulate metal forms. Similarly, riveted coil handles and tabbed lugs found on many of the Red Washed wares rely on production techniques more commonly seen in metalwork. In this way, handles were joined to a molded vessel when fairly dry, making them less likely to slump out of place after being set into the wall. These production strategies probably provided useful means for working with difficult earthen materials.

The extensive use of a red wash in this assemblage is consistent with most Iron Age pottery across this region and throughout the neighboring Xinjiang province [23]. Often, only trace amounts of this wash remains on many of these potsherds. This is because when the red wash was applied to the unfired pottery, it made surfaces more susceptible to cracking. The combination of flying wing or horns handles and lugs, red surface washes that are presumably more prestigious than the tan to orange plain pottery color (at least in view of burial ceramics), and the presence of small medium and large versions of pottery with these characteristics, all together indicates that aesthetic concerns for the provision of sets of pottery was a production goal. One suggestion has been that this pottery was the equivalent of china patterns repeated throughout discrete sets used as feasting regalia (Fig. 40). Ethnographic evidence and museum collections of tableware from southeastern Kazakhstan have wooden or ceramic sets used for serving koumiss to guests and present in household assemblages (pitcher and cups).

Figure 40. Similarities in handles of different sizes of vessel

Our current understanding of poor quality resources, the considerable difficulty of achieving thin-walled forms and complex rim and body profiles, the rarity of pottery at the site, the presence of extremely worn rims leads to the conclusion that these ceramics are high-status goods valued not solely for their roles in feasting activites, but for the labor and skill required to produce them. The wooden tableware historically used by agro-pastoralists throughout the Semirechye region, including the Talgar, Issyk and Almaty alluvial fans, is probably a more sensible and lightweight option than the gritty, fragile and heavy ceramics that characterize the Tuzusai assemblage. They had to really desire pottery as the appropriate material. Another line of evidence consisting of repair holes reinforces this view. Repair holes are found in many of these ceramics can be interpreted in a new light. Whereas these repairs were once viewed as a necessity due to the short length of stay and limited scale of ceramic production among highly mobile communities, in reality, they may have been the result of efforts to conserve important or valuable objects that were difficult to make. In this respect, the pattern of ceramic repairs bears closer resemblance to those observed in collectors pieces in which either rare or expensive objects are preserved. Despite their sometimes ordinary appearance, the pottery at Tuzusai seems not intended for the mundane uses of everyday cooking and consumption, but rather reserved for serving food and beverage from pitchers into bowls and cups.

ACKNOWLEDGMENTS

This work was conducted with assistance from the National Science Foundation (grant number 112359) awarded to Drs. Claudia Chang and Irina Panyushkina. Heinsch and Vandiver are especially grateful to Chang and Tourtellotte for inviting them to contribute to their work on Tuzusai. Information provided by Feodor Grigoriev through his years of archaeological expertise in the region provided insight into aesthetic parallels with ceramics in the Syr Daria region and in regional kurgan inventories. We would also like to thank Olga Kuznetsova at the Kazakhstan Institute for Archaeology and Ethnography for her photographic documentation of special finds and consultation on regional ceramic finds of the Early Medieval Period.

REFERENCES

1. L. Yablonsky, in *Nomads of the Eurasian Steppe in the Early Iron Age*, edited by J. Davis-Kimball, V.A. Bashilov, and L.T. Yablonsky (Zinat Press, Berkeley,CA, 1995), p. 200.
2. K.A. Akishev and G.A. Kushaev, *Drenmaya Kul'tura Sakov i Usurtei Doliny Reki Ili.* (Nauka, Alma-Ata, 1963).
3. C. Chang, in *The Archaeology of Power and Politics in Eurasia: Regimes and Revolutions*, edited by C.W. Hartley, G. B. Yazicioglu, A.T. Smith (Cambridge University Press, New York, 2012) pp. 122-142.
4. C. Chang, I.P. Panyushkina, J. Fassbinder, L. Kuhne, M. Heinsch, K. Lyublyanovics, M. Macklin, R. Spengler, W.H.J. Toonen, and P.A. Tourtellotte, in press.
5. J. Fassbinder, personal communication.
6. I.P. Panyushkina, in press.
7. R.N. Spengler, C. Chang, P.A. Tourtellotte, J. Field Arch. 38 (1) 68-85 (2013).
8. A.M.Rosen, C. Chang, F.P. Grigoriev, Antiquity. 74, 611-23 (2000).
9. R.N. Spengler, C. Chang, P.A. Tourtellotte, J. Field Arch. 38 (1) 68-85 (2013).
10. C. Chang, N. Benecke, F.P. Grigoriev, A.M. Rosen, P.A. Tourtellotte, Antiquity. 77 (296) 298-312 (2003).
11. C.Chang and P.A. Tourtellotte, (field report for the 2013 season).
12. O.V. Kuznetsova, Izvestia, Seriya Obshestvenni Nauk. 1 (224) 161-168 (2000).
13. O.V. Kuznetsova, pers. comm.
14. Environment Research Center, *Characterization of Ecosystems of Archaeological Excavations Area in Talgar River Basin.* (ENVIRC, Almaty, 1995), p. 7.
15. B.Machalett, M. Frechen, U. Hambach, E.A. Oches, L. Zoller, S.B. Marković, Quarternary Intl. 152-153, 192-201 (2006).
16. B.Machalett, M. Frechen, U. Hambach, E.A. Oches, L. Zoller, S.B. Marković, Quarternary Intl. 152-153, 192-201 (2006).
17. D.V. Abdrakhimov, E.S. Abdrakimova, and V.Z. Abdrakhimov, Glass and Ceramics. 56 (5-6) 190-193 (1999).
18. http://www.almatyceramics.kz/ru/page/6
19. R.E.Grim and N. Guven, *Bentonites: Geology, mineraology, properties, and uses*. (Elsevier, New York, 1978).
20. G. A. Kudailulaova, H. Strauss, V. Koeckritz, Acta Geodyn. Geomater. 2 (2)[138], 87-93 (2005).
21. D.Verbeken, S. Dierckx, K. Dewettinck, K., Appl. Microbiol. Biotechnol. 63, 10–21 (2003).

22. L. Yablonsky, in *Nomads of the Eurasian Steppe in the Early Iron Age*, edited by J. Davis-Kimball, V.A. Bashilov, and L.T. Yablonsky (Zinat Press, Berkeley,CA, 1995), p. 233.
23. M. Jianjun, in *The Golden Deer of Eurasia: Perspectives on the steppe nomads of the ancient world*, ed. J. Aruz, A. Farkas, E. V. Fino (Yale University Press, New Haven, 2006) 132-145.
24. F.P. Grigoriev, *Pozdnakasoi poselenia Tuzusai* (Shymkent and Almaty, 1995).

Characterization of Metal and Stone Characterization, and Metal Nanoparticle Corrosion

Mater. Res. Soc. Symp. Proc. Vol. 1656 © 2014 Materials Research Society
DOI: 10.1557/opl.2014.825

Benefits of the Complementary use of Archaeometry Investigations and Historical Research in the Study of Ancient Airplanes: the Breguet Sahara's Rivets.

Audrey Cochard[1], Joël Douin[1], Bénédicte Warot-Fonrose[1], Julitte Huez[2], Luc Robbiola[3], Jean-Marc Olivier[4], Philippe Sciau[1]

[1] CEMES, CNRS, Université de Toulouse, 29 rue J. Marvig, 31055 Toulouse, France .
[2] CIRIMAT-ENSIACET-INPT, 4 allée Emile Monso, BP 44362 Toulouse Cedex 4, France.
[3] TRACES-UTM, 5 allées Antonio Machado, 31058 Toulouse Cedex 9, France.
[4] FRAMESPA-UTM, 5 allées Antonio Machado, 31058 Toulouse Cedex 9, France.

ABSTRACT

The aim of this paper is to show that historical technical archives and complementary physico-chemical studies can be combined to obtain relevant information on the materials and processes used in the manufacturing of a Breguet 765 Sahara airplane. This will be useful both in history of sciences and technology and in the renovation of this more than fifty years old airplane.

The Breguet 765 Sahara plane is the last version of a family of French double-deck transport aircraft produced by Breguet between 1948 and 1960. The gathering of multi-disciplinary information from the literature of the period of production with laboratory investigations has revealed that a "new" aluminum-copper-magnesium alloy was used in the rivets of the Breguet 765. The A-U3G alloy was developed to meet properties requirements of the aeronautical industry for joining sheets of aluminum and was used in the Breguet 765 Sahara to strengthen the joints. Analytical techniques included TEM, EPMA microprobe and metallography.

INTRODUCTION

In the field of cultural heritage, chemical and structural information concerning aluminum alloys used in aeronautic industry is required. Indeed it is important to realize that the developments of aeronautical and aluminum alloys are connected. Interest in lightweight and high strength material stimulated research to develop new aluminum alloys as progress in materials development allowed lightening of the plane structure. However, although duralumin, the first strengthened aluminum alloy, was processed less than a century ago in an industrial context, it is not easy to find pertinent chemical and physical data concerning parts used in airplanes. Actually, the main aluminum alloys used for aeronautic industry are mentioned in various technical reviews but without direct links with airplanes made at the same period [1]. This missing information is required for a better understanding of the development of aeronautic material and for the renovation of the airplanes. In order to recover these data, we propose a double approach based on the study of indirect archival information and laboratory analyses on original aeronautic pieces. This type of study could also give interesting information in material sciences concerning the long-term aging of the Al-alloys.

To illustrate the interest in this double approach, we present the results obtained for some aluminum alloys used in the manufacturing of the Breguet Sahara airplane. We have chosen to limit our investigation to rivets because as they are a key point to maintain aircraft structural integrity, they were subjected to standardization processing ensuring constant well-defined mechanical properties.

HISTORICAL CONTEXT

Louis Breguet (1880-1955) was a French aircraft manufacturer. Fascinated by planes, he decided to create his company after the first flights of the Wright brothers in 1905. His company, *Société Anonyme des Ateliers d'Aviation Louis BREGUET*, started in 1911 and was finally merged with *Dassault Aviation* in 1971 [2]. In 1946, after the World War II, the Breguet Company developed a large-carrier airliner designed by the engineer Georges Ricard in 1936. The airplane, called Breguet 761, was aimed to replace the Douglas DC-4 but was sub-motorized. Then, engineers modified it and created the series of Breguet *Deux-Ponts* airplanes from the 761 to the 765. The novelty for those cargo planes was to present two flexible levels, or decks, with a removable floor inspired by the Breguet 730 seaplane. It could transport civil or military passengers on the upper deck and vehicles or bulky equipment on the lower deck. This series of aircraft began to be replaced in 1969 by the French-German Transall C-160 [2, 3].

During the conflict in Indochina, the French army looked for new troop transporters and ordered 27 airplanes from Breguet in 1951. Then the firm began to develop the Breguet 765 inspired by the civil airplane Breguet 763. The new aircraft had, amongst other enhancements, additional fuel tanks that extended the duration of its flight (up to 2485 miles without stopover). Because of budgetary constraints, the army bought only 4 aircrafts out of the 27 anticipated. The model studied flew from 1958 till 1972 and was neglected and stored on the base of Evreux in France. It is presently being renovated by the association *Ailes Anciennes Toulouse*, a work that started in 1987 [3].

ARCHIVAL INFORMATION

It was our intent to cross-correlate historical information with the results of laboratory investigations. This helped us to precise the nature of the alloy(s) used for the rivets and the chosen thermo-mechanical treatment(s). However, there is a difficulty to retrieve pertinent chemical and physical data on the elemental composition and processing of old aeronautical material as companies throw files away when the data are considered as out of date. This has been the case for most of archives regarding the Breguet 765 Sahara aircraft after the merger of Breguet and Dassault companies. Also, the access to the information may suffer from the selling or acquisition of a company by another industrial group: in our case, the access to Pechiney (manufacturer of raw aluminum alloys at the time of the airplane construction) archives was subjected to prior authorization from the Constellium group, which now owns these archives.

The documentary research focused on two topics: the aircraft itself (Breguet 765 Sahara) and the materials (aluminum alloys) used for the rivets.

When direct information is not available, it is often possible to retrieve indirect information about history of aircraft's use, for instance in advertising brochures, blueprints (used for airplane reproduction at different scales), aerodynamic calculations, and test flight reports. These data are found as primary sources, as archives of aircraft manufacturers, but also as secondary sources, in books about aeronautic history, periodicals or in the writings of enthusiastics and members of airplane associations. For the Breguet 765 Sahara, we have found some information in the documentation center of the *Bourget aeronautical and aerospace museum* as well as in the writings of the members of the association *Les Ailes Anciennes Toulouse* [4].

Concerning the material topic, we have compared old and recent documents about the aluminum industry and the processing methods. With these sources, it is possible to investigate the type of aluminum alloys and their properties according to their use in the airplane and to historic periods. Up to now, the principal institution where we can find this information is the *Institut pour l'Histoire de l'Aluminium (Paris)* and the Library of History of Science and Technology in *la Cité des Sciences et de l'Industrie (Paris).*

Duralumin

The alloy mainly used for rivets in the Breguet Sahara is Duralumin. In 1906, Alfred Wilm (1869-1937), a German metallurgist, invented Duralumin (Al-3.5-5.5%Cu-Mg-Mn) by discovering the age or precipitation hardening of aluminum alloys. This discovery was made while he worked on military research in Neubabelsberg near Berlin to make better cartridge and shell alloy. According to the legend, an increase in hardness was identified after measurements were interrupted for a weekend [5].

The age hardening is an increase of hardening of alloys related to the presence of fine precipitates (nm to µm in size) that provide barriers to the motion of dislocations, the defects in crystal structure that are responsible for plastic deformation. The precipitation process began after immersing in a heated solution, held at a temperature lower than the eutectic temperature to avoid melting at grain boundaries but above the solvus temperature in order to homogenize the material, and finally quenching into water to stabilize the microstructure [6, 7].

Rivets

We chose to concentrate our efforts on the rivets because they remain an essential element in the aircraft structure, even after the beginning of the use of adhesive or welding for structural joining. The use of electric welding appeared only during the World War II thanks to the needs of intensive production for fighter planes.

Classical alloyed aluminum wires were not of sufficient quality for rivets production to resist from in-use solicitations or breaking during a difficult riveting. Norms, tests and regulations were established. Wires had to be delivered cold worked after annealing, so that the heat treatments of rivets just before the riveting does not cause swelling of the grains that decreases its shear resistance and prevents good riveting [8]. In the journal, *Revue de l'Aluminium* [9], a 1955 article reported the necessity of an unification of the rivet alloys in the Duralumin class in order to end up with a defined grade suitable for each application. A new aluminum alloy for rivets was then developed and authorized by the French administration (*Aéronautique Française*) at the same time that the conception of the Breguet 765: the A-U3G alloy.

Before this development, duralumin A-U4G and A-U2G were used for rivets production. A-U4G is an aluminum alloy with 4 wt% of Cu and less than 1 wt% of Mg. With this material, rivets have a good shear resistance (26 kg/mm^2 minimum) but for a limited period of time after quenching (8 hours) before being too hard to be used. When the rivets became too hard, they were sent back to heat treatment workshop to be subjected to a solution treatment for 15 to 30 minutes at 923°F and quenched in water. A-U2G is an aluminum alloy with 2 wt% of Cu and less than 1 wt% of Mg. With this material, there is no limit of time for using the rivets but they have a lower shear resistance (theoretically 20 kg/mm^2 minimum and between 22 and 24 kg/mm^2 in practice).

Finally, the A-U3G is an aluminum alloy with 3 wt% of Cu and less than 1 wt% of Mg. Rivets made with this alloy do not have a limited time for use in riveting, and they have a shear resistance of 24 kg/mm^2 in theory and 26 kg/mm^2 in practice. The designers of this alloy hoped that after some improvements this value would be a guaranteed minimum. This happened to be the case as A-U4G and A-U3G alloys where still cited as alloys used for rivets manufacturing in the 1958 encyclopedia of the work of aluminum [8].

LABORATORY INVESTIGATIONS

The objective is to characterize directly alloys used for rivets of the Breguet 765 Sahara by concentrating on microstructure and elemental composition, in order to compare these data with the indirect information found in archives.

Sampling and experiments

The analyzed rivets come from two (L1 and L2) damaged and replaced parts during the restoration. The L1 part was localized in the front of the internal structure and the L2 part from lower deck as indicated in Figure 1. Usually, a rivet is constituted by a cylindrical shaft, a factory head (matrixed at the fabrication) and a shop head (formed during the setting up). As shown on the figure 1, the studied parts do not exhibit a significant corrosion but unfortunately the factory heads of all rivets were lost during the extraction step.

The metallographic observations were performed with an optical inverted microscope (Nikon MA200) and a Scanning Electron Microscope (LEO 435 LV operating at 15 kV) to attain higher magnifications. The studied rivets were cut in two parts along their length and subsequently coated cold with epoxy resin (Figure 1). Then the samples were mechanically polished with silicon carbide discs up to P4000 (5μm particles size) and with diamond polishing paste from 3 to 1/4 μm particles size. Finally they were etched with a Keller etchant for 30 seconds (Keller reagent: 1 mL of HF (48%), 1.5 mL of HCl (conc.), 2.5 mL of HNO_3 (conc.), 95 mL of water) [10].

The nanostructure was observed using Transmission Electron Microscopy (Jeol 2010 operating at 200 kV). To carry out the TEM observations, samples were polished mechanically (silicon carbide disc up to P2400) and thinned by an electrochemical process to obtain areas of appropriate thickness, around or less than 100 nm.

The elemental composition of the alloy was determined using the Cameca SX50 electron microprobe of Géosciences Environnement Toulouse (GET) laboratory. The measurements were led with a current of 50 nA and an accelerating voltage of 20 kV. The depth of analysis in the aluminum is 2 μm maximum, and the diameter of analyzed area is around one micrometer. 60 points of measurement were performed for each sample.

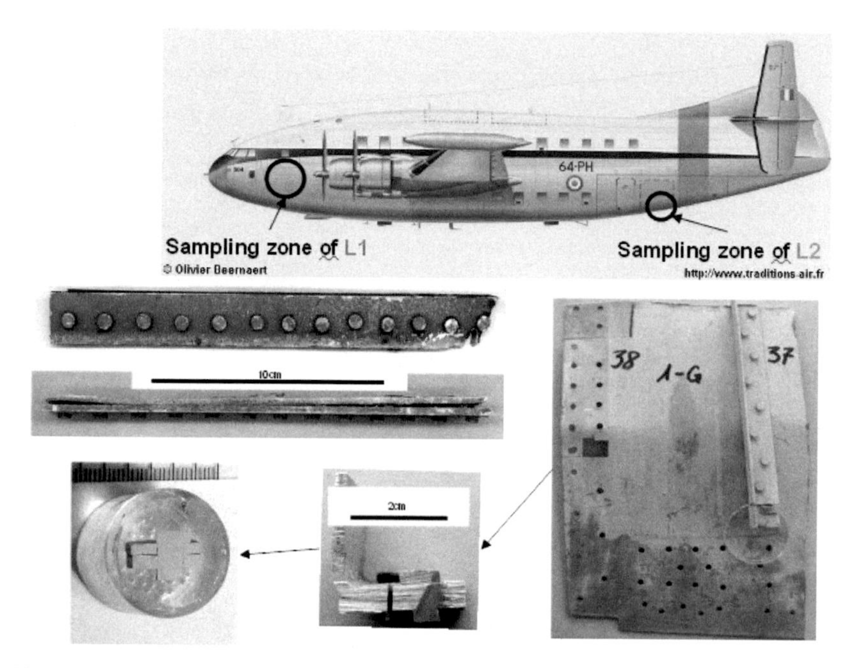

Figure1. Sampling sites (on the top), the extracted L2 part (bottom right) and an example of sample prepared for elemental analysis (bottom left)

Elemental composition

The elemental analyses (Table1) show that the alloys used for the rivets of parts L1 and L2 are different. L1 rivets contain around 3 wt% of Cu while the Cu rate is around 4 wt% for the L2 rivets. Mg and Mn contents are also different; L2 rivets contain twice as many Mg and Mn as L1 rivets.

In the paper of *Revue de l'Aluminium* [9], rivet compositions are given and the composition of L1 and L2 rivets are very similar to the A-U3G and A-U4G compositions respectively (Table 1).

Electron Probe Micro-Analyzer Weight % (this study)	Cu	Mg	Mn	Si
L1 (3 rivets analyzed)	3.02 ± 0.05	0.25 ± 0.01	0.25 ± 0.01	0.47 ± 0.02
L2 (2 rivets analyzed)	4.09 ± 0.15	0.61 ± 0.20	0.60 ± 0.30	0.55 ± 0.30
Revue de l'Aluminium (1955) [1]	**Cu**	**Mg**	**Mn**	**Si**
A-U4G alloy	4.10	0.70	0.65	0.55
A-U3G alloy	3.30	0.35	<0.10	0.50
A-U2G alloy	2.10	0.50	0.45	<0.20

Table 1. Microprobe analysis performed at Géosciences Environnement Toulouse (GET) laboratory compared to chemical compositions from reference [9]. Elemental compositions are given in wt.%.

Microstructure

Observations by optical microscopy were used to characterize the microstructure of the different parts of the rivets at a micrometer scale. In order to reveal the grain boundaries, the samples were etched in order to make them clearly visible. As shown in figure 2, the grains in the shaft of the rivets extracted from L1 and L2 parts are significantly different. The rivets of L1 (Figure 2a) have grains with a polygonal form that measures around 50 µm in diameter, while the rivets of L2 (Figure 2c) have grains less regular with a diameter ranging from 10 µm to 100 µm. There is no particular grains orientation in the shaft (Figures 2a and 2c), while in the rivet shop head, a large deformation of grains crushed by the riveting process can be evidenced (Figures 2 b and 2d).

The microstructure of shafts (Figure 3) is very close to the one presented in the paper of *Revue de l'Aluminium* [9]. As for the L2, the microstructure of the A-U4G is less regular and with smaller grains compared with the A-U3G, which is very similar to L1. This result confirms the use of these two alloys in the manufacturing of Breguet 765.

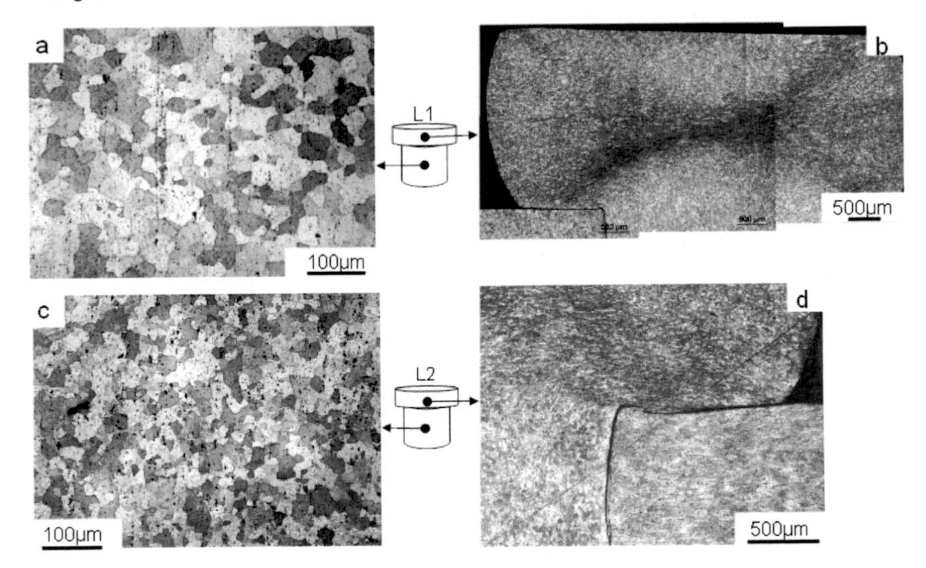

Figure 2. Microstructure views of L1 (on the top) and L2 (on the bottom) rivets by optical microscope after etching.

Precipitates and nanostructure

The two rivet types contain several types of precipitates of various sizes (from nanometers to a few tens of micrometers). The larger ones (5-10 µm) distributed randomly are essentially composed of Al, Mg and Cu. No significant different was observed between the two alloys concerning these precipitates. Moreover, they are too wide and the distances between them are too large for they play a role in hardness. On the other hand it is worth pointing out that at the micrometer scale, the observed microstructure depends essentially on the thermo-mechanical treatments and not on the composition, while the difference in the mechanical properties is to be found in the different populations of small precipitates that can be evidenced at higher magnification (Figure 4). At this scale the difference between the two alloys is obvious, confirming the difference of thermo-mechanical treatments used in the elaboration process. The nanoprecipitates concentration is lower in the L1 rivets in accordance with a less shear resistance. Note that the two rivet types exhibit a singular contrast structure (Figure 4), in the form of darker streaks that are believed to be related to a periodic variation of the content of Cu in the matrix. Further work is needed to investigate more properly this effect.

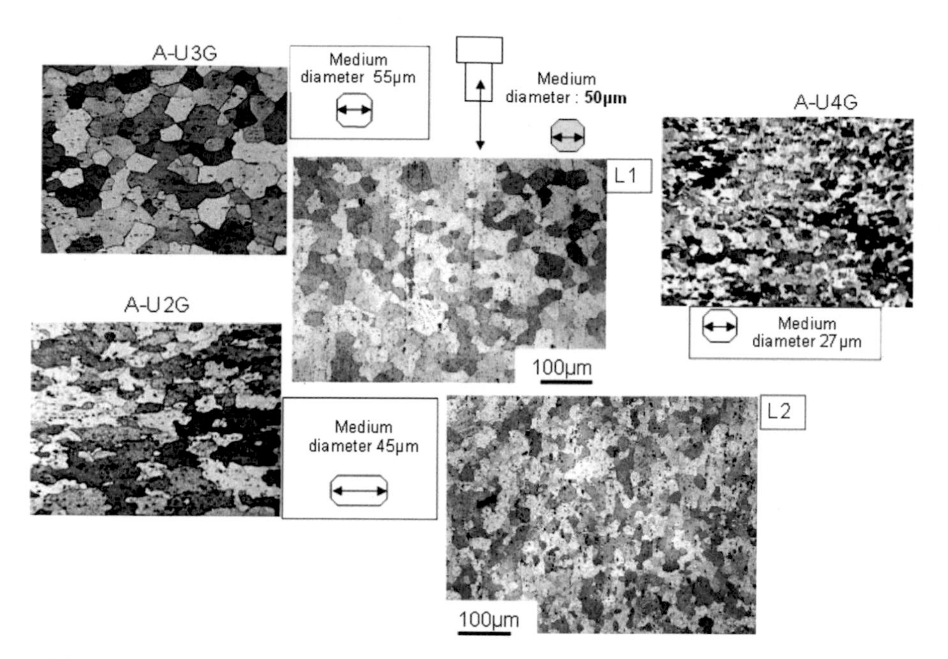

Figure 3. Comparison between microstructures presented in the Journal de l'Aluminium, January 1955 [9] and microstructure of L1 and L2 rivets obtained by optical microscopy after etching.

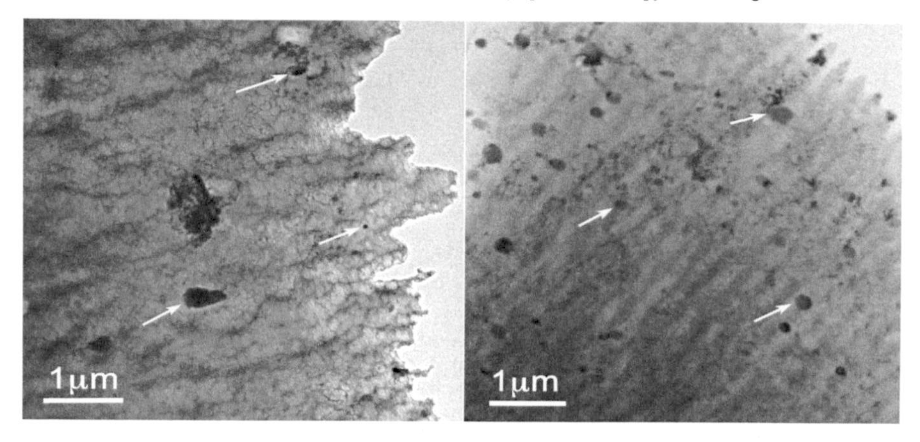

Figure 4. TEM observations of the microstructure showing the size and density of precipitates (arrowed) in a L1 rivet (left) and in a L2 rivet (right). L2 rivets show a larger density of small precipitates. In both materials, there exists a singular contrast in form of parallel darker streaks.

DISCUSSION

Electron microprobe and microstructural analyses have revealed a different composition between rivets from L1 and L2. L1 and L2 rivets can be related to the A-U3G and A-U4G compositions mentioned in the article published in January 1955 in *Revue de l'Aluminium* concerning the development of new alloys for rivets [9]. In another of this same review published in 1958, it is indicated that the rivets in A-U3G and A-U4G involved solution treating (495±5°C) and quenching before the riveting. The absence of texture for the grains in the shafts could be due to this treatment. The A-U3G composition was developed by the French Pechiney Company for rivet application, at the time of conception of the Breguet 765 airplane. This study established clearly that this plane has benefited of these new rivets. However as "old" A-U4G composition was also used, we wonder if these two alloys were chosen according to the requisite mechanical properties. The A-U3G was developed to replace A-U4G, which requested a solution treating just before the riveting while the thermal treatment could be done several days before the setting up for A-U3G. However the shear resistance of A-U3G was lower than A-U4G [9], and maybe the last one was still used for this reason. In the other hand, the Breguet 765 was only built in small numbers at the beginning of the A-U3G rivet production, one cannot exclude that the two types of rivets were used indifferently up to the end of A-U4G stocks. Without direct archival information about it, it is difficult to answer to this question on the base of these only analyses.

CONCLUSION

Despite difficulties in archives retrieval, we have shown that information concerning the Al-alloys used in the manufacturing of rivets of the Breguet 765 Sahara aircraft can be recovered using a double approach based on indirect archival information and laboratory analyses. Links have been established between data found in the contemporaneous scientific literature and present laboratory analysis. We have shown that rivets of L1 part have the elemental and structural characteristic of the A-U3G developed specially for rivet application at the same time that the Breguet 765 program. Future work will encompass learning more about the rivets, in particular the different processes of age hardening for the two alloys, that influence rivet performance, strength and durability.

ACKNOWLEDGEMENTS

The authors would like to thank the *Ailes Anciennes Toulouse* association for providing the samples and the *Institut pour l'histoire de l'aluminium* and *the center of documentation in aeronautical and aerospace museum of Bourget* for their responsiveness and help. They would also like to thank Dominique Lamirault for the TEM samples preparation. This work was supported by the BQR2012 INPT-INSA-ISAE "AAAdurable".

REFERENCES

1. Dollfus C., *Les divers domaines d'applications : L'Aviation*, Revue de l'Aluminium, **211** (1954) pp.201-209.

2. Breguet, E., *Breguet, un siècle d'aviation,* édition Privat, pp. 91-97, (coll. Aviation, Toulouse, 2012).

3. Hartmann, D., *Sauver Brigitte ou le sauvetage et la récupération du Breguet 765 « SAHARA » d'Évreux à Toulouse, pp.37-39,* (http://www.calameo.com/subscriptions/57493).

4. Service Technique de l'Aéronautique, *Clauses techniques du Breguet 765*, édition 2bis (1956).

5. Hardouin Duparc Olivier, *Alfred Wilm et les débuts de Duralumin*, Cahiers d'histoire de l'aluminium **64**, (Institut pour l'Histoire de l'Aluminium, Paris, 2005) p.62.

6. Saulnier, A., *Le durcissement structural des alliages aluminium-cuivre 4%*, Revue de l'Aluminium **217**, pp.41-46 (1955).

7. Dubost, B., Saintfort, P., *Durcissement par précipitation des alliages d'aluminium*, Dossier Techniques de l'ingénieur **M240**, pp.2-4 (1991).

8. Penel, P., *Le rivetage- première et deuxième partie*, Encyclopédie du travail de l'Aluminium **fiche G81,** pp.1125-1127 (1958).

9. Tournaire, M., Renouard, M., *Alliages pour rivets de la famille du Duralumin*, Revue de l'Aluminium **217** (1955).

10. American Society for Metals Handbook committee, *Metals Handbook vol. 8: Metallography, Structures and Phase Diagrams*. Edited by American Society for Metals, p.124, (Metals Park, Ohio, 1973)

Mater. Res. Soc. Symp. Proc. Vol. 1656 © 2015 Materials Research Society
DOI: 10.1557/opl.2015.2

Non-Invasive Characterization of Stone Artifacts from the Great Temple of Tenochtitlan, Mexico

Mayra Dafne Manrique-Ortega[1], Pieterjan Claes[1], Valentina Aguilar-Melo[1], Malinalli Wong-Rueda[1], José Luis Ruvalcaba-Sil[1], Edgar Casanova-González[1], Emiliano Melgar[2], Reyna Solis[2]

[1]Instituto de Física, Universidad Nacional Autónoma de México, Circuito de la Investigación Científica s/n, Ciudad Universitaria, Mexico, DF 04510, Mexico. e-mail: sil@fisica.unam.mx
[2]Museo de Templo Mayor, INAH. Seminario 8, Centro Histórico, Mexico, DF 06060, Mexico.

ABSTRACT

The Museum of the Great Temple of Tenochtitlan in Mexico City holds a collection of several thousands of polished stone artifacts that were excavated and identified as temple offerings. These can stratigraphically be related to the sequential construction stages (II-VII) of the ceremonial area of the Aztec capital from the foundation of the city in 1325 to 1521, when the Spaniards conquered the city. A non-destructive investigation of the elemental and chemical composition of these archaeological artifacts helps us to understand the provenance of these pieces, their use and the specific mineralogical choice for these artifacts as well as more information regarding trade routes relevant to the development of the Aztec empire. A mineralogical analysis of, in total, 450 stone artifacts was carried out using infrared spectroscopy (FTIR) and X-ray Fluorescence (XRF). From this, eighty-five pieces were selected according to their excavation location, either in the Great Temple itself or in the surrounding buildings, as well as to represent the different construction stages of the area (this is part of a World Heritage Site). The resulting mineralogical and chemical information was related to possible mineral resources that were controlled and used as the empire expanded. Artifacts made from high-status semi-precious minerals, like jadeite and turquoise, are found to be concentrated in the central buildings and in the Great Temple itself, but also in the later construction periods of the area.

INTRODUCTION

Although Tenochtitlan was already founded around 1325 A.D. on a small island in the Mexico Basin as a result of a territorial concession made by the lord of Azcapotzalco to the Aztecs, it lasted a century before the city finally became the Aztec capital [1]. In 1428, the Aztecs were released from the power of Azcapotzalco and could begin their expansion throughout the Aztec heartland. Eventually, they conquered and controlled about 340 ethnic groups located in a vast region covering of Mesoamerica from the southwest coast to the Gulf of Mexico and Soconusco (Southwest coast) in the Maya area (Figure 1). Besides military benefits, the Aztec expansion also resulted in the acquisition of materials which were not present in the capital region but that were needed in daily and ritual life in the city. Conquered cities were forced to pay tributes, or taxes, to Tenochtitlan. Not only merchandises, such as agricultural products, raw materials and handicrafts, each of them characteristic of their region of origin, made their way to the capital, but also laborers, farmers, builders, craftsmen, and servants were recruited and employed in the city. This resulted in the introduction of new material processing and craft techniques into Tenochtitlan. The tax system established by the Aztecs during their dominion is the main reason why in the offerings found at the Great Temple of Tenochtitlan, and in other excavations in the city, many materials and manufacturing techniques have been identified that are not characteristic for the center of Mexico, but are related to other parts of their territory.

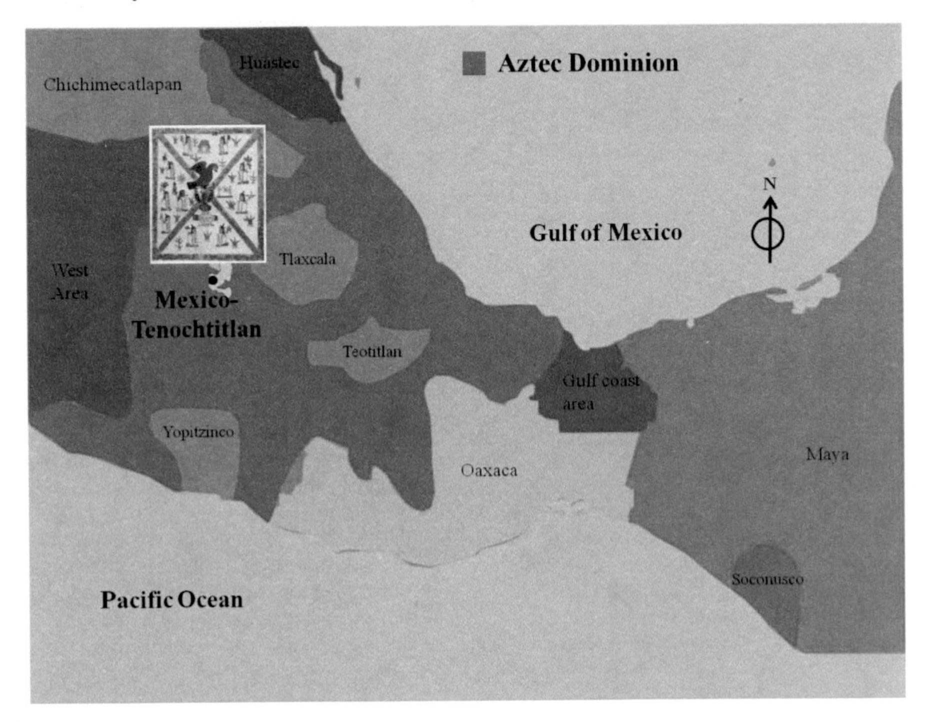

Figure 1. Regions controlled by the Aztecs in Mesoamerica, highlighted in red. Regions related to other cultures are marked in a purple, yellow, brown, and green color.

As a result of the growth and prosperity of Tenochtitlan, the Aztecs were able to enlarge and embellish The Great Temple of Tenochtitlan. This can be seen in the seven construction stages of the temple, one under each Aztec ruler, called Tlatoani (Table 1). Its architecture and geographical position represented the sacred nature of the temple. It was built at the intersection of the four basic divisions of the city and represented the levels of the Aztec universe. As the main temple, the Great Temple was the most important place for burials and religious offerings [2]. Archaeological records indicate that these offerings are extremely diverse, both in the main and surrounding buildings. Not only were the objects found in the Aztec's offerings in the seven stages of construction diverse, but also the materiality, presence and quality of the objects were varied. With regard to minerals, lithics and lapidary objects, hundreds of items made from raw materials foreign to Tenochtitlan, such as serpentine, turquoise, jadeite and marble, were discovered in these burials. Also foreign and non-contemporary manufacturing styles, such as Olmec, Teotihuacan, Mezcala, Mixtec, Mayan and Toltec styles, can be distinguished among the thousands of artifacts [3]. Upon the Spanish arrival in 1519, Tenochtitlan covered about 13 km^2 and with a population of more of 100,000 inhabitants, it was one of the largest cities in the New World. After the Spanish invasion, the city was almost completely destroyed and now its remains can still be found under the colonial buildings in the historical center of Mexico City. In recent decades, archaeologists have made many discoveries thanks to the large number of excavations that were performed as a result of the expansion of the modern city. However, many of the primary contexts were lost or unknown and a wealth of historical and archaeological information disappeared.

In the case of lithics and lapidary materials, previous archaeological studies of the majority of these artifacts were restricted to a visual investigation of the main materials and their division into typographic classifications, leaving aside the possibility of imitations by other groups or by the Aztecs themselves. Consequently, without further analysis the artifacts originally manufactured in Tenochtitlan cannot be separated from those handcrafted elsewhere [4].

Due to the recent increase in characterization studies of lithics and lapidary materials from Mesoamerica using non-destructive and non-invasive techniques [5-10], we initiated a study of a large quantity of lapidary objects found at the Great Temple of Tenochtitlan [11-12]. The objectives of this study are not limited to the mineralogical characterization of the different pieces but also focus on the provenance of the pieces and their archaeological significance. The analyses are useful to understand the different trade routes established by the Aztecs but also to determine which of the materials were commercialized and which were offered during the various stages of the Great Temple. Additionally, we compared the mineralogical information with the archaeological data concerning the location where these pieces were excavated. A representative selection of 87 pieces was chosen based strategically on the archaeological problems and other information to enable further interpretation (chronology, style, color). Further analysis involved the comparison of this information with technological studies of the artifacts [3-4, 13-14].

Table 1. Construction stages of the Great Temple of Tenochtitlan, their chronology and the offerings studied of each stage.

Stages of Construction	Chronology	Ruler (Tlatoani)	Studied Offerings
I	1325-1375	Tenochtitlan Foundation	-
II	1375-1426	Acamapichtli, Huitzilíhuitl, Chimalpopoca	39, 45
III	1427-1440	Itzcóatl	25, 26, 28
IVa	1440-1469	Moctezuma I	18, 19, 48, 85, 97
IVb	1469-1481	Axayácatl	C2, 1, 5, 6, 9, 17, 22, 82
V	1481-1486	Tízoc	27, 54, 77
VI	1486-1502	Ahuízotl	65, 70, 87

EXPERIMENTAL METHODS

The non-invasive mineralogical identification for the artifacts from The Great Temple of Tenochtitlan was performed using IR spectroscopy as it has demonstrated efficient data acquisition in previous studies [8-9]. Compositional information was obtained using X-ray Fluorescence (XRF) measurements. From the relative intensities of specific elemental information, an understanding of the artifacts was gained.

Infrared reflectance spectra of the artifacts studied in this article were acquired *in situ* using a Bruker Alpha FTIR spectrometer without an optical fiber but with an external reflection module. The IR spectra have a resolution of 4 cm^{-1} and are measured in the mid-IR spectral range, namely from 400 to 4000 cm^{-1}. Objects are placed near the focal point, 15 mm from the spectrometer, and an area of 5 mm^2 is illuminated.

Our portable X–Ray Fluorescence system SANDRA (Sistema de Análisis No Destructivo por RAyos X or System for Non Destructive Analysis using X-rays) [15] was used to collect elemental information. This SANDRA system consists of a detector fixed at an angle of 45° with respect to the direction of the excitation X-ray beam. The X-ray tube has a molybdenum anode and, for this analysis, the power of the tube was set to 35 kV but the current ranged from 0.08 to 0.250 mA with an acquisition time of 90 to 180 s, depending on the materials and conditions required to avoid the saturation of the system. Relative intensities of the XRF spectra were measured using the AXIL program [16].

CORPUS OF OBJECTS

To obtain high-quality information about the lapidary objects found at the Great Temple of Tenochtitlan, an analysis of 450 artifacts was performed. Not only were these objects found in distinct locations around the temple, they can also be linked to different building stages of it. Here, we will discuss a selection consisting of 85 artifacts, from which 47 originate from the temple, while the other 38 pieces were found in the surrounding buildings, so called buildings A, B, E and F. (Figure 2 and 3).

This selection was based on archaeological classifications, considering the different styles and temporalities of the objects as well as the abundance of certain materials and their representativeness for the ancient Aztecs. An overview of the obtained information can be found in Tables 2 and 3 for the offerings at the adjacent buildings and the Great Temple itself, respectively.

DISCUSSION

Mineralogical identification

Via infrared spectroscopy we were able to identify various minerals as is presented in Figure 4 (a-e) where an overview is shown of the precious minerals found among the artifacts. These results were confirmed by the elemental compositions obtained via XRF measurements.

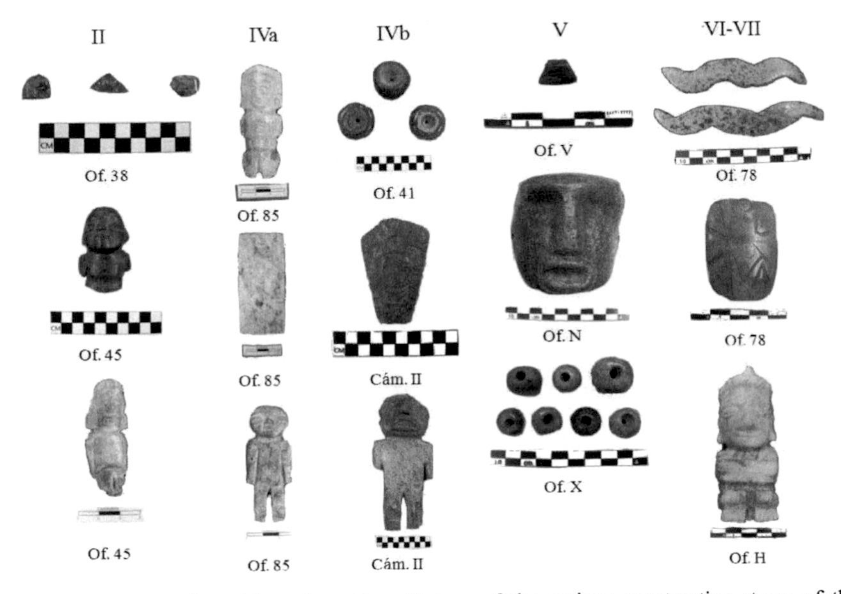

Figure 2. Representative objects from the offerings of the various construction stages of the Great Temple of Tenochtitlan, includes beads, tablets, sculptures, Offering number is indicated.

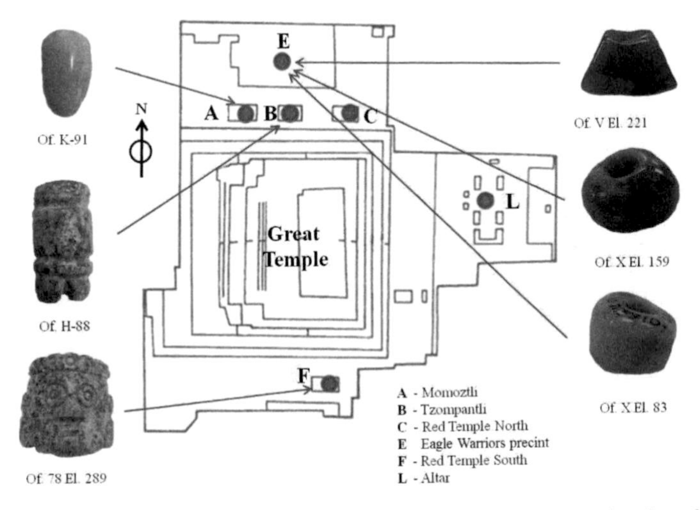

Figure 3. Overview and relative locations of the different buildings at the site of the Great Temple of Tenochtitlan and some representative offerings from these buildings.

Table 2. Overview of the 38 artifacts found at the adjacent buildings of the Great Temple of Tenochtitlan. Mineralogical identification was performed by FTIR.

Construction stage	Building	Offering	Mineral identification / number of artifacts identified
IV b - V	E	T, V, X	Albite + alumino-silicate / 6
VI	B	H-75, 76, 86, 87, 97, 98, 135	Calcium carbonate /1 Calcium carbonate + serpentine / 6 Calcium carbonate + unidentified /1
	F	78	Calcium carbonate / 1 Calcium carbonate + unidentified /1 Calcium carbonate + serpentine / 10 Serpentine / 4
VII	A	K-55, 86, 93	Jadeite / 1 Albite + alumino-silicate / 6 Quartz / 1

Here we identified serpentines, quartz, turquoises, and jadeites. Albite mixed with another alumino-silicate was also present; however, positive identification of this unidentified mineral was not possible without X-ray Diffraction information (XRD). Also calcium carbonates, such as calcite (Figure 5a) and aragonite (Figure 5b) were found among the offerings of the main temple as well as in the adjacent buildings. Although both minerals are calcium carbonates with similar infrared spectra, it is possible to differentiate between them. This is done using the shape of the broad infrared absorption band in the region 1350 – 1600 cm^{-1} as well as by comparing the relative intensity of the bands at 858 and 872 cm^{-1} for aragonite and calcite, respectively. In addition, other spectra show traces of minerals mixed with these calcium carbonates. In some cases, it is possible to identify these extra minerals present in an artifact. However, it is not trivial to identify the mixture of minerals, as is shown in Figures 5(d-e) and 5(g-h). Comparing the other infrared bands with the infrared spectra of serpentine and quartz, see figure 4a and 4b respectively, the majority of the remaining bands can be identified as such. For instance, combinations of calcium carbonates and quartz (Figure 6 a-b) or serpentine (Figure 5 e and h; Figure 6 c-d) were resolved. In Figure 6, the contribution of the calcium carbonate is highlighted in light gray.

Table 3a. Overview of 30 artifacts found at the Great Temple of Tenochtitlan listed. The artifacts are listed according to the construction stages of the temple for stages II to IV. Archaeological identification is contrasted with its mineralogical identification by FTIR.

Construction stage	Offering	Archaeological identification	Mineral identification / number of artifacts identified
II	39	Tecali*	Calcium carbonate / 1 Serpentine / 1 Quartz / 1
	45	Stone	
III	25	Green stone	Calcium carbonate + quartz / 1 Quartz / 1 Albite + alumino-silicate / 1
	26		
	28		
IV a	18	Green stone	Albite + alumino-silicate / 1
	19	Green stone	Albite + alumino-silicate / 1 Quartz / 1 Serpentine / 1
	48	Stone	Calcium carbonate + serpentine / 1
	85	Stone	Quartz / 1 Calcium carbonate + serpentine / 1
	97	Stone	Serpentine / 2
	Room III	Tecali*	Calcium carbonate / 3
IV b	Room II	Stone	Serpentine / 1
	1	Turquoise	Turquoise / 2
	2		
	5	Stone	Calcium carbonate + quartz / 1
	6	Jadeite	Jadeite / 1
	9	Stone	Calcium carbonate + serpentine / 1
	11	Tecali*	Calcium carbonate / 1 Calcium carbonate + unidentified /1
	17	Green stone	Calcium carbonate + serpentine / 1
	22	Tecali*	Calcium carbonate / 2
	82	Stone	Serpentine / 1

*Tecali is the pre-Hispanic name of a Travertine mineral, rich in calcium carbonates with a layered deposition structure [17]. The main sources in central Mexico are in a location in Puebla region with the same name (Tecali de Herrera).

Table 3b. Overview of 17 artifacts found at the Great Temple of Tenochtitlan. The artifacts are listed according to the construction stages of the temple for stages V to VII. Archaeological identification is contrasted with its mineralogical identification by FTIR.

Construction stage	Offering	Archaeological identification	Mineral identification / number of artifacts identified
V	27	Green stone	Calcium carbonate + quartz / 1
	54	Green stone	Jadeite / 1 Calcium carbonate + serpentine / 1
	77	Stone	Jadeite / 1 Calcium carbonate + quartz / 1
VI	65	Green stone	Albite + alumino-silicate / 1
	70	Stone	Jadeite / 2
	87	Green stone	Jadeite + unidentified /1 Albite + alumino-silicate / 1
VII	50	Green stone	Calcium carbonate + serpentine / 1
		Stone	Calcium carbonate / 1
	64	Green stone	Albite + alumino-silicate / 1 Jadeite / 1
	98	Green stone	Jadeite / 3

Apart from the artifacts mentioned here, many other pieces from the Great Temple and its surrounding buildings were studied. The mineralogy indicates an even more extensive variety than presented here. Many of these objects have infrared spectra similar to jadeite, but with shifted bands and/or intensity changes, and they are listed as an alumino-silicate type mineral. Further analyses will be conducted using XRD combined with XRF. Currently, we are modifying a commercially available XRD spectrometer into a portable one for this purpose.

Comparison among the different construction stages of the Great Temple of Tenochtitlan

Figure 7 and Table 3 illustrate the different minerals identified during the seven construction stages of the Great Temple. In the earliest stages (II – IVa), the minerals found in the burials are very similar to those identified in the surrounding buildings, such as calcium carbonates, quartz, serpentines and albite mixed with an unidentified alumino-silicate. During the intermediate and late construction stages (IVb – VII) of the Great Temple, the mineralogical identification shows an increase of more precious minerals, such as jadeite and turquoise. This phenomenon can be related to the growth and expansion of Tenochtitlan, as has been proposed earlier in [3, 4]. Over the years, Tenochtitlan expanded, strengthened and enriched itself with materials found in the conquered regions. Starting from the reign of Axayacatl (1469 – 1481 C.E. corresponding to stage IVb of the temple), jadeite samples were found in the offerings of the temple. However, this lasted until Ahuízotl (stage VI) before the Aztecs were able to expand their territory as far as Soconusco in Chiapas and gained more direct access to regions near to jadeite sources.

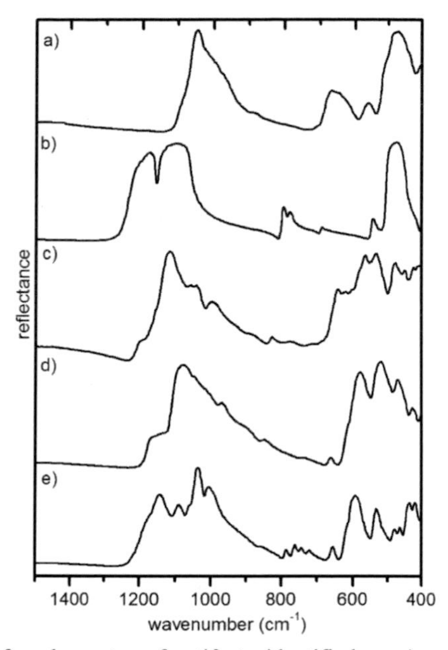

Figure 4. Reflectance infrared spectra of artifacts identified as a) serpentine, b) quartz, c) turquoise, and d) jadeite. The lower spectrum e) shows the combination of albite with an alumino-silicate that has not been identified yet.

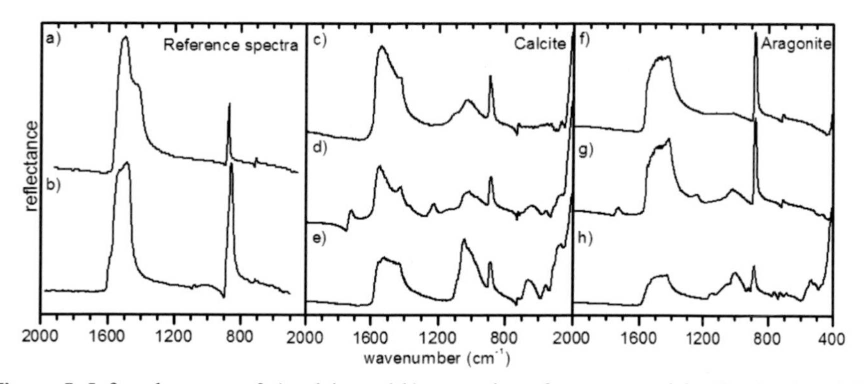

Figure 5. Infrared spectra of a) calcite and b) aragonite reference materials. Graphs c) and f) show the infrared spectra of artifacts identified as calcite and aragonite, respectively. Spectra (d-e) and (g-h) show mixtures of these minerals (respectively, calcite and aragonite) with serpentine (traces e and h) and additionally also an unidentified mineral (traces d and g).

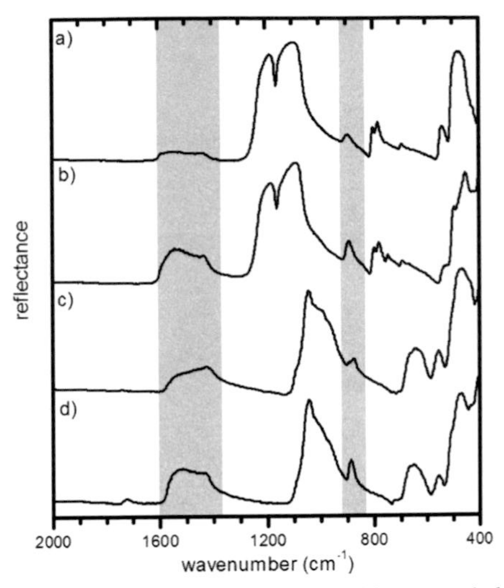

Figure 6. Infrared spectra of calcium carbonate mixed with quartz (a-b) and serpentine (c-d). Possible contributions of other minerals cannot be excluded. The contribution of the calcium carbonate is highlighted in light gray.

Therefore, the first jadeite artifacts found among the offerings of the Great Temple cannot be linked directly to specific mines, but originate from indirect access to raw jadeite materials or artifacts. This indicates that trade routes existed or that conquered tribes were forced to pay their taxes using jadeite, or a combination of both. For the later stages (Figure 7), after the conquest of Soconusco, a wide variety of jadeite artifacts can be found in the burial offerings of the Great Temple, indicating a more direct availability of the mineral, as well as a cultural preference for green jadeite, as was proposed in a previously smaller study of offerings from the Great Temple [12].

Despite the importance of the pieces found in the burials of the Great Temple of Tenochtitlan, the archaeological information, as presented in Table 3, based on a visible inspection is scarce, superficial, and inaccurate in some cases. Therefore, it is of utmost importance to obtain more detailed information via analytical techniques to establish deeper insight into the relationship between Tenochtitlan and other (conquered or friendly) states.

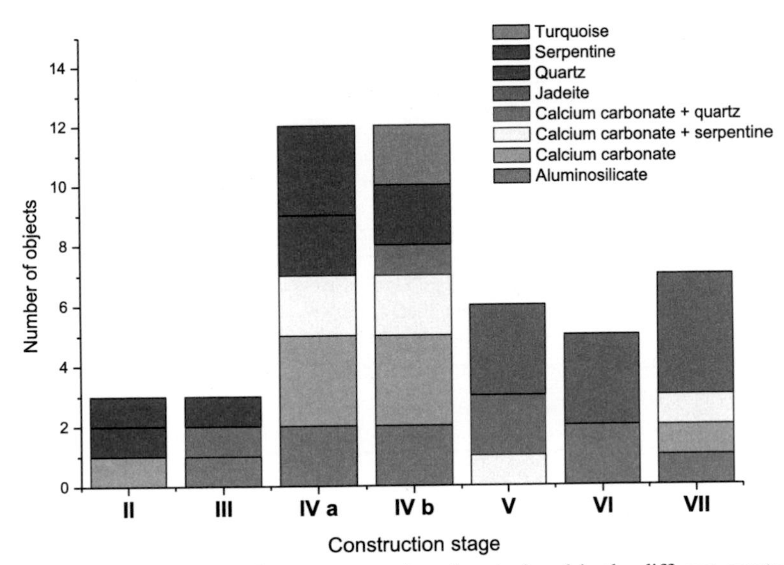

Figure 7. Histogram showing changes among the minerals found in the different construction stages of the Great Temple of Tenochtitlan.

Comparison of Minerals in the Offerings from the Great Temple to those in the surrounding buildings

Objects from the Great Temple were compared to those deposited in the adjacent surrounding buildings using the mineralogical identification based on FTIR and XRF. As shown in Tables 2 and 3, the offerings found in the main temple were handcrafted using higher value minerals, such as jadeite and turquoise. The majority of the pieces collected in the neighboring buildings were identified as calcium carbonates, serpentines, quartz and albite mixed with alumino-silicates phases. These minerals are also considered as precious stones, however with less value and a minor significance for the ancient Aztecs [18]. Only one exception was observed, namely in building K (offering 55, piece 10-262619 22/54) where a single piece of jadeite was identified. The Great Temple was a special holy place and reserved for important noblesse in Aztec society. Therefore only the best, most precious minerals were used in the offerings at the temple, while slightly less precious minerals can be found aside, at the burials in the surrounding buildings.

The jadeite from building K corresponds to the construction stages (VII), where as in the main temple, the use of jadeite was more commonly used in the offerings. Certain minerals occur more frequently in certain buildings. For example, calcium carbonate and mixtures are abundant in buildings H and F. Building K has greater amount of albite mixed with an alumino-silicate. An analysis of more artifacts might help to obtain a clearer pattern of this behavior.

Elemental compositional information

XRF measurements provided elemental compositional information and aided in the determination of possible resource areas. Additionally, we studied the homogeneity of the minerals by comparing the relative intensities of Fe and K with respect to the Ca intensity for the serpentine artifacts excavated from buildings B (offerings H-75, 76, 86, 97, 98) and F (offering 78), corresponding to the VI construction stage, and offering 48 from stage IVa. Figure 8 shows a relationship between the relative intensities of Fe/Ca and K/Ca suggesting that the elemental composition of serpentines are more diverse in IVa than in the later stage VI. The Aztecs in the later periods were more aware of the different minerals and may have had preferences of their use for status in the hierarchy of elite materials. In earlier stages not much importance was given to the uniformity and selection of the serpentines and most likely several sources were used. Moreover, standardization may have been introduced and practiced, resulting in a more thorough selection of the materials after stage IV, as supported by the results of previous technological studies [3, 19-20].

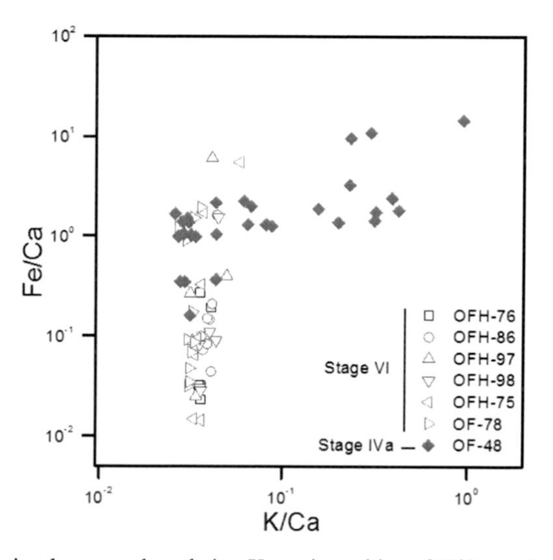

Figure 8. Correlation between the relative X-ray intensities of K/Ca vs Fe/Ca for serpentines. Offering 48 is related to the construction stage IVa of the temple, while the other offerings correspond to the stage VI.

CONCLUSIONS

In the mineralogical study of archaeological artifacts, a combined molecular and elemental spectroscopic investigation allow us to obtain the most relevant information. Elemental compositional information can be achieved via XRF measurements while infrared spectroscopy is the most direct manner to unambiguously identify the mineral according to its molecular structure. In certain cases, the mineralogical identification is hindered mainly because of the complexity of the composition, and other techniques such as XRD may provide complementary data.

In this article, we presented a strategically chosen selection of pieces originating from the Great Temple of Tenochtitlan, covering the different construction stages of the temple as well as the different excavation locii (both in the temple and in the adjacent buildings). Firstly, using FTIR and XRF *in situ* we were able to identify the majority of the pieces as serpentines, quartz, turquoises, and jadeites, as well as albite mixed with an alumino-silicate and mixtures of calcium carbonates with other minerals. With this information a correlation was found between the different construction stages of the temple and the minerals that were selected as offerings. A clear change in the choice of minerals is found during stage IVb, coinciding with the reign of Axayacatl. Starting in this period, more precious minerals were used for the artifacts found at the Great Temple of Tenochtitlan. However, it lasted until Ahuizotl, stage VI, before Tenochtitlan had more direct access to jadeite sources within its territory. Secondly, the presence of certain minerals in the temple or in the adjacent buildings can be related to the importance of the buildings. For instance, the most precious stones were found in the offerings of the temple, while less precious minerals were excavated in the surrounding buildings. Thirdly, by comparison of the relative intensities of Fe/Ca and K/Ca, we were able to relate the provenance and variability between the different serpentines. This shows the importance of Tenochtitlan and the Aztec dominion in Mesoamerica.

ACKNOWLEDGMENTS

The authors wish to express their gratitude to Fernando Carrizosa of the curatorial area and the direction of Museo de Templo Mayor for the possibility to study this wide variety of stone objects in the Museum. P. Claes is grateful for financial support via a personal postdoctoral fellowship given by the DGAPA-UNAM. This research has been supported by grants from CONACyT 131944 MOVIL II, PAPIIT IN402813 ANDREAH II and ICyTDF PICCO10-57.

REFERENCES

1. M. Matos Moctezuma, *The Great Temple of the Aztecs* (Thames and Hudson, London, 1988).
2. L. López Luján, *Las ofrendas del Templo Mayor de Tenochtitlan*, (Instituto Nacional de Antropología e Historia, Mexico, 1993).
3. E. R. Melgar-Tisoc, "Análisis Tecnológico de los objetos de piedra verde del Templo Mayor de Tenochtitlán", in *El jade y otras piedras verdes. Perspectivas interdisciplinarias e interculturales*, coordinated by W. Wiesheu and G. Guzzy (Instituto Nacional de Antropología e Historia, Mexico, 2012) pp. 181 – 195.
4. E. R. Melgar-Tisoc and R. B. Solis-Ciriaco, "Manufacturing Techniques of the Turquoise Mosaics from the Great Temple of Tenochtitlán, Mexico" in *2nd Latin – American Symposium on Physical and Chemical Methods in Archaeology, Art and Cultural Heritage Conservation. Selected papers* 2009, edited by J.L. Ruvalcaba-Sil, J. Reyes-Trujeque, J. Arenas-Alatorre and A. Velázquez-Castro (Universidad Nacional Autónoma de México, Universidad Autónoma de Campeche and Instituto Nacional de Antropología e Historia, México, 2010) pp.119 – 124.

5. M.D. Manrique Ortega, *Análisis no destructivo por técnicas espectroscópicas de las piedras verdes del ajuar funerario del rey Maya Pakal*, Dissertation thesis on physics, Universidad Nacional Autónoma de México, Mexico, 2012.
6. P.L. Hauff, "The enigma of Jade, with Mineralogical Reference to Central American Source Materials", in *Precolombian Jade: New Geological and Cultural Interpretations*, edited by Fredrick W. Lange (University of Utah Press, Salt Lake City, 1993) pp. 82 – 103.
7. J.L. Ruvalcaba-Sil, A. Daneels, M. Vaggi and M. Aguilar-Franco, "Non–destructive Characterization of Green Stone Pieces from La Joya Site, Veracruz", in *2nd Latin – American Symposium on Physical and Chemical Methods in Archaeology, Art and Cultural Heritage Conservation. Selected papers* 2009, edited by J.L. Ruvalcaba-Sil, J. Reyes-Trujeque, J. Arenas-Alatorre and A. Velázquez-Castro (Universidad Nacional Autónoma de México, Universidad Autónoma de Campeche and Instituto Nacional de Antropología e Historia, México, 2010) pp. 49 – 55.
8. J.L. Ruvalcaba-Sil, L. Filloy, M. Vaggi, L. H. Tapia-Gálvez, R. Sánchez-Becerra, "Estudio no destructivo in situ de la Máscara de Malinaltepec", in: *La Máscara de Malinaltepec*, edited by S. Martínez del Campo-Lanz (Instituto Nacional de Antropología e Historia, Mexico, 2010) pp. 153 – 167.
9. L. Filloy Nadal, D. Magaloni Kerpel, J.L. Ruvalcaba Sil, R. Sánchez Hernández, "Las materias primas utilizadas para la manufactura de las figurillas y las hachas de la Ofrenda 4 de La Venta: caracterización y fuentes de origen", in *La Ofrenda 4 de la Venta*, (Museo Nacional de Antropología, Instituto Nacional de Antropología e Historia, México, 2013) pp. 103 – 127.
10. F. Casadio, J.G. Douglas and K. T. Faber, *Anal. Bioanal. Chem.* **387**, 791 (2007).
11. J.L. Ruvalcaba, E. Melgar, Th. Calligaro, *Manufacturing Analysis and Non Destructive Characterisation of Green Stone Objects from the Tenochtitlan Templo Mayor Museum, Mexico*, in *Proceedings of the 37th International Symposium on Archaeometry*, 13th - 16th May 2008, Siena, Italy. Edited by I.Turbanti-Memmi (Springer, Heildelberg, 2011) pp. 299-304.
12. J.L. Ruvalcaba-Sil, E.R. Melgar-Tisoc, J. Curado, K. Laclavetine and T. Calligaro: "Caracterización y procedencia de piedras verdes de las ofrendas del Templo Mayor de Tenochtitlán" in *Técnicas analíticas aplicadas a la caracterización y producción de materiales arqueológicos en el área maya*, edited by A. Velázquez-Castro and L.S. Lowe (Universidad Nacional Autónoma de México, 2013) pp. 163 – 177.
13. E. R. Melgar Tisoc, J.L. Ruvalcaba, "Technological Analysis of the Calcite Beads from the Great Temple of Tenochtitlan", *Cultural Heritage and Archaeological Issues in Materials Science II*, edited by J.L. Ruvalcaba Sil, J. Reyes Trujeque, A.Velazquez Castro, M. Espinosa Pesqueira, Materials Research Society Proceedings, Cambridge University Press, in press.
14. R.B. Solis Ciriaco, J.L. Ruvalcaba Sil, "Provenance and Manufacture of Mixtec Style Objects Found on the Surrounding Structures of the Precinct of the Great Temple of Tenochtitlan", *Cultural Heritage and Archaeological Issues in Materials Science II*, edited by J.L. Ruvalcaba Sil, J. Reyes Trujeque, A.Velazquez Castro, M. Espinosa Pesqueira, Materials Research Society Proceedings, Cambridge University Press, in press.
15. J.L. Ruvalcaba-Sil, D. Ramírez-Miranda, V. Aguilar-Melo, F. Picazo, *X-Ray Spectrom.* **39**, 338 (2010).
16. B. Vekemans, K. Janssens, L. Vincze, F. Adams, P. Van Espen, *X-Ray Spectrom.* 23 (1994) 278-285

17. O. H. Jiménez Salas, R. Sánchez Hernández, J. Robles Camacho, *Arqueología*, 24 (2000) 129-143.
18. C. McEwan, C. Cartwright, R. Stacey, *Turquoise mosaics frome Mexico*, (The British Museum, 2006), 8-23.
19. A. Velazquez Castro, *Ancient Mesoamerica*, 22 (2011) 437–448.
20. N. Schulze, "For whom the bells tolls? Mexican cupper bells from the Templo Mayor offerings: Analysis of its production process and its cultural context", in *Materials Issues in Art and Archaeology VIII*, edited by P. Vandiver, B. Mc Carthy, R. Tykot, J.L. Ruvalcaba-Sil, F. Casadio, (Materials Research Society Proceedings Vol. 1047, Warrendale, 2008), 195-204.

Mater. Res. Soc. Symp. Proc. Vol. 1656 © 2014 Materials Research Society
DOI: 10.1557/opl.2014.709

Multiscale characterization of limestone used on monuments of cultural heritage

Mandana Saheb[1], Jean-Didier Mertz[2], Estel Colas[2], Olivier Rozenbaum[3], Anne Chabas[1], Anne Michelin[1], Aurélie Verney-Carron[1], Jean-Pierre Sizun[4]
[1] LISA, UMR CNRS 7583, Université Paris-Est Créteil and Université Paris-Diderot, 61 avenue du Général de Gaulle, 94010 Créteil Cedex France
[2] LRMH, USR 3224, 29 rue de Paris, 77420 Champs-sur-Marne, France
[3] ISTO, UMR 7327, 1A rue de la Férollerie, 45100 Orléans, France
[4] LCE, UMR 6249, 16 route de Gray, 25030 Besançon, France

ABSTRACT

In the context of the preservation of the cultural heritage, it is important to understand the alteration mechanisms of the materials constituting historical monuments and architecture. Limestone especially is widely used in many French monuments exposed to an urban aggressive atmosphere affecting their durability. To better understand the alteration mechanisms, the first step is to characterize at different scales the stone material properties. In one hand, the pore network that drives the fluids transfer inside the materials was characterized. And on the other hand, the alteration layer formed on several decades aged materials was studied. Results on this fine-scale characterization are discussed.

INTRODUCTION

The preservation of the built heritage constitutes an environmental, economic, and cultural challenge. In France 52% of the stone buildings are made out of limestone, so that their preservation is of primary and immediate importance. The climate plays a role on the natural ageing of the building constructions. Moreover atmospheric pollution directly affects the evolution of the materials due to the chemical reactions induced by dry and wet deposition and their alteration kinetics. For over 200 years, the increased production of energy has caused high atmospheric emissions in the form of gases (SO_2, CO_2, NO_x) and particulate matter. The gases cause an acidic deposit that could accelerate the materials dissolution and/or the formation of crusts. The particulate matter blackening the surface of the materials often increases the water sorption mechanism or catalyzes the sulphation reaction. The present work belongs to an ongoing study aiming at understanding the alteration processes on limestone exposed to an urban polluted area. The first step is the characterization of the relevant parameters of the stones that could influence the further alteration processes and/or evolve with the alteration. Thus, the pore network needs to be characterized because it allows the water transfer (that is the main alteration factor) inside the materials and because it can be modified by dissolution, pressure solution or crystallization processes. The alteration zones need to also be characterized by a multiscale approach (macro- to micrometric) as they play a significant role in the further alteration (changes of the kinetics) and in order to evaluate their evolution, physical pathways and chemical mechanisms as a function of time. To this purpose, non-destructive methods are favoured.

EXPERIMENT

Materials

The selected calcareous material is the so called "Saint-Maximin roche fine" limestone widely used for the restoration of historical monuments in France, especially in the Parisian basin. This limestone from the Lutetian period (45 My) is relatively homogeneous from a chemical point of view but heterogeneous regarding its physical properties that we demonstrate are linked to the fluid transfer parameters. Materials of different alteration types and stages were studied: pristine materials from quarries (Q) and samples of replacement blocks from various monuments in Paris and Paris suburbs. Three stones from the 'Basilique of Saint-Denis' (SD, 12th c.), and pre-Haussmannian monuments such as the 'Tribunal Administratif' (TA, 17th c.) and the 'Comédie Française' (C, 17th c.) have been selected. These stones are part of restoration stone blocks of the 2nd part of the 20th century, and no surface treatment has been applied since the replacement.

Samples preparation

A block of each sample of 10x10x10 cm^3 was extracted from the stone block on the buildings or from the quarry. Cores were drilled in the blocks and analyzes were performed directly on the bulk cores. Mercury intrusive porosimetry, X-ray diffraction and Raman microspectroscopy were performed on 13 mm diameter and 10 mm high rod samples. Macroscopic petrophysical measurement was performed on 30 mm diameter and 80 mm high rod samples. X-ray tomography was performed on the surface of 6 mm diameter and 10 mm high cores. The location of the sampling zones (from the same part of the block) and the size of the analyzed zones ensure the representativeness of the different measurements.

Pore network characterization

All samples were characterized at different scales (from micrometric to nanometric) using complementary analytical tools.

Classical petrophysical measurements allow determination of several parameters linked to storage and fluid transfer properties of the stones, such as the total porosity (Nt) and the water capillary coefficient (A).

Qualitative visualization of the porosity is investigated using an original staining method of pore network in thin section. The method is based on the imbibition of the stone by two different stained resins. The first one, a red stained resin plays the role of a wetting fluid and fills the 'free' or porosity accessible to wetting fluid by capillary imbibition [1,2]. The fraction of the porosity which is not accessible to water during rainwater absorption, represents the trapped or closed porosity; this pore volume is blue stained in thin sections.

Quantification of the porosity was performed by mercury intrusion porosimetry (MIP) using a Micromeritics® Autopore IV system in standard conditions [3]. Because mercury is a non-wetting liquid, progressive saturation of the pore network with the mercury is obtained by applying a controlled pressure on the mercury in the range of 0.035 to 206 MPa. MIP technique allows for the determination of the opened pore access corresponding to the pore size distribution of the limestone in the range from 0.005 μm to 180 μm in radius.

Non-destructive microtomography analysis was performed using an industrial CT device Nanotom 180NF (GE Phoenix X-Ray, Wunstorf, Germany). The 6 mm-diameter rod samples were used to enable collection of 3 µm voxel size images. Multiple scans allow imaging the rod on a large depth: from the top surface down to 16 mm under this surface.

Alteration characterization

To identify the structural composition of the alteration phases, macroscopic X-ray diffraction is performed using an Empyrean diffractometer from PANalytical Co equipped with a copper anode. Identification of the phases was performed based on the comparison of the diffractograms with JCPDS files (Joint Committee on Powder Diffraction Standards). The Highscore software allows the production of semi-quantitative data.

Micro-Raman measurements are performed on an Invia reflex spectrometer from the Renishaw Company equipped with a frequency-doubled Nd YAG emitting at 532 nm. A 50× optical microscope LEICA objective is employed to focus the laser beam and to collect the scattered light, providing a laser spot size on the samples less than 2 µm. A 2400 l/mm grating induces a spectral resolution of about 1 cm^{-1}. The laser power on the sample is about 100 µW to avoid any phase transformation due to the heat.

RESULTS AND DISCUSSION

Pore network characterization

Microscopic observations on thin sections (Figures 1 and 2) show that the limestone is mainly composed of miliole clasts and crinoid fragments, quartz grains and micritic shell clasts in different proportions. The main petrographic difference between the stones is a lower proportion of quartz grains in the quarry stone (Q) than in the SD, C and TA types. The white, gray and brown zones correspond to the grains of the stone (quartz grains and calcite) and the difference of the coloration lies in the grain size.

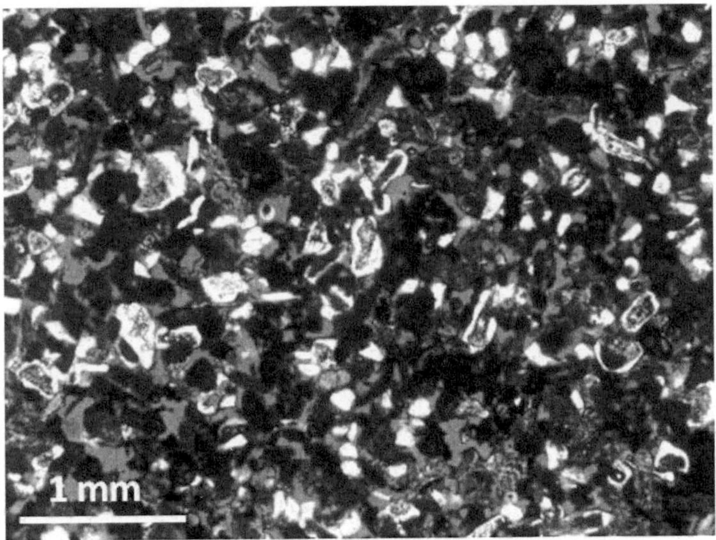

Figure 1. General free connected and trapped pores distribution.

Figure 2. (Weathered SD limestone) Exterior surface on the left. Sulphation zone mainly contains trapped porosity sometimes associated with emerging cracks.

The high porosity of the stones (27-40%) is induced by diagenetic conditions and most of the pore spaces are accessible to capillary water (Table 1). Pore network is composed by most of the macropores (mode 1) and all the micropores (mode 2) (< 7μm). The trapped porosity corresponds to spherical macropores located in the central part of the largest pore spaces of the network and the macropores sheltered inside the foraminifera tests (Figures 1 and 2). In addition, epigenetic phenomena resulting from weathering can locally create a new porosity.

In the altered samples from monuments, the exposed stone surface is modified. For TA and C limestone, alteration of the exposed surface leads to a moderate increasing of the original roughness due to dissolution. Considering the SD limestone type (Figure 2), the upper part-located at few millimeters from the surface is heterogeneous because it has been modified during the weathering. This alteration zone leads to a significant reduction of the initial porosity of the limestone. The residual pores are mainly trapped pores (Figure 2) and could reduce or inhibit the fluid transfer. Such structural modification and pore blocking resulting in a gypsum enriched layer at around 2 mm under the outer surface is related to the sulphation mechanism.

Mercury intrusion porosimetry reveals a bimodal pore size distribution (Table I). The main mode is in the range 10-15 μm in radius and a secondary mode is present in the range 0.1-8 μm. The first one is clearly in close relation with the size of the macropore network. The difference between the stones concerns the amount of microporosity and the position of the secondary mode. It probably means that the modification of the structural parameters of the samples occur at this micrometric to sub-micrometric scale.

The well-connected pore network results of the main mode and explains the high capillarity parameters of the stones. Nevertheless, local porosity and additional transfer parameters as water capillary kinetic coefficient confirm the effect of the sulphation layer on the fluid transfer properties, especially for the SD samples (Table I).

Table I. Petrophysical parameters of the materials (*: measurements on 3 samples; Nt: total porosity (EN1936); A: water capillary coefficient (EN1925); PHg: porosity open to mercury; D: bulk density; SSA: surface specific area)

Limestone type	Nt* (%)	A* ($g.cm^2.h^{-0.5}$)	PHg (%)	D ($g.cm^{-3}$)	SSA ($m^2.g^{-1}$)	Mode 1 r_1 (μm)	Mode 2 r_2 (μm)
Quarry Q	38.1±0.4	3.7±0.1	36.2	2.174	0.59	20	0.15
C	41.7±2.1	1.9±0.7	40.4	1.903	0.24	10	3
TA	43.0±0.2	4.7±0.5	39.8	2.019	0.22	15	0.1
SD	30.7±1.0	0.5±0.1 to 2.4±0.4	27.6 to 33.9	2.120 to 2.140	0.31 to 0.49	10	3.5

As expected, X-ray tomography shows that the porosity of the pristine stone (not shown here) is homogeneous and the average porosity is 24 % for this resolution (i.e. 3 μm). On the altered samples, decrease of porosity is observed due to re-crystallization processes (Figure 3). The porosity of the C sample gradually decreases from 20 % on the surface to 15 % between 6 and 9 mm under the surface. After this range, the porosity slightly increases to a stable value around 21 % at 11 mm up to 16 mm. The porosity of the SD sample rapidly decreases from 21 % on the surface to 4 % between 1.8 and 3.5 mm under the surface. After this range, the porosity increases and reaches a stable value (around 21 %) at 11 mm up to 16 mm. This observation on the SD stone is in good agreement with the ones based on the other (destructive) characterization methods. However, for the TA sample, porosity increases in the 1.5-3 mm distance under the surface and decreases in the 0-1.5 mm range. Images (not shown here) show that re-crystallization phases close the porosity and replace some missing grains.

To sum up, the main differences between the quarry limestone and the limestone from monuments are linked to the heterogeneities in the pore network generated by the alteration (i.e. dissolution and precipitation processes). This morphological evolution is important as it will influence the fluid transfer in the materials and thus the further alteration.

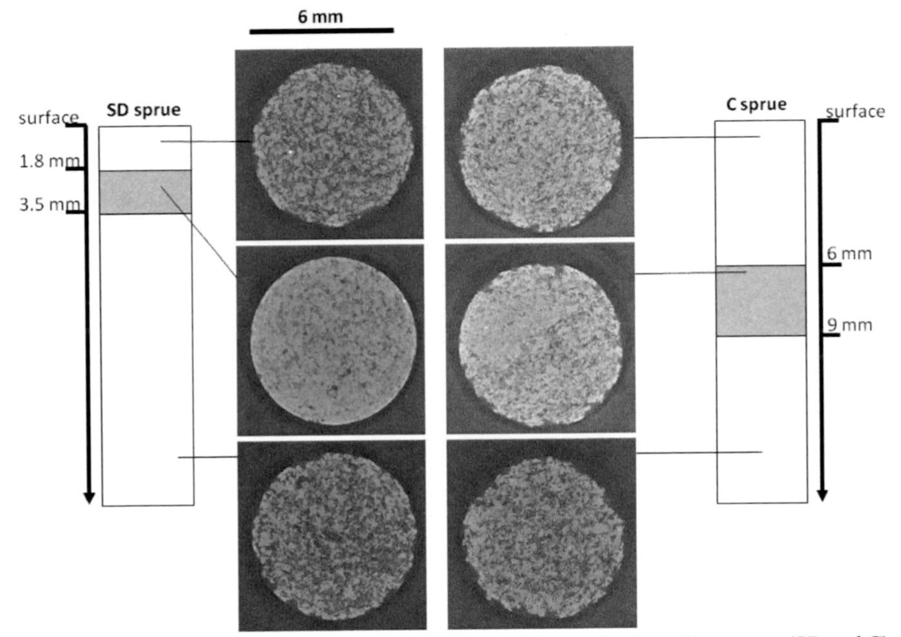

Figure 3. Images obtained using X-ray tomography at different depths on limestones (SD and C)

Alteration layer characterization

X-ray diffraction has been performed on a black crust of a sample located on part of monuments that is sheltered from rainfall. The comparison with JCPDS files shows that it is mainly composed of carbon (graphite-60%), gypsum ($CaSO_4.2H_2O$-25%) and calcite ($CaCO_3$-15%) (Figure 4).

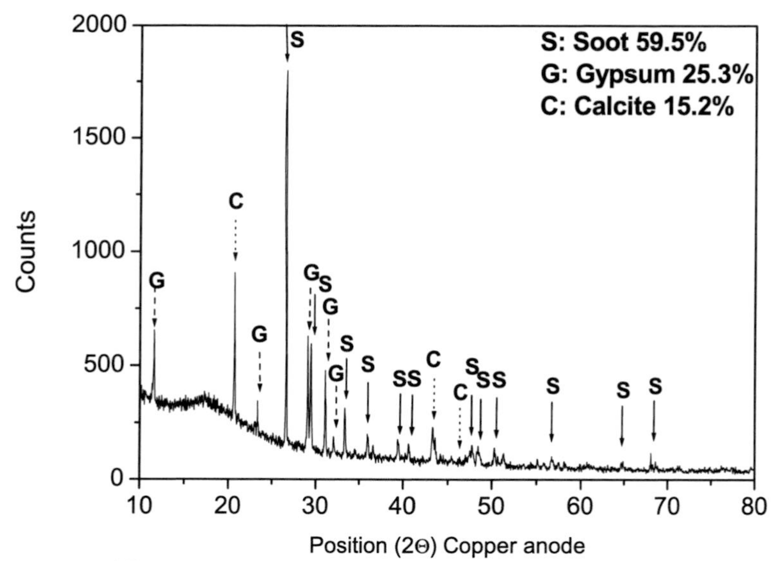

Figure 4. Diffractogram of a black crust formed on the limestone (SD) (copper anode)

This composition is in good agreement with literature data on such stone samples exposed to a polluted atmosphere [4]. The presence of graphite corresponds to the crystalline core of the soot emitted from diesel exhaust and absorbed on limestone. The presence of gypsum has also been evidenced using optical observations on thin sections. Its formation is due to the interaction between the atmospheric SO_2 and the calcite. SO_2 is oxidized to SO_3 in condensed water at the surface of the calcite which is dissolved by the product H_2SO_4 and that leads to the gypsum formation. The presence of calcite inside the crust could come from the substrate itself (during sampling) or correspond to the dissolution / re-crystallization processes happening during the water cycles. Actually, water transiting inside the pore network can dissolve part of the calcite of the materials and then re-precipitate somewhere else during the drying. That leads to modification of the porous network and consequently to the modification of the further alteration.

On some localized zones, Raman microspectrometry has identified scytonemin (Figure 5). The high luminescence of the Raman spectrum is linked to the organic carbon. This aromatic organic molecule has been detected in other studies on limestone and is linked to a biological activity exposed to UV radiation or a polluted environment [5]. The urban area of the Paris suburbs is a suitable environment for the activity of such bio-organisms.

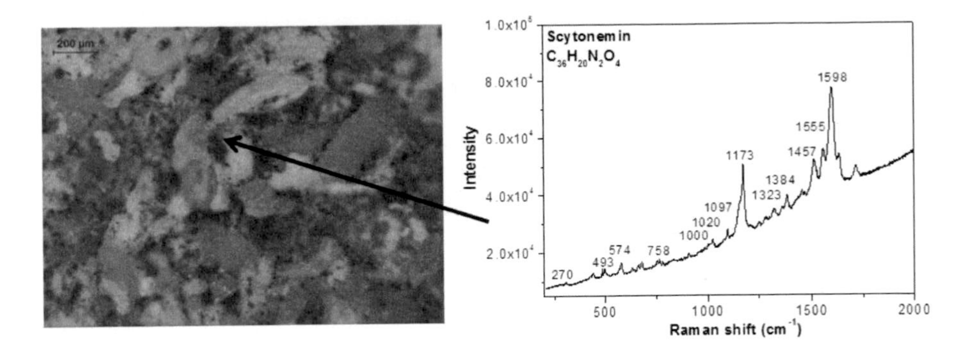

Figure 5. Optical micrograph on the limestone from Saint-Denis and associated Raman spectrum (λ=532 nm)

CONCLUSIONS

This study is part of an ongoing work aiming at understanding the alteration processes occurring on limestone exposed to an urban area. The fine characterization of the pore network shows a macroporous limestone where the water can transit. Moreover open micropores could also play a role in the alteration processes. The alteration layer is mainly composed of gypsum and soot which is consistent with a polluted atmosphere. Markers from a microbiological activity have also been detected.

In parallel, another part of the project aims at reproducing in laboratory the alteration of limestone exposed to an urban area in controlled and realistic environments. Associated with the characterization of the stone, it will allow determining the modifications of its properties. This will help studying specific questions on the alteration processes (identification of the reaction zones), the kinetics of the alteration (quantification of the modified zones) and the role of the alteration layer (evaluation of the kinetics).

ACKNOWLEDGMENTS

The authors would like to thank the French Ministry for Culture and Communication for the financial support. Special thanks to Mr Chauvelier from Lanfry Company and Mr Debray from Dubocq Company for their decisive help during the sampling on the monuments. For the characterization measurements, the authors thank Sophie Nowak from the ITODYS for the XRD and the SIS2M/LAPA for the Raman analyses.

REFERENCES

1. B. Zinszner and C. Meynot, Rev. Inst. Fr. Pétrole, 37, 337 (1982).
2. F.M. Pellerin and B. Zinsner, A Geoscientist's Guide to Petrophysics, Technip Eds, 384 p. (2007).

3. ASTM D4404-10, 2010 - Standard Test Method for Determination of Pore Volume and Pore Volume Distribution of Soil and Rock by Mercury Intrusion Porosimetry, 7p. (2010).
4. C. Sabbioni, Air Pol. Rev., 2, 63-106 (2003).
5. N. Prieto-Taboada, I. Ibarrondo, O. Gómez-Laserna, I. Martinez-Arkarazo, M.A. Olazabal and J.M. Madariaga, J. Hazard. Mat., 248–249, 451-460 (2013).

Mater. Res. Soc. Symp. Proc. Vol. 1656 © 2014 Materials Research Society
DOI: 10.1557/opl.2014.706

Characterization of a Surface Tarnish Found on Daguerreotypes Revealed under Shortwave Ultraviolet Radiation

Aaron Shugar[1], Krista Lough[1], Jiuan Jiuan Chen[1].
[1]Art Conservation Department, SUNY Buffalo State, Buffalo, New York, USA.

ABSTRACT

A characteristic fluorescent tarnish can be observed on some daguerreotypes under shortwave ultraviolet radiation. The fluorescence can be seen in several distinct patterns: edge tarnish, rings, and continuous films. Dispersive Raman spectroscopy, scanning electron microscopy (SEM), and X-ray diffraction (XRD) were applied to characterize and identify the fluorescent compound. Raman spectroscopy identified the characteristic peak for copper cyanide, CuCN, at 2172 cm^{-1}. Elemental k-ratio maps of the SEM analysis indicated an increase in copper, sodium, carbon and nitrogen in the area of fluorescence. XRD confirmed the identification of a copper cyanide compound. Shortwave ultraviolet radiation can be used in a monitoring program of daguerreotypes to further characterize the fluorescent tarnish and its effect on the deterioration of daguerreotypes.

INTRODUCTION

Ultraviolet radiation is a common technique employed in conservation to examine a variety of art and artifacts including paintings, objects, paper, and photographs [1]. Shortwave ultraviolet radiation (UVC, 200 – 280 nm with the peak at 254 nm) was first used to examine daguerreotypes in 1992 [2]. It was observed that some daguerreotypes produced a bright green fluorescence examined under UVC. Early research on this phenomenon documented the occurrence of this fluorescence in daguerreotype collections, and made advancements in its initial identification. The extent at which this phenomenon was identified was extensive with 50% of the collection investigated showing fluorescence [2, 3]. Although initial attempts to characterize the cause of the fluorescence were made, they were by no means conclusive.

Non-destructive and non-invasive analysis is essential when working with daguerreotypes. This is due to the sensitivity of the image surface. Even though direct sampling can potentially provide additional information, it is not recommended, nor would it be reproducible on daguerreotype plates not deemed sacrificial. Therefore, this research focuses on in-situ analysis to avoid removing the tarnish, limiting the damage to the plate itself, and suggests a non-invasive method of identification for other researchers.

This research focuses on the characterization of the fluorescence through different analytical techniques, including dispersive Raman spectroscopy, scanning electron microscopy (SEM) using elemental k-ratio maps, and X-ray diffraction (XRD). Several mockups were produced and the daguerreotype selected for analysis exhibited strong fluorescence, possibly indicating a greater amount of the fluorescent compound.

Background Research

The use of UVC for the investigation of daguerreotypes was first discussed by Daffner, Kushel, and Messinger [2]. Their research includes a description of a tarnish that occurs on some daguerreotypes, its analysis and preliminary identification. They investigated the occurrence of the fluorescent tarnish on a group of 110 daguerreotypes where it was observed on 50% of the plates. The bright fluorescence observed with UVC occurred in primarily three forms, edge tarnish, rings, and continuous films. The areas that fluoresce do not always show a visual difference under normal illumination, but when they do, they correspond with a brownish color.

The methods of analysis used in their research included scanning electron microscopy (SEM), Fourier transform infrared analysis (FTIR), and solvent tests. The solvent tests showed that the fluorescence was eliminated by a xylene, and N,N-dimethyl formamide treatment. The fluorescence was not affected by water, ethanol, hydrochloric acid, and ethylemedianimetetraacetic acid (EDTA). Their SEM and FTIR analysis detected copper and nitrogen suggesting that the fluorescent material may be a copper cyanide compound but they were not conclusive in this determination.

Tragni [3] conducted an extensive research project on the use of ultraviolet-induced visible fluorescence for the examination of photographs and a component of this study involved the continuation of the earlier research by Daffner, Kushel, and Messinger [2]. Tragni examined an additional 180 daguerreotypes using UVC. It was noted that the fluorescent tarnish was observed on a large number of the daguerreotypes; however the percentage was not listed. New observations made during this research included the finding of fluorescent tarnish on the back of many daguerreotypes. The fluorescence on the back was observed on electroplated and roll-clad plate (Sheffield plate). ATR-FTIR analysis of the back of a roll-clad daguerreotype, where a strong fluorescence existed, indicated the presence of cyanide. Cyanide-based cleaning compounds were once used on daguerreotypes, and may be the source of the current identified cyanide. Electroplated procedures often included the use of cyanide but the roll-cladding technique does not involve a cyanide compound [4] suggesting that the cyanide was deposited by other means.

Modern daguerreotypes produced by Grant Romer were also observed and fluorescence was visible on the back of all ten of the modern plates. This indicated that in some situations, the fluorescent compound formed soon after the creation of the daguerreotype. Fluorescence was also observed on two of the brass mats. Two daguerreotypes that exhibited the fluorescence were treated with ammonium hydroxide. After the treatment the fluorescence was no longer visible under UVC. The cleaning solution was examined under UVC and the fluorescence was not detected. Fluorometric readings were performed on the cleaning solution of one of the plates and an absorption peak in the shortwave ultraviolet spectrum was observed. This suggests that the fluorescent compound was removed from the plate in the cleaning solution.

The decision was made to continue this research with the prospect of conclusively identifying the fluorescent tarnish. After the tarnish is identified, its source and its long-term effect on daguerreotypes can hopefully be determined. Potentially, a treatment protocol can be established based on this research as well.

Daguerreotypes

Daguerreotypes are a one-of-a-kind image on a hightly polished, silver-plated sheet of copper first presented at the meeting of the French Academy of Sciences in Paris in 1839 [4]. The polished plate is exposed to iodine vapors to create a light-sensitive surface on the silver. After the plate is exposed it is developed over hot mercury. The image is fixed with sodium thiosulfate and then gilded with gold chloride. The polished silver surface of the daguerreotype is subject to tarnish. Daguerreotypes were usually placed in a protective packaging consisting of the daguerreotype, paper or metal mat, and glass, all bound together with paper tape. The sealed package was then placed in a small case. The history, conservation and preservation of daguerreotypes is extensively covered in other liturature [4-7].

The most common form of deterioration found on daguerreotypes is a smooth tarnish film exhibiting interference colors. This deterioration is formed with the plates exposure to air and thus occurs: at the edges of the plates, at the interior edge of the mat window, below cracks in broken cover glass, adjacent loose edges of the sealing tape, and along air gaps between the plate and the mat. One analytical study of the colored tarnish films suggests they consist of silver sulfide [8].

The daguerreotype surface is inherently fragile, so there is a great chance that physical damage, such as fingerprints, smudges, scratches, and gouges, may occur on the plate. Other common forms of deterioration are copper corrosion products erupting from the plate and localized debris: from the glass, other case components, and residues from polishing compounds [4].

Ultraviolet radiation

Ultraviolet radiation induced visible fluorescence refers to the emission of visible light from a substance being exposed to ultraviolet radiation and is a common non-destructive technique used in art conservation for the examination of objects. Ultraviolet radiation is not visible to the human eye. It falls at 10 to 400 nanometers, between visible light and X-rays, on the electromagnetic spectrum. Ultraviolet radiation can be broken down into four regions: UVA, also known as longwave ultraviolet radiation, near-ultraviolet, black light, or Wood's light (between 320 and 400 nm); UVB, also known as middlewave ultraviolet radiation (between 280 and 320 nm); UVC, also known as shortwave ultraviolet radiation, far-ultraviolet, or germicidal UV (between 200 and 280 nm), and; Vacuum UV (between 10 and 200 nm).

UVA and UVC are the most useful for the examination of museum objects. In the fluorescence process a material temporarily absorbs energy, which is then reemitted as lower-energy radiation in the visible light region [1, 9]. The color and intensity of the fluorescence is a response to several factors including: the nature and amount of the material, the extent of degradation, and the excitation source (wave-length intensity). Therefore, a material may show no fluorescence under one region of the ultraviolet spectrum and show a strong fluorescence under a different region. UVC has been shown to be a useful tool in the examination of daguerreotypes. The emission peak of the ionized mercury for UVC is at 254 nm. Overtime, long exposure to ultraviolet radiation can potentially change the chemical and physical properties of an object [10]. The amount of exposure of ultraviolet radiation should be limited to the object. It is important to insure the object being examined with UVC is not sensitive to ozone as ozone can be generated when the UVC lamps are turned on.

METHODOLOGY

Daguerreotypes are inherently fragile and conducting in-situ analysis on their surface should be completed only after safe methods are determined. Therefore, a series of mock-ups were created in order to gain a greater understanding of the fluorescent tarnish and to develop secure procedures for any *in-situ* analysis of the daguerreotypes.

Based on research by Tragni [3] two compounds were suggested as being the cause of the fluorescence, sodium cyanide (NaCN) and potassium cyanide (KCN), with sodium cyanide providing a higher fluorescent yield. Mock-ups included both silver electroplated samples and pure copper samples that were treated with both sodium cyanide and potassium cyanide for comparative purposes. For both compounds, plates were dipped in a 0.2 molar solution in distilled water for 30 seconds, rinsed with distilled water and left to air dry.

Documentation and Shortwave Ultraviolet Radiation

Normal and specular illuminated photographed were taken with a UV-VIS-IR modified Nikon D700 camera with a Nikon 105 mm lens using a PECA 918 filter. The light source was one GTI graphic light box, with 2 fluorescent bulbs (F20T12, 20W), D5000. For normal illumination white foam core was used to fill the light onto the far side of the plate. For specular illumination a sheet of glass at a 45-degree angle was used to create a specular light and block the cameras reflection from appearing on the reflective surface. The settings on the camera were as follows: ISO 200, with white-balanced taken with the gray card, Aperture Priority, with an aperture set to F8 or F11. Daguerreotypes photographed with specular illumination results in a negative image that highlights the area of tarnish [11].

The UVC-induced visible fluorescence photographs were taken with a UV-VIS-IR Modified Nikon D700 camera with a Nikon 105 lens with PECA 918 and a Kodak 2E filters. Visible fluorescence was induced with one UV Systems SuperBright II UVC lamp peak/254 nm. The settings on the camera for the UVC-induced visible fluorescence images were as follows: ISO 200, the white balance set to "Shade," Aperture Priority, with an aperture set to F8 or F11. The temperature and tint settings were adjusted individually to match the appearance of the visible fluorescence observed in each daguerreotype.

Dispersive Raman Spectroscopy

Dispersive Raman spectroscopy was used to examine: green fluorescent areas and blue fluorescent areas on two different test plates, powdered samples of potassium cyanide and sodium cyanide, and the green fluorescent areas on the front and back of two daguerreotypes. The spectra were collected with a Bruker Senterra Raman Spectrometer. Spectral resolution was 3-5 cm^{-1} across the spectral range analyzed. Analysis was performed using Opus 6.5 software.

The spectra of the green fluorescent areas on the test plate and the daguerreotypes were collected using a 633 nm excitation laser operating at a power of 0.83 mW at the sample. A 100x ultra-long working distance objective was used to focus the excitation beam to an analysis spot of approximately 1 μm directly on the surface of the objects under study.

The spectra of the blue fluorescent area on the test plate and of sodium cyanide were collected using a 532 nm excitation laser operating at a power of 1.2 mW at the sample. A 50x ultra-long working distance objective was used to focus the excitation beam to an analysis spot of approximately 2 μm directly on the surface of the test plate and on a powdered sample of sodium cyanide placed on a microscope slide.

The spectrum of potassium cyanide was collected using a 785 nm excitation laser operating at a power of 4.4 mW at the sample. A 50x ultra-long working distance objective was used to focus the excitation beam to an analysis spot of approximately 2 μm directly on a powdered sample placed on a microscope slide. The resulting Raman spectra for all samples are the average of 5 scans at 2 second integrations each.

Scanning electron microscopy

Scanning electron microscopy, energy-dispersive spectroscopy (SEM-EDS) was used to examine a green fluorescent area on one plate. The mapping was performed with a TESCAN tFEG with four 30 mm^2 SDD-EDS detectors. Elemental k-ratio maps were collected at beam energy of 5 kV to maximize sensitivity to nitrogen and other light elements. The background and peak-interference corrected k-ratio maps are presented in the logarithmic three-ban color encoding which enables sensible inter-comparison of the images. The data cubes were compiled with the image analysis program Lispix [12].

X-ray diffraction

X-ray diffraction (XRD) was collected using a Rigaku Ultima X-ray Diffractometer to analyze the green fluorescence on a test plate and on a daguerreotype. The plates were placed directly in the diffractometer, so no sample was taken. The measurement conditions for the test plate were 40 kV at 44 mA with a .5 deg/min scan speed, a sampling step at 0.03 deg, and a measurement axis of 2*f*Æ/*f*Æ. The scan range was ~ 20-40 deg.

DISCUSSION

Tarnish Visible in Normal Illumination and Short-Wave Ultraviolet Radiation

The fluorescent tarnish described in the early literature [2] was clearly observed and documented. The tarnish was observed to occur primarily in three forms; edge tarnish, rings, and continuous films. The research performed by Tragni [3] documented the fluorescent tarnish in the three forms described previously, the back of electroplated and roll-clad plates, and on the brass mat. The daguerreotypes observed and documented in this experiment exhibited all of the forms of fluorescence described in early research. In addition to these forms of fluorescence, a fluorescent compound was also observed on the bevel of the brass mat.

Several of the plates exhibited a combination of the forms of tarnish. A plate with a strong concentration of fluorescence along the edges could also have several fluorescent rings in the center. A plate with several rings could also display areas of fluorescence in continuous film. On a couple of the plates a clear connection between the fluorescence and brown spots or stains was observed (Figure 1). The connection between the brown stains and the fluorescence are more easily identifiable on plates with less visible tarnish. The fluorescence was also observed on the back of electroplated and roll-clad daguerreotypes. The back of several plates had a stronger concentration of tarnish around the taped edges. Fluorescence was observed on the back of 5 brass mats. The tarnish on the back of the mats was primarily a continuous film. The fluorescence was also seen on the bevel of three of the brass mats.

Figure 1 Front of 19th-century daguerreotype under normal illumination (left), specular illumination (center), and UVC-induced visible fluorescence (right) showing the characteristic brown staining and green fluorescence.

An additional observation was made on a plate that had been electro cleaned in 2009. Even after cleaning, a few areas of the fluorescent tarnish was still observed. One of the daguerreotypes observed in this research has a significant amount of smooth tarnish film of silver sulfide. The plate, when viewed under UVC, exhibited a few small rings of fluorescent tarnish seen in some but not all of the holes in the tarnish film.

Dispersive Raman Spectroscopy

The Raman spectrum for the green fluorescence on one of the test plates is presented in Figure 2a. The peak at 2172.40 cm^{-1} is close to the literature value for copper cyanide, which for solid CuCN occurs at a wavelength of 2174 cm^{-1} [13]. The significant intensity of the peak to background could be a result of the metallic surface of the daguerreotype creating a surface-enhanced Raman scattering effect [14]. The significant increase of peak to background is enhanced on the daguerreotypes due to the mirrorlike surface of the plate. The fine silver surface is inherently a SERS substrate. In another study of the use of Raman on daguerreotypes the natural SERS surface allowed the detection of some deterioration products that were well below the detection limit of SEM-EDS [14]. There are some potential problems that should be considered which may limit the SERS effect on daguerreotypes. This includes the level of deterioration or tarnish on the surface, which can reduce the enhancement provided by the silver substrate.

Copper cyanide was found on the front and back of both of plates tested. The front of the plate showed a peak from CuCN at 2174.50 cm^{-1} (Figure 2b). The back of the plate had a peak for CuCN at 2175.00 cm^{-1} (Figure 2c). The Raman spectra of the fluorescence on the front and back of a significantly tarnished daguerreotype were also collected. The front of the plate showed a peak for CuCN at 2175.203 cm^{-1} (Figure 2d). The back of the plate showed a peak for CuCN at 2172.00 cm^{-1} (Figure 2e).

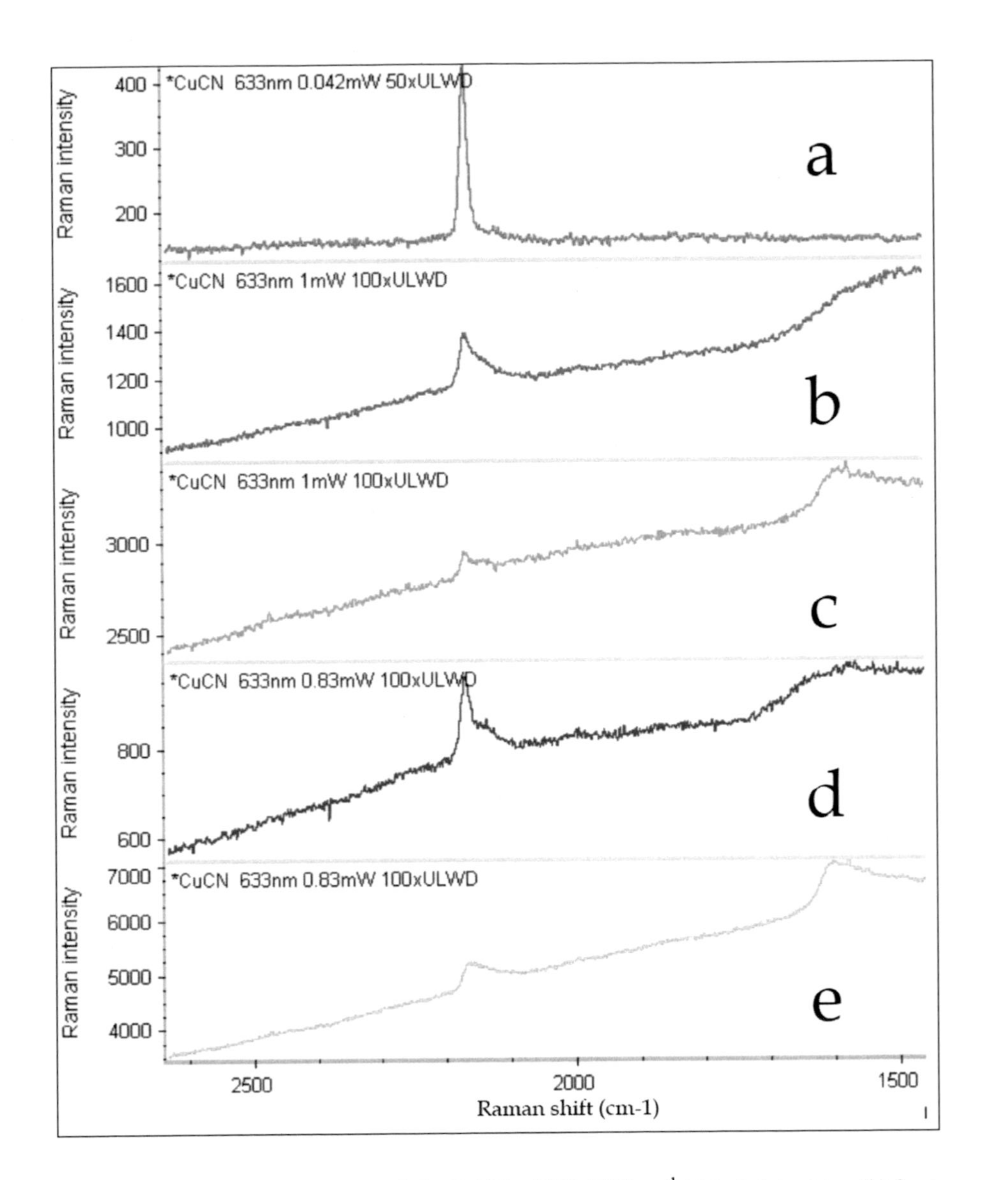

Figure 2 Raman spectra showing peak for CuCN at 2172-2174 cm^{-1} (a) test plate at top, (b) front of first plate, (c) back of first plate, (d) front of second plate, and (e) back of second plate at bottom.

Raman peak intensity has been reported to directly correspond to the amount of CuCN present in samples [15]. A linear response was found in the concentration range of CuCN from 250 ppb to 10 ppm with a detection limit around 100 ppb. The peak intensity for each of the CuCN peaks was determined using the peak height tool in the Omnic program (Table 1). The peak intensity for CuCN was significantly reduced on the daguerreotypes when compared to the test plate, which could be a result of the deterioration on the daguerreotypes limiting the SERS effect. The intensity of the CuCN was reduced on the back of the daguerreotypes compared to the front. It is difficult to make a direct comparison between the peak height determined in this experiment and the peak height presented in another study [15] due to the different substrate, but it appears that the peak height determined corresponds to a concentration of between 250 to 500 ppb.

Location of Raman spectra	Raman shift (cm^{-1})	Peak height determined using Omnic peak height tool
(a) Test Plate	2172.40	250.504
(b) Front of Daguerreotype	2174.50	134.847
(c) Back of Daguerreotype	2175.00	41.481
(d) Front of second daguerreotype	2175.203	139.342
(e) Back of second daguerreotype	2172.00	37.165

Table 1 Raman shifts and peak heights for the spectra in Fig. 2 a-e.

Scanning electron microscopy

Scanning electron microscopy-energy dispersive spectrometry (SEM-EDS) was used to analyze several fluorescent spherolites that appeared as brown stains under normal illumination (Figure 3). Elemental analysis is encoded in a logarithmic multiband color scale with three distinct color bands: trace (blue), minor (green), and major (red) (Figure 4). The color bands run from pure color at the low end of the scale to pastel color to the high end of the scale [16]. A k-ratio map shows an increase in copper, carbon, and nitrogen, in the location of the fluorescence, which are the elemental components of copper cyanide. The k-ratio maps also showed a decrease of silver and gold indicating degradation of the image material. In the same region, the map also shows an increase of oxygen and sodium concentrations near the ridge. The sodium increase is seen as a ring around the tarnish and is possibly related to a previous sodium cyanide treatment.

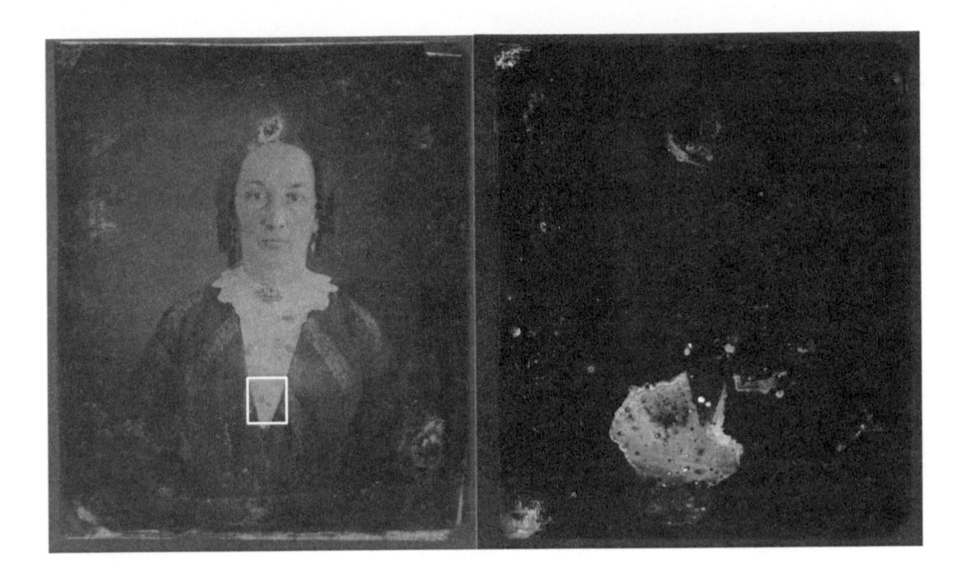

Figure 3 Location of the SEM analysis indicated in the white square on the 19th-century daguerreotype under normal illumination (left) and shortwave ultraviolet radiation (right).

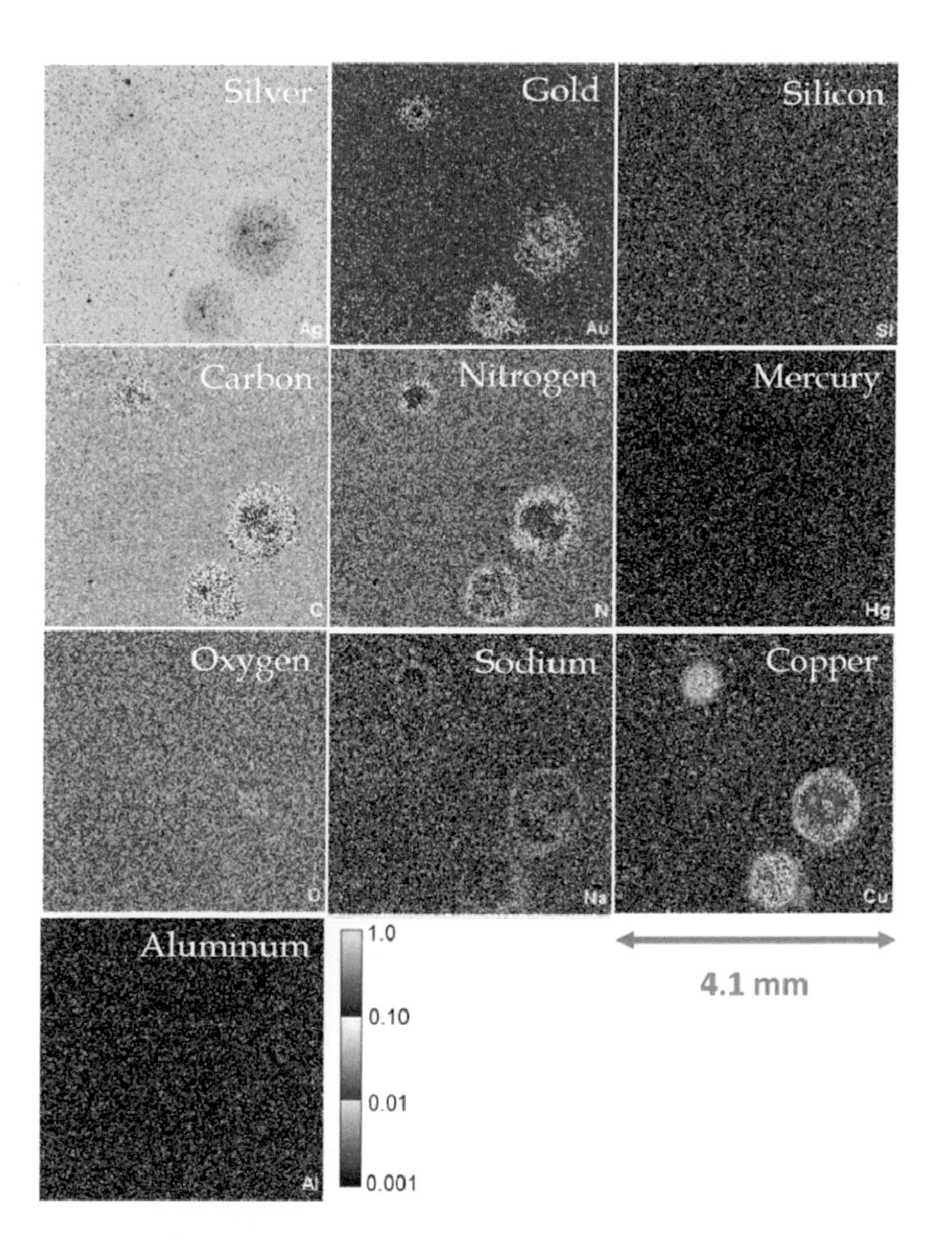

Silver
Ag
Gold
Au
Silicon
Si
Carbon
C
Nitrogen
N
Mercury
Hg
Oxygen
O
Sodium
Na
Copper
Cu
Aluminum
Al
1.0
0.10
0.01
0.001
4.1 mm

Figure 4 Elemental k-ratio maps of 19th-century daguerreotype, Figure 3. Collected at beam energy of 5 kV. The background and peak-interference corrected k-ratio maps are presented in a logarithmic three-band color encoding scale.

X-ray Diffraction

The X-ray diffraction analysis for the test plate is presented in Figure 5. The spectrum indicated one peak at 2-theta (deg) 29.63. This peak corresponded to the most intense peak of copper (I) cyanide, seen in the ICSC pattern. The most intense peak for copper (I) cyanide has a 2-theta at 29.90. Copper (I) cyanide has a trigonal crystal system with cell parameters of: Alpha=90.000, Beta=90.000, and Gamma 120.000 [17].

The XRD analysis of the daguerreotype was not successful. The amount of fluorescence on the plate is probably in the parts per billion range based on the intensity of the Raman peaks [15] and is below the detection limit of the powder XRD. It might be possible to extract a sample for micro-XRD to confirm the presence of CuCN as was found on the test plate.

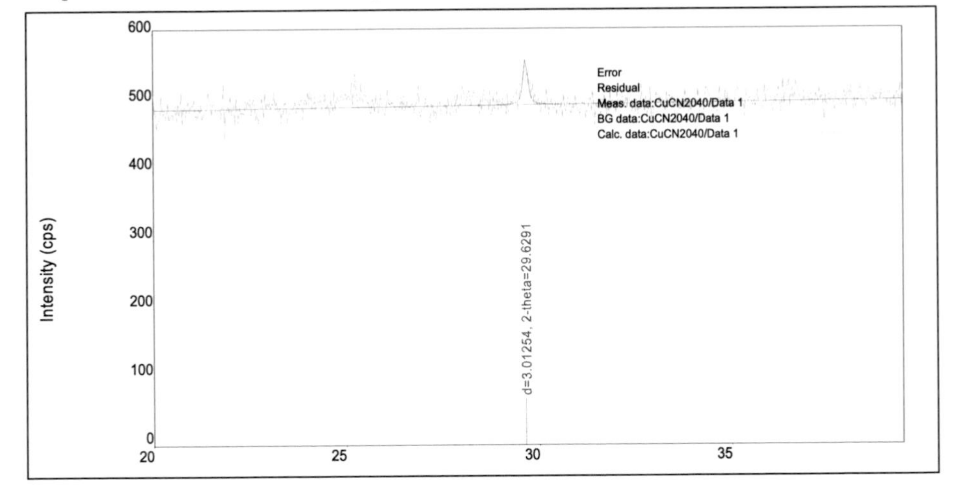

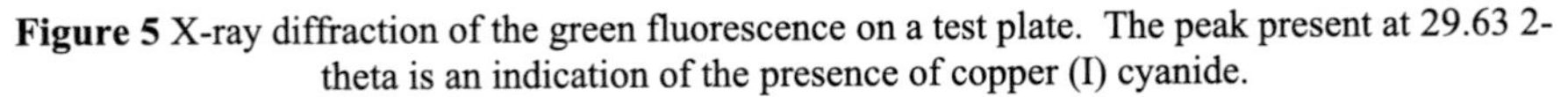

Figure 5 X-ray diffraction of the green fluorescence on a test plate. The peak present at 29.63 2-theta is an indication of the presence of copper (I) cyanide.

CONCLUSIONS

Shortwave ultraviolet radiation (UVC) is a useful tool for the examination of daguerreotypes and can yield new information about their tarnish patterns. Under UVC a characteristic bright fluorescence can be observed that appears to indicate the presence of copper cyanide. Early research into the phenomenon by Daffner, Kushel, and Messinger [3] indicated a high concentration of approximately 50%, of the fluorescent tarnish when observing a large daguerreotype collection.

Dispersive Raman spectroscopy identified copper cyanide in the areas of fluorescence on the front and back of two daguerreotypes. The characteristic peak at 2172 cm^{-1} for copper cyanide was observed even when only a small amount of the compound was present, in the order of parts per billion. The presence of copper cyanide in the fluorescence was confirmed with scanning electron microscopy. Elemental k-ratio maps were collected in an area of fluorescence and a logarithmic multiband color scale was used to analyze the data. The analysis indicated an area high in copper, carbon, and nitrogen that corresponds to fluorescent areas and their presence helps to confirm the Raman data. Powder X-ray diffraction showed the presence of copper cyanide on the test plate but was not sensitive enough to detect its presence on the daguerreotype. A methodology using a micro-beam XRD may be more successful and will be attempted in the future.

Certain questions about the fluorescent tarnish need to be addressed, such as, whether the fluorescent tarnish will increase in time or if the regions stabilize and become unreactive over time. A program to monitor daguerreotypes using UVC at intervals may solve this question. A long-term plan including UVC-induced visible fluorescent documentation of daguerreotypes can help to create a greater understanding of the nature of the fluorescent tarnish. Through the examination of large collections of daguerreotypes with UVC it may be possible to observe trends in the fluorescence to help determine if the copper cyanide is formed from a particular treatment and or time period. At this moment it would be difficult to determine exactly what cyanide procedure used in the daguerreotype process aided in the formation of copper cyanide.

UVC is a powerful tool and its proper use is crucial in ensuring the safety of the object and the examiner. UVC cannot penetrate glass therefore daguerreotypes observed under UVC will have to be removed from their packages. Bare plate daguerreotypes are extremely fragile and require extra care when handled. Gloves should always be worn when handling bare plate daguerreotypes not only for the objects safety but for the examiner as well. There may be the presence of highly toxic compounds, such as copper cyanide or silver cyanide, on the daguerreotype surface and it is important to protect oneself from these toxins. UVC will cause significant damage to eyes and can also be damaging to skin. It is important to wear protective UV-filtering goggles and cover exposed skin when working with this ultraviolet radiation.

After the initial discovery of the fluorescent tarnish in 1996 there has been little advancement into understanding and characterizing the fluorescent compound. From this research a few new developments were made to aid in the understanding of the nature of the compound. It is hopeful that research into this phenomenon can continue and provide a greater understanding of the first commercially successful photography process.

ACKNOWLEDGMENTS

The SEM analysis was completed by Dale Newbury and Nicholas Ritchie at the National Institute of Standards and Technology. The X-ray diffraction analysis was carried out by Biswajit Sarkar at the South Campus Instrument Center, University at Buffalo, State University of New York. The authors would like to thank Daguerreotypist Rob McElroy for his generous donation of daguerreotype plates to use in the construction of mock-ups. The author would also like to thank Patrick Ravines, director, SUNY Buffalo State Art Conservation Department, for his donation of a copper plate to use in the construction of mock-ups.

REFERENCES

[1] M. S. Grant, "The Use of Ultraviolet Induced Visible-Fluorescence in the Examination of Museum Objects, Part I," *US National Park Service - Conserve o gram,* vol. 1, 2000.

[2] L. A. Daffner, D. Kushel, and J. M. Messinger II, "Investigation of a Surface Tarnish Found on 19th-Century Daguerreotypes," *Journal of the American Institute for Conservation,* vol. 35, 1996, pp. 9-21.

[3] C. B. Tragni, "The Use of Ultraviolet-Induced Visible Fluorescence for Examination of Photographs," Rochester Institute of Technology, Advanced Residency Program in Photograph Conservation, George Eastman House International Museum of Photography and Film and Image Permanence Institute 2005, pp. 43-56.

[4] M. S. Barger and W. B. White, *The Daguerreotype Nineteenth-Century Technology and Modern Science*. Washington, D.C.: Smithsonian Institution Press, 1991.

[5] J. Buerger and A. Swan, *French Daguerreotypes*. Chicago: The University of Chicago Press, 1989.

[6] B. Newhall, *The Daguerreotype in America*. New York: Dover Publications, 1976.

[7] A. Swan, C. E. Fiori, and K. F. J. Heinrich, "Daguerreotypes: A Study of the Plates and the Process," *Scanning Electron Microscopy,* vol. I, pp. 411-23, 1979.

[8] A. Swan, "The Preservation of Daguerreotypes," in *AIC Preprints*, ed Washington D.C.: The American Institute for Conservation of Historica and Artistic Works, 1981, pp. 164-172.

[9] J. J. Rorimer, *Ultra-Violet Rays and Their Use in the Examination of Works of Art*. New York: Metropolitan Museum of Art, 1931.

[10] G. Tompson, *The Museum Environment*, 2nd ed. Oxford: Butterworth-Heinemann, 1986.

[11] J. J. Chen, "A Photodocumentation Method: Taking the Negative Image of a Daguerreotype," 2003. unpublished Report

[12] D. Bright. (2012, 11/13). *Lispix: An Image Processing and Analysis Tool for the PC*. Available: http://www.nist.gov/lispix/doc/contents.htm

[13] G. C. Lukey, J. S. J. v. Deventer, S. T. Huntington, R. L. Chowdhury, and D. C. Shallcross, "Raman Study on the Speciation of Copper Cyanide Complexes in Highly Saline Solutions," *Hydrometallurgy* vol. 53, 1999, pp. 233-244.

[14] C. S.A., T. Meller, N. Kennedy, and M. Wypyski, "The Daguerreotype Surface as a SERS Substrate: Characterization of Image Deterioration in Plates from the 19th Century Studio of Southworth & Hawes," *Journal of Raman Spectroscopy,* vol. 39, 2008, pp. 914-921.

[15] C. V. G. Reddy, F. Yan, Y. Zhang, and T. Vo-Dinh, "A Highly Sensitive Raman Method for Selective Cyanide Detection Based on Evaporated Cuprous Iodide Substrate," *The Royal Society of Chemistry: Analytical Methods,* (5) 2010, pp. 458-460.

[16] D. E. Newbury and D. S. Bright, "Logarithmic 3-Band Color Encoding: Robust Method for Display and Comparison of Compositional Maps in Electron Probe X-ray Microanalysis," *Microscopy and Microanalysis,* vol. 5, 1999.
[17] O. Reckeweg, C. Lind, A. Simon, F. J. Disalvo, and Z. Naturforsch, "ICSD Pattern copper (I) cyanide," 2012.

Method Development in Image Processing, Analysis and Proof of Concept

Mater. Res. Soc. Symp. Proc. Vol. 1656 © 2014 Materials Research Society
DOI: 10.1557/opl.2014.711

Quantitative Porosity Studies of Archaeological Ceramics by Petrographic Image Analysis

Chandra L. Reedy[1], Jenifer Anderson[1], Terry J. Reedy[2]

[1]Center for Historic Architecture & Design, University of Delaware, Newark, DE 19716, USA
[2]Independent Statistician, 3 Stage Road, Newark, DE, 19711, USA

ABSTRACT

Pores in archaeological ceramics can form in a number of different ways, and reflect both deliberate choices and uncontrollable factors. Characterizing porosity by digital image analysis of thin sections holds a number of advantages as well as limitations. We present the results of experiments aimed at improving this method, focusing on high-resolution scans of entire thin sections. We examine the reproducibility of pore measurements by petrographic image analysis of ceramic thin sections using laboratory-prepared specimens of clay mixed with sand of known amount and size. We outline protocols for measuring Total Optical Porosity, using the Image-Pro Premier software package. We also briefly discuss use of pore size and pore shape (aspect ratio and roundness) in characterizing archaeological ceramics. While discerning reasons for observed amounts, sizes, and shapes of pores is an extremely complex problem, the quantitative analysis of ceramic porosity is one tool for characterizing a ware and comparing a product to others. The methods outlined here are applied to a case study comparing historic bricks from the Read House in New Castle, Delaware; the porosity studies indicate that different construction campaigns used bricks from different sources.

INTRODUCTION

Porosity has long been recognized as an important feature to characterize in any study of ceramics [1]. Porosity in ceramic materials results from choices in raw materials, clay processing and object fabrication methods, drying and firing regimes, and use, burial, or deterioration factors [2-4]. For example, during clay processing and vessel fabrication, air bubbles can become trapped. Shrinkage during drying and firing can enlarge those pores. Long linear pores with parallel alignment, often wavy with tapering ends, can appear as a result of shrinkage of the clay as excess water is released during firing; this alignment may also emphasize patterns of pressure placed during fabrication. At higher firing temperatures these linear pores may be less interconnected than at lower firing temperatures [5]. As carbonates dissociate and organics burn out or char during firing, additional porosity can be created. If firing temperatures are high enough, porosity can decline if vitrification occurs [1, 6]. Temper additives such as sand or grog can keep porosity higher, as clay tends to shrink away from those particles during drying and firing, creating additional porosity. Round secondary pores can be produced by trapped gases as the clay matrix and silica minerals begin to melt, off-gas, and vitrify [7]. If temperatures become too high, the round pores can become bloated and expand in number, indicating overfiring.

Some ceramics are deliberately designed to be porous for certain functions, while others are designed with low porosity. Vessels intended for cooking need to be porous enough to expand and contract over a fire without cracking, and storage ceramics intended to keep contents cool need a pore structure that allows evaporation; during use these pores can become enhanced. Another complication is that porosity changes can occur due to burial of a ceramic, with microcracking or dissolution from groundwater, so what is being measured may not always reflect the original material. Porosity variation within a ware, or even within areas of a single vessel, is likely to be higher for traditional ceramics compared with modern industrial products.

With so many factors affecting ceramic porosity, cultural interpretation of observed pore systems is a complex problem. Nonetheless, quantitative analysis of ceramic porosity is one tool for characterizing a ware and comparing a product to others [8]. But, with so many potential causes of porosity, relying on a bulk porosity analysis would not be as informative as being able to examine the variety in size, shape, and distribution of the pores. Being able to quantify and better understand pore characteristics are important because porosity is a direct reflection of the choices that people make in selecting and processing their materials and in fabricating and firing their ceramic products.

Archaeologists have quantitatively characterized porosity in ceramics (shape, size, number, and volume percentage) for decades [1]. A primary method has been point counting of thin sections under a transmitted light microscope. This involves superimposing a grid over the thin section and counting or measuring what is under each grid intersection point (usually with 300-400 points counted per thin section). An eyepiece micrometer is used to measure each pore. In addition to thin section analysis, traditional techniques for measuring volume porosity of ceramic materials include liquid immersion, water absorption, liquid nitrogen, or mercury intrusion porosimetry. Most of these methods are limited to measuring open pores connected to the surface and cannot access closed pores sealed off from the surface. This can sometimes lead to inaccurate or incomplete results [8, 9]; and too much sample material may be required.

Digital image analysis (DIA) of thin sections has more recently been used to quantify porosity in ceramic materials. Utilizing thin sections for this task is crucial because these can also be used for qualitative mineral identification, quantitative analysis of non-plastics, and the study of structural aspects that relate to fabrication and decorative choices. With archaeological ceramics we may not want to sacrifice the amount of sample material often required for performing tests found in ISO or ASTM standards for assessing porosity, so the relatively small sample size needed for thin sections, and the usefulness of those thin sections for addressing so many other research questions, are important considerations [10]. While some techniques for porosity measurement focus on impregnation of the material with a liquid or gaseous substance and quantify the volume of open pore spaces, DIA is a direct observation method that measures both open and closed pores [8] although it cannot distinguish between the two pore types [11]. Image analysis of petrographic thin sections has been demonstrated to provide comparable results to optical point counting and micrometer measurements, but is much faster, allowing for inclusion of a larger number of areas and specimens in quantitative work [12].

Thin section DIA focuses on measuring macropores, although the definition of these versus micropores varies considerably. In IUPAC terminology, macropores are wider than 0.05 μm; mesopores have diameters between 0.05 – 0.002 μm; and micropores are <0.002 μm diameter [13]. The limit of resolution of pores in thin sections varies depending upon the image capture system used. According to some researchers [14], pores of diameter <8 μm are not measurable with optical microscopy. Some petrographers define the boundary between macropores and micropores as a pore area of 500 μm^2, which they say translates to a pore length of about 20 μm, and close to the resolution of optical microscopy [15]. Others define microporosity as including any pores not easily detected in thin section images, and consider these to be pore diameter of <30 μm [16]. Some researchers [9] define macropores as those about 62.5 μm, the size of fine sand [17], even with an image acquisition resolution that would allow detecting smaller pores.

Macroporosity has been found sufficient for quantifying total porosity, size, and shape of pores for most soils and ceramic materials, although this of course depends upon the research questions. Sedimentary petrographers [18] use DIA of thin sections made from undisturbed soil samples to measure pores, defining macropores as >50 μm. With our image acquisition system for scanning whole thin sections, described below, the limit of resolution we work with is generally 7 μm in length (feret diameter), with smaller pores measurable if higher scanning resolutions are used (down to about 1 μm). We are also currently experimenting with use of ultra-thin sections (15 μm thickness) for measuring smaller pores.

If adding measurements of micropores is important, and enough sample can be sacrificed, mercury porosimetry can be added. DIA of scanning electron microscope-generated images can also provide microporosity data, although some questions have been raised about their reliability for porosity studies due to variations in parameters (magnification, voltage, working distance, and detector type) leading to variable results [19], so some additional experimental work is probably also called for with these types of images; and the overall area size being examined is also an important consideration if one hopes to characterize porosity of the material as a whole.

We have been exploring ways of improving the measurement of macroporosity to characterize ceramics through DIA of thin sections. Here we first test the ease and reproducibility of DIA of pores in ceramic thin sections using 96 laboratory-prepared specimens of known recipes. We found that high-resolution scans of entire thin sections are preferable to analysis of many fields of view under the microscope. We outline our preferred protocols for measuring Total Optical Porosity using the Image-Pro Premier software package. We present data for replicate analyses of individual thin sections and for multiple specimens separately made from the same batch of raw materials. We also analyze the effect of varying amount and size of sand additives on amount of porosity after firing the experimental specimens. We conclude with a discussion of the application of this protocol to archaeological ceramics and of additional size and shape parameters that can be used to help characterize ceramic pore systems.

THIN SECTION POROSITY STUDIES

Ceramic thin sections are slices of a three-dimensional network of pores, some interconnected and others isolated. Some researchers have reported a consistent underestimation of porosity in thin sections compared with three-dimensional analysis of larger core samples [20], attributed to the presence of complex pore geometries in some specimens and to the inability to view and measure the smaller micropores; and, image analysis procedures that involve conversion to binary images may also result in the loss of some small pores [21]. However, it has been demonstrated that porosity and pore size distributions of a 3D block can be adequately predicted using 2D images [22, 23]. DIA has also been shown to be a reliable method for characterizing porosity, in agreement with results from techniques using 3D images [11]. Because thin sections are relatively low cost, easily accessible, useful for a variety of other analyses [9], and because improvements continue to be made in image analysis programs, much research has gone into DIA of thin sections.

Since sands and sandstones have similarities to ceramic materials, work on DIA performed by sedimentary geologists is relevant to ceramic studies. J. Layman [12] compared

DIA of thin sections (using Image-Pro Plus software) to other methods traditionally used by sedimentary geologists to analyze porosity. Working with carbonate rocks and averaging five fields of view for each thin section, he found it to be a reasonable substitute for determining porosity by other petrographic methods such as point-counting of thin sections; and found that pore size obtained by this method for macropores was a useful substitute for mercury porosimetry. Other sedimentary petrographers expanded the number of fields of view analyzed to ensure coverage of the entire thin section, up to 20-30 [20]. Below, we show how consistent results can be obtained by using one high-resolution scan of the entire thin section.

In the past, most ceramic thin sections were prepared using a clear epoxy resin as the mounting medium. This presents problems, since in plane polarized light the pores are difficult to differentiate from clear, low-relief quartz that is ubiquitous in most ceramic materials; and as the epoxy turns dark in crossed polarized light the empty pores are then difficult to discern from clay and quartz or other minerals in extinction position. Some ceramic petrographers tried to get around this problem by obtaining an image in both plane polarized and cross polarized light, subtracting the second image from the first, then calculating a new gray level for each pixel corresponding to the difference in value of the two images [24]. A faster and less complex approach is to impregnate the thin section with a dyed resin, so that the pores easily stand out in plane polarized light [25].

Perhaps the most useful measure for characterizing pores using image analysis is Total Optical Porosity (TOP), the ratio of the sum of all pore areas to the area of the entire image (area percentage of pores) [26] (Figure 1). Sedimentary petrographers have found other pore parameters useful as well; an advantage of DIA is that it provides simultaneous measurement of as many different parameters as one selects. In our experiments for developing a fast and reliable protocol for characterizing pores in ceramic thin sections, we focused on Total Optical Porosity.

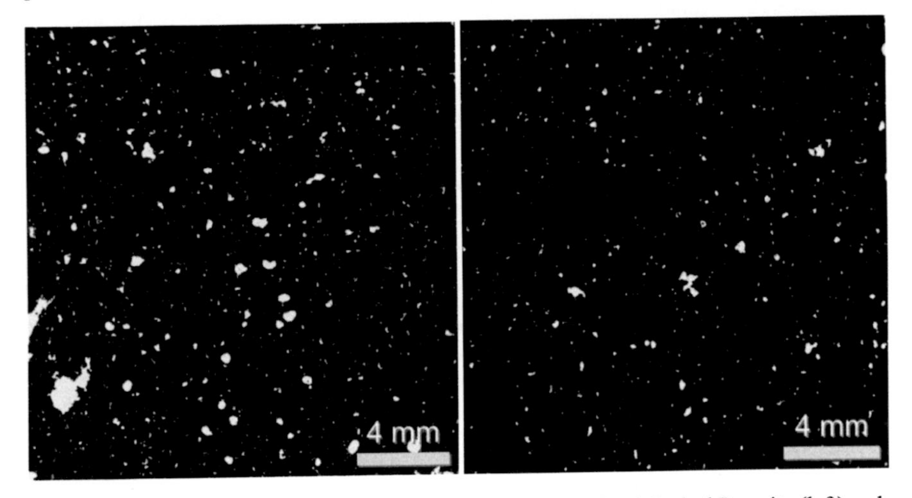

Figure 1. Pores (light areas) of two specimens, one with 4.6% Total Optical Porosity (left) and another with 2.0% Total Optical Porosity (right).

EXPERIMENTAL PROCEDURES

Experimental Units

To test image analysis protocols, we prepared a set of specimens comprised of clay and a quartz-rich sand additive. Two types of clay were used, a red earthenware that fires to a deep orange red (Earthen Red from Clay-King) and a gray earthenware that fires to a white color (White Earthenware from Clay-King), to give two very different visual appearances for the image analysis work. We picked three size ranges of sand (from Mile Hi Ceramics): coarse (16 mesh), medium (30 mesh), and fine (70 mesh). Each was then added to the clays in three different proportions: 10, 25, and 40% by volume. After kneading well, each clay-sand mixture was rolled out to an even thickness and specimens of a standard size were cut out with a cookie cutter-like mould. We deliberately chose a preparation method that would mimic how traditional potters might mix temper into a clay so that the resulting level of homogeneity would be similar to what might be encountered with ancient ceramics. Specimens were first dried in air indoors for a week, then in an oven at 150°C for another week. They were then fired with gradual increase in temperature over the course of a day, to the temperature recommended for the clays (900 - 1000°C). This careful drying and firing regime was intended to minimize cracking or loss of grains during firing, to keep macroporosity to a minimum.

For each sand percentage, size, and clay combination, five replicate specimens were made for thin sections, for a total of 90 specimens. Another six contained clay with no sand additive (three replicates of each clay). A thin section was prepared for each of the 96 specimens.

We also investigated how much information DIA can provide on whole samples examined under a stereomicroscope, without the extra work of preparing thin sections. An advantage would be that the time and expense of thin-section cutting, mounting, and grinding would be eliminated if sherds could be characterized directly. For this research, additional replicate specimens were prepared and cut using a diamond-edge saw to obtain a fresh surface. This possibility was considered because sherds themselves, especially fresh edges or cleaned surfaces, have been used to obtain some quantitative data supplementing thin-section microscopy [27]. If DIA could be used to quantify porosity, this would allow data to be collected on even larger numbers of sherds. However, while sand grains were readily visible and with some effort could be used for image analysis, the pores in the whole specimens were difficult to see even with various processing methods. An additional problem is that only those pores open to the surface can be quantified by this method. Thus we chose to focus on thin sections.

The technique of impregnating ceramics with a dyed epoxy in thin-section preparation for DIA has been used by a number of researchers. For example, historic brick samples were mounted in a blue-dyed epoxy resin for a hue-saturation-intensity segmentation procedure (using Image-Pro Plus software) to separate the blue pores from other components of the thin sections [10]. The goal was to characterize macropores, focusing on pores at or above an area of 400 μm^2. Pottery specimens have also been prepared as blue epoxy-resin-impregnated thin sections to quantify porosity [28]; in this case a plug-in for a freeware image analysis program (Scion Image) was employed to automatically recognize the blue hues representing porosity and calculate a percentage pore abundance of the thin section. Results were contrasted with those for optical point counting and found to be comparable, but required much less time that the older method. We used a blue-dyed epoxy resin to impregnate and mount all thin section specimens.

We experimented with applying measurement protocols to multiple fields of view of a thin section under a transmitted light microscope, and then obtaining an average. However, we found this to be problematic. It is time consuming, and if there are some large pores, each field of view may be quite unrepresentative of the material as a whole, meaning that many fields of view will then have to be measured in order to be representative. An additional problem is that pores touching a border have to be discounted from any measurements other than area percentage, since they are at least partially cut off from view, so cannot be measured accurately. This reduces the number of pores available for many analyses, and affects the statistics.

Some image analysis packages have a tiling capability – so images can be taken of multiple fields of view with overlap, and then tiled and stitched together to make a single large image for analysis. Then only the grains on the border of that one large image need to be eliminated from measurements. The Image-Pro Premier software we use allows live tiling, where one can move through adjacent areas of a thin section under the microscope and a larger image is automatically tiled and stitched together. However, this too takes time and care, and creates a very large image file, so also is not ideal. As a result, we continued to search for an alternative method for processing large numbers of samples more efficiently, leading to our decision to use entire scanned thin sections for image analysis.

Images Produced in a High-Resolution Film Scanner

The solution of scanning an entire thin section has been tried by several researchers studying archaeological ceramics, historic mortars, sands, stone, and other materials [29, 30, 31]. A single image can then be used for DIA. This procedure makes quantitative analysis practical for a larger number of thin sections, and gives results that apply to the entire thin section rather than to small fields of view. One method is to use a flat-bed scanner to take high-resolution scans [9, 32-35]. For example, a flatbed scanning system was used to study macroporosity (defined as pores larger than 62.5 μm) in archaeological and historical mortars [9]. An advantage that was noted was the ability to obtain an image of the entire thin section with homogeneous illumination, something that can be difficult to achieve with multiple microscope fields of view. Color images were then converted to binary ones for analysis by ImageJ software. Flat-bed scanners have also been used to conduct DIA of pores in a variety of cementitious materials. In one case a high-resolution flat-bed scanner (3175 x 8000 dpi) was used for analysis of binary images (with pores white against a black background) for quantitative characterization of pores [36]; it was shown how much better these images were for DIA than images taken with a typical office flat-bed scanner (4800 x 4800 dpi maximum optical resolution capability).

Rather than a flat-bed scanner, thin-section scanning in a film scanner (at a resolution of 9.4 μm/pixel, incorporating polarizing filters, was successfully used in the study of texture of granitoid rocks [37]. After trying out several systems, we also chose to pursue the use of a film scanner. For this work we used a low-cost Plustek OpticFilm 7600i film scanner with maximum optical resolution of 7200 x 7200 dpi. However, scanning at maximum resolution creates very large file sizes, which can take longer for the image analysis processing, and for most image analysis work (especially on coarser ceramics) scanning at a lower optical resolution will usually suffice. A 10-mm scale bar scanned under identical conditions was used to spatially calibrate the system. If there are any glaze layers to be excluded from measurements, or areas off the edge of the thin section (showing just the epoxy) in the image, these can be excluded from measurements either by cropping the image or by using a Region of Interest mask during analysis.

Image Analysis Protocols

After experimenting with a variety of DIA software programs, we chose Image-Pro Premier software distributed by Media Cybernetics. It is intuitive and easy to use, while incorporating a wide range of operations pertinent to ceramic thin section analyses. Whichever program is used, issues that need to be considered include calibration of the image capture setup so that measurements are in a specific unit rather than in number of pixels counted, image quality (most analyses require uncompressed file formats, such as TIFF, and images must be in excellent focus), and statistical validity of the sample size for the material under analysis [25].

A major strength of the Image-Pro Premier package is its Smart Segmentation protocol. This allows one to more effectively define the reference areas for background and objects of interest in an image; the reference areas are then used to create segmentation masks to count and measure objects of interest. Smart Segmentation allows one to include a wide variety of channels for segmentation including monochrome intensity level, RGB, YIQ, and HIS color spaces and morphological filters, and to automatically correct for uneven background. It then classifies every pixel by distance to the closest reference area, using distance in multidimensional space where every segmentation channel is a dimension. Pores are highlighted by clicking on a selection of them as reference points, ensuring that the full range of colors and intensities are marked. Background is set by clicking on clay and sand (or other temper components). If the material has very small pores, zooming in for marking reference points is helpful, to ensure that the reference marks cover only the pores and do not spill over into background, and vice-versa.

For our experimental specimens, the best segmentation was achieved by RGB color channels, although other parameters were important for other archaeological ceramic products. The "fill holes" option was selected, so that if any pixels within a pore are missed in the selection of reference color ranges, they will be filled in. The parameters of the background and object areas are then analyzed, and a segmentation recipe is created which includes the transformations with maximum degree of difference between the objects and background. Multiple images can be used to construct the recipe, to ensure that the full range of variation is captured. The recipe can be saved to be applied to other similar images in a project, eliminating the need to manually mark reference areas for particles, pores, and background for many thin sections. This procedure generally works well, is fast, and is reproducible. However, when the clay is gray or white, it can sometimes have a blue tint where thin that makes it more difficult to distinguish from the blue of the pores; in this case, prior to segmentation the image can be adjusted to create more contrast between the matrix and the pores. We find that lowering the brightness and increasing the contrast makes the segmentation easier in this case.

We also found success with two alternative protocols using a different version of Image-Pro (Image-Pro Plus, also by Media Cybernetics) [38]. However, these protocols require more manual steps in image adjustment and application of filters than does the Smart Segmentation process. Whichever approach is followed, the resulting segmented areas can be carefully examined prior to going forward with measurements, to ensure that the pores have been correctly marked. Sometimes this is best done by viewing the original image side-by-side with the segmented image. Once a segmentation protocol is found to work well for the images within a research project, this process can move forward quickly.

Once the pores are correctly highlighted (segmented) (see Figure 1), any desired measurements can be performed simultaneously. The analysis can be further refined by using the classification options in Image-Pro, to identify whether or not there are clear categories of pore diameters or shapes, and the relative abundance of each; it may then be useful to quantify these categories separately. Or, the highlighted pores can be displayed in a new image where they are sorted (Figure 2) according to length, area, aspect ratio, roundness, etc. to get a sense of the range and to inform data collection and analysis strategies. Data for each individual pore, as well as summary statistics, can be saved into a file for statistical analysis.

Figure 2. Here pores are sorted by size (length/ferret diameter).This type of image is useful for examining the range (here we see that most pores are very small), for deciding on cut-off points for certain measurements, and for making decisions about data collection and analysis strategies.

Once the saved segmentation recipe is correctly identifying pores, and the measurement choices have been selected, macros and batch processing can automate the procedure of opening up an image file, segmenting and measuring pores, and saving data to a file, so many scanned thin sections can be analyzed rather quickly. Data on individual pores and/or summary data can be statistically analyzed and combined with data on various particles, each segmented and analyzed with their own saved segmentation recipes and measurement parameters.

RESULTS AND DISCUSSION

We first analyzed the scanned thin sections by selecting reference points for pores and background individually for each image, choosing about 20 reference points total and trying to select the full range of colors for each component. It is possible to include multiple phases in the same analysis, by marking separate reference points for pores, sand grains, and clay background, for instance. But, given the variation in appearance within each, we found that it was easier to focus on one component at a time. Since different measurement parameters are also often selected and saved for each phase, this procedure seems more efficient.

However, it is very tedious to mark each thin-section image separately. We moved to the option in the Smart Segmentation procedure to save a recipe (parameters that give the maximum difference between the pores and the background), to reuse on other images. Not having to spend time marking each and every image allowed us to focus instead on spending more time being very careful on one representative image (or set of images), selecting 50-60 reference points. While some types of materials would not need this high number of reference points for successful segmentation, with ancient ceramics there is often much subtle variation, and this approach proved to more completely segment the full range of pore structures within the matrix. While the recipe derived from the red clay worked remarkably well for the white clay, we found that constructing a second recipe by separately marking reference points for the white clay specimens worked better. Once the recipes were developed and saved, analysis of the remaining images proceeded quickly, by simply opening up the image and applying the recipe. For this project we did, however, carefully examine each segmented image to ensure that the recipe worked well to identify the pores without incorrectly marking any background areas as pore.

Table 1 gives the results for this analysis by carefully-constructed recipe for each of the 96 thin sections. The five numbers for each clay type and added sand size/percentage combination (three numbers for clays with no added sand) represent separate objects, each made from the same batch of clay/sand mixture. These are analogous to sherds from separate vessels made from the same batch of clay plus sand temper additive, and illustrate the variation one might expect in multiple ceramic objects made from the same clay-temper batch.

Table 1. Total Optical Porosity (Area %)

Added Sand	Red Clay	White Clay	Added Sand	Red Clay	White Clay
None	0.6	0.0	Coarse, 25%	2.2	2.9
	0.4	0.1		3.2	1.5
	0.5	0.0		2.1	1.5
Fine, 10%	1.9	1.7		2.4	1.4
	2.0	1.5		2.2	1.9
	2.4	1.6	Fine, 40%	2.7	2.2
	2.1	1.6		3.1	4.4
	2.2	1.7		3.4	1.5
Medium,10%	2.2	1.2		2.9	3.9
	3.5	1.8		2.1	2.2
	2.5	2.5	Medium,	2.9	4.4
	2.2	1.3	40%	2.5	1.6
	2.4	1.7		3.2	2.5
Coarse, 10%	2.8	1.4		3.4	1.4
	3.5	1.8		3.4	0.8
	1.5	1.3		2.4	1.9
	2.9	1.5	Coarse, 40%	2.9	4.0
	2.8	1.9		2.6	2.5
Fine, 25%	4.1	2.3		6.6	1.3
	2.8	1.1		1.9	2.1
	2.2	1.5			
	2.8	1.4			
	2.6	1.7			
Medium,	4.6	1.9			
25%	3.1	2.0			
	3.8	1.5			
	3.0	2.0			
	4.0	1.4			

Replicate specimens sampled from each batch; TOP determined on an entire scanned thin section (optical resolution 7μm/pixel). Using Smart Segmentation (Image-Pro Premier) saved recipe for each clay type, 50-60 reference points. Fine = 70 mesh; medium = 30 mesh; coarse = 16 mesh.

Table 2 shows a selected group of images analyzed by individual reference marking of 10 points each for pores and background, repeating the procedure three times for each thin section. Analyses were in random order so that a different set of reference points would get selected each time. This table shows that reproducibility of segmentation is good, with about 0.1% standard deviation for the three analyses of each image. Results for the two measurement approaches are highly correlated (Pearson Correlation Coefficient .91). However, overall pore percentages for individually measuring are somewhat lower than for applying the saved recipe (Figure 3).

Table 2. Replication with 20 Reference Points, Total Optical Porosity (Area %)

Sand %	**Red**	**Clay**		Mean	σ	**White**	**Clay**		Mean	σ
Fine, 10	1.9	1.4	1.6	1.6	0.3	1.5	1.3	1.2	1.3	0.2
Medium,10	1.9	2.3	2.0	2.1	0.2	1.0	1.0	1.0	1.0	0.0
Coarse, 10	2.0	1.9	2.0	2.0	0.1	1.2	1.1	1.1	1.1	0.1
Fine, 25	4.1	3.9	4.1	4.0	0.1	1.9	1.8	2.0	1.9	0.1
Medium, 25	3.9	4.1	4.0	4.0	0.1	1.8	1.8	2.0	1.9	0.1
Coarse, 25	1.6	1.7	1.6	1.6	0.1	2.6	2.5	2.4	2.5	0.1
Fine, 40	1.3	1.3	1.4	1.3	0.1	2.0	2.1	1.9	2.0	0.1
Medium, 40	1.7	1.7	1.7	1.7	0.0	3.5	3.6	3.8	3.6	0.2
Coarse 40	1.4	1.3	1.2	1.3	0.1	1.7	1.7	1.7	1.7	0.0

Process repeated three times for each scanned thin section.

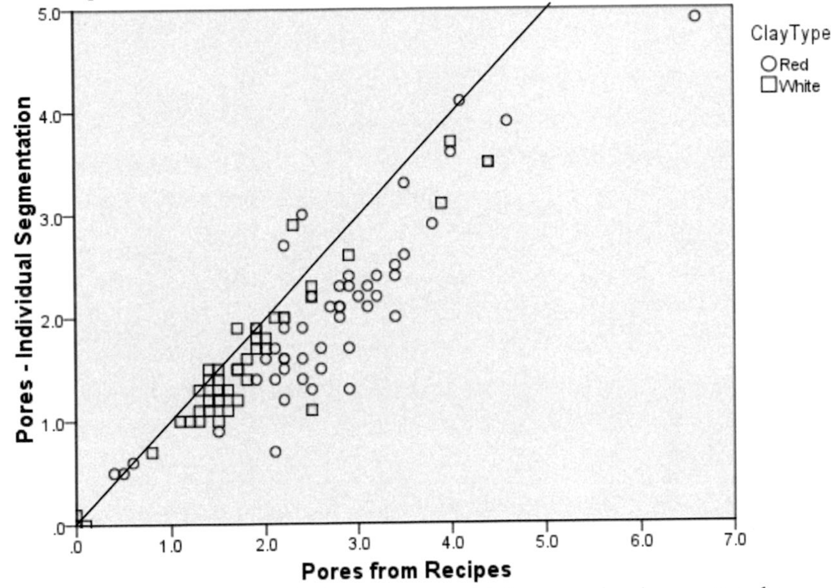

Figure 3. Individually-segmenting images versus applying a saved recipe are strongly correlated, but individual results are lower implying that some pores get missed.

To examine this comparison more closely, we computed a new variable, Pore Difference, as % pores identified via recipe minus % pores identified via individual measurement. This showed that most of the percentages derived from the recipe are higher, with only four cases where the individually-measured thin sections had significantly higher results. The difference between the two methods is significant, with a mean difference of 0.45% (significantly different than the 0.1% difference between individual measurements). So the approach of selecting about 10-20 new reference points on each thin section produces relatively consistent results, but gives somewhat lower percentages than obtained by using one carefully constructed (50-60 reference points) segmentation recipe in a standard way on all samples of that clay color. Examination of the images with all identified pores highlighted shows that this is not due to over fitting (marking some non-pore areas as pores), but does in fact produce a more accurately segmented image, with more of the smaller and thinner pore areas that tend to have a greenish-blue tinge being correctly identified as pores.

Why are 25-30 each pore and background points better than 10 of each? The bluish pore points form a cluster in three-dimensional space. Four points will define a tetrahedron that will exclude many pore points. As more points are added, those outside the existing bounding polyhedron will expand it. After several pore and background points have been chosen, the Image-Pro Premier software we use makes it possible to see and monitor which pore points will not be included (by using a semi-transparent mask for the pore segmentation step-by-step as it occurs) so that one may add new boundary points one at a time on a stepwise basis. A selection of 25-30 points seems to be enough to include any pores encountered, whereas 10 are not.

Some materials that are more uniform and with clearly distinct phases may require only a few points for successful segmentation. But, with the range of variation present in ancient and historic ceramic samples, the 25-30 points seem to be required consistently. For both the red and white clay specimens in this study, we did not set out to choose 25-30 points, but chose just the right number to ensure that pore areas were fully selected but no background areas incorrectly marked as pore. With the brick shown in Figure 6, where an existing recipe did not work so a new one had to be constructed, again 25 points each for pores and background produced the best result. In that case, with a brownish clay and a different blue epoxy, a different mix of RGB, HSI, and morphological characteristics formed the recipe. The ability to save the segmentation recipe and reuse it on similar images makes this careful and more time-consuming procedure more practical than it would be if we had to separately segment each and every image in a research project.

To examine the relationship between porosity and clay type, sand size, and sand amount, we did an analysis of variance (ANOVA), using the data derived from segmentation recipes. All three factors initially appeared to have a significant effect on Total Optical Porosity. However, sand size and amount might only appear to have an effect because those with no added sand have almost no porosity. Hence we redid the ANOVA with data removed for the clay with no added sand. This showed that clay type continued to be significant, but once at least 10% sand was added to the clay the sand size and amount were not significant determiners of porosity (Figure 4). This plot also clearly shows that the means for the red clay specimens are definitely higher than the means for the white clay specimens.

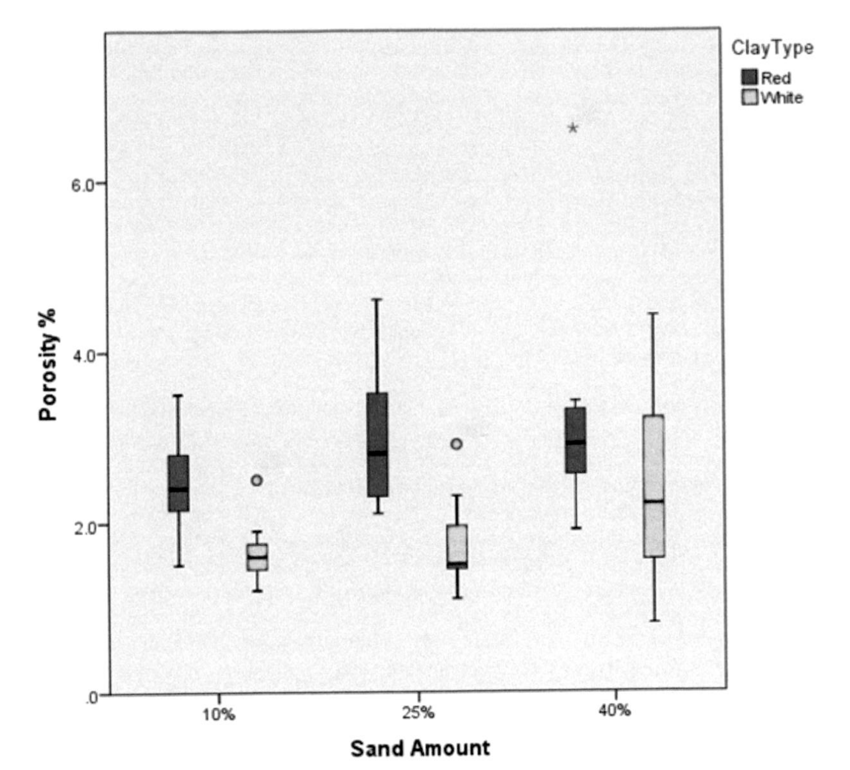

Figure 4. Box plot showing the relationship between sand amount and porosity percentage; the means for red clay specimens are higher than those for the white clay specimens, but once at least 10% sand is added, the amount has no significant effect on porosity. Brackets mark the upper and lower ranges, shaded areas the interquartile range occupied by 50% of the numbers. Outliers are separately marked.

For comparison, we analyzed a subset of thin sections using images taken at the microscope, with five fields of view in plane polarized light under 50x magnification. Since we were measuring only Total Optical Porosity, and not size and shape parameters, pores touching borders did not have to be excluded. Results of the microscope views were problematic, because there was often much variation from one field of view to another. If one or more large pores happen to be in the field of view, the area percentage is quite high; if the field of view happens to be in an area with few or only very small pores, the percentage is quite low. Since ancient ceramic materials tend not be uniform, this may always be the case. To achieve a representative number, one would likely have to follow procedures recommended for sedimentary materials [12, 20], including as many as 20-30 images to be averaged, and possibly using an even lower power objective (to give 25x magnification). While using a saved segmentation recipe on these images could speed up analysis, it would still be time consuming to systematically mark out 20-30 fields of view, capture the digital images, and analyze them all. Using an entire scanned thin section instead makes characterizing a large number of thin sections more practical.

We have even found that one recipe can work well for multiple research projects, if the ceramic phases are similar in appearance. For example, the image in Figure 5 is a porous historic brick thin section. The pore network includes many long, stringy, narrow pores which cut across, and are embedded in, or are closely adjacent to sand grains or charred organic materials. As a result, even by zooming in closely it was often difficult to mark reference points on either pores or background without accidently touching onto the other. We decided to try the recipe already constructed for the red clay in our experimental study. Fortunately, it worked quite well at correctly highlighting the pore network.

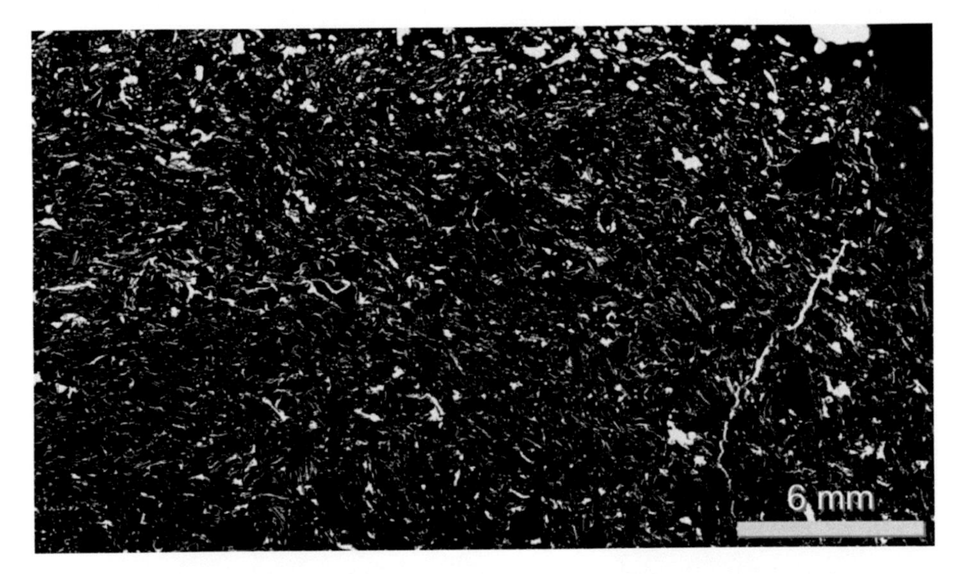

Figure 5. Historic brick (from the Read House in New Castle, Delaware) with a very stringy network of many (over 6,500) small pores.

Finally, once a pore network has been correctly segmented, we can collect data on many parameters in addition to area percentage. Ones that have been found most useful include those related to pore size and shape. Pore size is defined by either maximum length (feret diameter) or pore area [6, 10, 39]; or by number or percentage of pores falling into various size categories, defined according to the type of material and the research questions. Many petrographers have also measured pore shape, usually determined by measuring aspect ratio and roundness. Aspect ratio (ratio between the major and minor axis) is a measure of pore elongation; ratios between 1 – 1.5 are considered equidimensional, larger ratios indicate elongation. Roundness is defined as perimeter^2 / 4 π a, where a is equal to pore area [12]. Roundness measures between 1.0 and 3.0 indicate a round object, those greater than 3.0 an angular one. A round object with low aspect ratio is considered circular; with high aspect ratio it is an ellipse. Non-round objects with low aspect ratio are square; with a high one, rectangular.

As an example, we can compare the historic brick specimen in Figure 5 (from the Read House, New Castle, Delaware) with a specimen from another brick found at a different location within the same site (Figure 6). Table 3 compares the data for the two bricks, including Total Optical Porosity, size (length/ferret diameter), aspect ratio, and roundness, and shows that there are some significant differences between the bricks. While both have the same Total Optical Porosity, the brick in Figure 6 has a much higher amount of that accounted for by large pores, fewer in number than the brick in Figure 5 (under 1,000 versus over 6,500).The pores of the brick in Figure 5 are often smaller, significantly more elongated (higher aspect ratio), and more angular. Removing from analysis pores that touch borders had little effect on the results, indicating that this issue is not as significant when an entire thin section is analyzed compared with its effect on microscopic fields of view. The results of this example from the Read House bricks indicate that different construction campaigns used bricks from different sources. This hypothesis is supported by a more extensive case study.

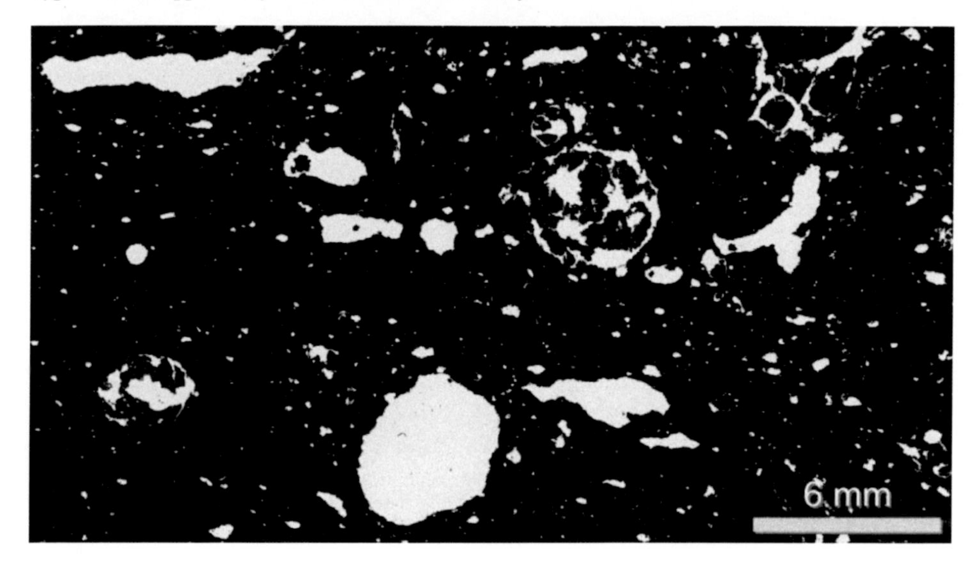

Figure 6. Historic brick (from Read House, New Castle, Delaware) excavated from a different area of the site than the brick in Figure 5. Total Optical Porosity of the bricks are identical, but a much higher amount of the porosity here is accounted for by large pores that are significantly less elongated (lower aspect ratio) and much more round.

Table 3. Porosity Characteristics of Bricks Excavated From Different Locations at Read House, New Castle, DE

	Brick 1	**Brick 2**
Total Optical Porosity (%)		
Pores ≥ 0.007 mm	14	14
Pores ≥ 0.063 mm	12	13
Pores ≥ 0.2 mm	7.6	13
Pores ≥ 0.63 mm	2.2	11
Length (feret diameter) (mm)		
Pores ≥ 0.007 mm, maximum	3.5	7.8
Pores ≥ 0.007 mm, mean	0.07	0.11
Pores ≥ 0.063 mm, mean	0.15	0.24
Pores ≥ 0.2 mm, mean	0.34	0.58
Pores ≥ 0.63 mm, mean	1.0	1.8
Aspect Ratio, mean		
Pores ≥ 0.007 mm	2.4	2.1
Pores ≥ 0.063 mm	3.1	2.1
Pores ≥ 0.2 mm	3.8	2.2
Pores ≥ 0.63 mm	4.0	2.6
Roundness, mean		
Pores ≥ 0.007 mm	2.4	2.1
Pores ≥ 0.063 mm	3.5	2.3
Pores ≥ 0.2 mm	4.8	3.0
Pores ≥ 0.63 mm	9.3	5.7
Perimeter, mean (mm)		
Pores ≥ 0.007 mm	0.18	0.34
Pores ≥ 0.063 mm	0.41	0.79
Pores ≥ 0.2 mm	1.0	2.1
Pores ≥ 0.63 mm	3.7	7.0
Area, mean (mm^2)		
Pores ≥ 0.007 mm	<0.01	0.03
Pores ≥ 0.063 mm	0.01	0.08
Pores ≥ 0.2 mm	0.03	0.27
Pores ≥ 0.63 mm	0.17	1.3
Sand (% ≥ 0.063 mm)	20	10

Analysis using Smart Segmentation (Image-Pro Premier); 0.063 mm is lower limit for fine sand, 0.2 for medium sand, and 0.63 for coarse sand [17] so were used for pore size categorization.

CONCLUSIONS

These experiments with laboratory-prepared specimens made with two different clays of very different visual appearances and with sand additions of varying sizes and amounts have demonstrated that digital image analysis of petrographic thin sections can be a fast, reliable, and reproducible method for quantifying macropore systems in ceramic materials. Key elements of the successful procedure include using a thin section impregnated with a dyed epoxy to highlight pores, use of a high-resolution scan of an entire thin section, and development of a carefully-constructed segmentation recipe that can be applied to all thin sections of a similar appearance. Total Optical Porosity and data quantifying the pore sizes and shapes can be used to characterize a ceramic product or ware and to compare it with others. While fully interpreting the reasons for the amount, shape, and size distribution of pores and pore networks in ancient and historic ceramics may be very complex and difficult, characterization of visible pores can provide a "porosity fingerprint" for a material [12]. Such porosity fingerprints may help in identifying similar materials, grouping materials, or, in conjunction with other petrographic and chemical data, in identifying workshops. Application of the methods discussed here to historic and archaeological case studies, such as the Read House bricks, are now helping us to better interpret and understand choices that workshops made in materials selection, processing, fabrication, and firing procedures, and how those choices affected the final ceramic products made and used by past societies.

ACKNOWLEDGMENTS

This work was supported by the National Science Foundation under Grant No. 1005992. Yimeng Liu in the University of Delaware's Center for Historic Architecture and Design helped with the fabrication and firing of the experimental specimens, and Dr. Lu Ann DeCunzo (University of Delaware Anthropology Department) provided the Read House brick specimens.

REFERENCES

1. P. Rice, *Pottery Analysis: A Sourcebook* (University of Chicago Press, Chicago, 1987).
2. C. L. Reedy, *Thin-section Petrography of Stone and Ceramic Cultural Materials* (Archetype Publications, London, 2008).
3. P. S. Quinn (editor), *Interpreting Silent Artefacts: Petrographic Approaches to Archaeological Materials* (Archaeopress, Oxford, 2009).
4. P. S. Quinn, *Ceramic Petrography: The Interpretation of Archaeological Pottery & Related Artefacts in Thin Section* (Archaeopress, Oxford, 2013).
5. O. S. Rye, Archaeology & Physical Anthropology in Oceania **11** (2), 106-137 (1976).
6. G. Cultrone, E. Sebastián, K. Elert, M. J. de la Toree, O. Cazalla, and C. Rodriguez-Navarro, Journal of the European Ceramic Society **24**, 547-564 (2004).
7. C. K. Koh Choo, Y. E. Lee, I. W. Shim, W. K. Choo, G. H. Kim, W. Y. Hug, and S. Chun, Archaeometry **46** (2), 247-265 (2004).
8. K. G. Harry and A. Johnson, Journal of Archaeological Sciences **31**, 1567-1575 (2004).
9. D. Miriello and G. M. Crisci, Journal of Cultural Heritage **7**, 186-192 (2006).
10. U. Mueller and E. F. Hansen, in *Proceedings of the 8th Euroseminar on Microscopy Applied to Building Materials*, edited by B. Georgalia and E. E. Toumbakari, (Cosmosware, Athens, 2001), pp. 603-610.
11. S. Deshpande, A. Kulkarni, S. Sampath, and H. Herman, Surface Coatings & Technology **187**, 6-16 (2004).
12. J. M. Layman, MS. Thesis, Texas A & M University, 2002.

13. IUPAC, Pure Applied Chemistry **31**, 578-638 (1972).
14. E. Borrelli, *Porosity* (ICCROM, Rome, 1999).
15. F. Anselmetti, S. Luthi, and G. P., Eberli, American Association of Petroleum Geologists Bulletin **82** (10), 1815-1836 (1998).
16. N. S. I. Bashah and B. Pierson, in *Quantification of Pore Structure in a Miocene Carbonate Build-up of Central Luconia, Sarawak and its Relationship to Sonic Velocity* (International Petroleum Technology Conference, Bangkok), DOI 0.1523/14625.MS.
17. D. Norbury, *Soil and Rock Description in Engineering Practice* (CRC Press, Boca Raton, 2010).
18. M. Pagliari, N. Vignozzi, and S. Pellegrini, Soil & Tillage Research **79**, 131-143 (2004).
19. M. von Bradke, M. F. Gritzhofer, and R. Henne, Scanning **27**, 132-135 (2005).
20. A. Adams, PhD. Thesis, Texas A & M University, 2005.
21. T. T. Mowers and D. A. Budd, American Association of Petroleum Geologists Bulletin **80,** 309-322 (1996).
22. E. Moreau, B. Velde, and F. Terribile, Geoderma **92**, 55-72 (1999).
23. V. Marcelino, V. Cnudde, V. Vansteelandt, and F. Caro, European Journal of Soil Science **58**, 133-140 (2007).
24. A. Schmitt, Archaeometry **40** (2), 293-310 (1998).
25. C. L. Reedy, Journal of the American Institute for Conservation **45** (2), 127-146 (2006).
26. R. Erlich, S. K. Kennedy, S. J. Crabtree, and R. L. Cannon, Journal of Sedimentary Petrology **54**, 1365-1378 (1984).
27. A. Nijboer, P. Attema, and G. Van Oortmerssen, Paleohistoria **47/48**, 141-205 (2006).
28. G. A. Thompson and D. A. Jerram, presented at the 2003 British Sedimentological Research Group General Meeting, Leeds, England (unpublished).
29. T. L. De Keyser, Journal of Sedimentary Research **69** (4), 962-964 (1999).
30. E. F. Hansen, PhD. Thesis, University of California, Los Angeles, 2000.
31. S. Tarquini and P. Armienti, Image Analysis and Stereology **22**, 27-34 (2003).
32. F. Casadio, G. Chiari, and S. Simon, Archaeometry **47** (4), 671-689 (2005).
33. P. C. Livingood, in *Plaquemine Archaeology*, edited by M. A. Rees and P. C. Livingood (University of Alabama Press, Tuscaloosa, 2007), pp. 108-126.
34. E. Frahm, M. Nikolaidou, and M. Kelly-Buccelati, SAS Bulletin **32** (2), 8-12 (2008).
35. E. H. Van Den Berg, V. F. Bense, and W. Schlager, Journal of Sedimentary Research **73** (1), 133-143 (2003).
36. J. Carlson, L. Sutter, T. Van Dam, and K. W. Peterson, Transportation Research Record (Concrete Materials) **1979**, 54-59 (2006).
37. S. Tarquini and M. Favalli, Computers & Geosciences **36**, 665-74 (2010).
38. C. L. Reedy, Studies in Conservation **57** (S1), 227-233 (2012).
39. C. M. Dicus, MS Thesis, Texas A & M University, 2007.

Mater. Res. Soc. Symp. Proc. Vol. 1656 © 2014 Materials Research Society
DOI: 10.1557/opl.2014.873

Dual-Beam Scanning Electron Microscope (SEM) and Focused Ion Beam (FIB): A Practical Method for Characterization of Small Cultural Heritage Objects

Matthew Carl, Chris A. Smith, and Marcus L. Young
Department of Materials Science and Engineering, University of North Texas, Denton, Texas 76203-5017, U.S.A.

ABSTRACT

Knowledge of the composition of many cultural heritage objects is limited, resulting in many unanswered questions in regards to the provenance, composition, and production methods. In this paper, our objective is to show that dual beam scanning electron microscope (SEM) and focused ion beam (FIB) can be used rapidly and non-destructively to determine the surface and bulk metal compositions in small cultural heritage objects. We show, for the first time, that this novel FIB technique can be successfully applied non-destructively to cultural heritage objects by examining three representative silver plated objects (Candelabra, "Century" spoon, and New York World's Fair spoon) from the Dallas Museum of Art's unparalleled collection of modern American silver. In each case, we successfully reveal and characterize the bulk metal as well as the Ag-plating, up to ~80 μm deep and show that there is no visual damage resulting from the milling process of the FIB. This novel characterization technique can be applied, due to its ease of availability and rapid use, to many other problems in addition to silver plated objects, making dual beam SEM/FIB a possible cornerstone technique in the study of cultural heritage objects.

INTRODUCTION

Scientific studies of historical and cultural heritage objects are focused on determining the provenance, condition, and manufacturing technologies of the people who produced them [1-13]. One major challenge of studying heritage objects is that they must be non-destructively and non-invasively examined, often requiring a combination of analytical techniques for full characterization. Analytical techniques, such as conventional X-ray diffraction (XRD) and X-ray fluorescence (XRF) allow for surface studies of the objects, but cannot properly examine the bulk material; furthermore, these surface studies can be misleading since many objects have surface alterations such as corrosion layers, tarnishes, patinas, plating, and post-excavation cleaning residues [4-11]. Studies of the bulk material can be performed non-destructively using synchrotron, neutron, and proton sources available at large user facilities; however, it is difficult to examine large bodies of work due to the nature of these facilities [12-15]. Our initial study is focused on the Jewel Stern American Silver Collection at the Dallas Museum of Art (DMA) which contains over 400 Ag and Ag-plate objects from the Modernist movement of the late 19th and early 20th centuries [16]. The DMA has an important collection of 20th century American silver and silver-plated objects and houses one of the most comprehensive collections of 19th and 20th century silver objects. The bulk metal and plating compositions and processing are not well documented for most objects in the DMA collection; however, evidence of changes in composition and processing have been periodically released by the manufacturers, e.g. through advertisements. As shown in Figure 1.a-b, these objects vary in size, shape, and surface flatness creating interesting challenges for scientific analysis. In this paper, we show that it is possible to non-destructively study, for the first time, both surfaces and bulk materials of cultural heritage objects by using a dual beam scanning electron microscopy (SEM) and focused ion beam (FIB) system with energy dispersive spectroscopy (EDS) and electron backscattered diffraction (EBSD) capabilities and that this technique, due to its availability and rapid use, can be applied to large bodies of objects. Although FIB systems have been used for TEM sample preparation in other cultural heritage studies [17], this new technique allows for rapid, *in situ* characterization of the objects. To demonstrate this technique, one sacrificial Ag-plated object, identical to objects found in the DMA's collection, and two representative Ag-plated objects from the DMA's collection were examined.

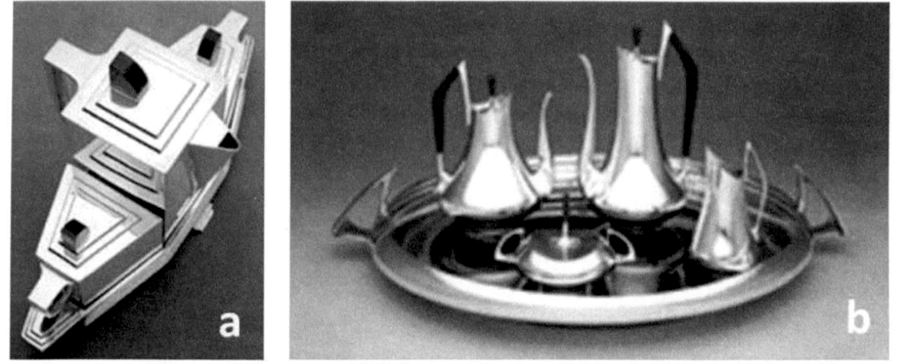

Figure 1. a) "Diament" dinette set, 1928 designed by Jean G. Theobald for the Wilcox Silver Plate Company. Dallas Museum of Art, the Jewel Stern American Silver Collection. b) "Circa '70" tea and coffee service, 1963. Dallas Museum of Art, the Patsy Lacy Griffith Collection.

EXPERIMENTAL METHOD

A FEI Nova 200 Nanolab dual column ultra-high resolution field emission SEM/FIB system equipped with EDS and EBSD was used to non-destructively examine the objects. The term "non-destructive" is used to illustrate the ability of the FIB system to perform *in situ* milling from the surface and bulk materials of the objects without resulting in visual or structural damage to the object. All objects studied in the FIB system were secured to the stage using carbon tape. For FIB operation, the objects were tilted 52° from horizontal, assuming a flat surface, to align the surface normal to the ion beam (i.e., milling perpendicular to the surface). Before milling starts, a thin strip, $20 \times 5 \times 5\ \mu m^3$, of Pt was deposited on the surface, using a gas-injection-system (GIS) needle, to reduce "curtaining" effects, i.e., uneven topography resulting from differential milling rates due to variations in the object's surface. During the FIB milling process, high-energy (30 keV) charged particles, e.g. Ga^+ ions, were focused on a small cross-sectional area, ranging between $20 \times 20\ nm^2$ to $100 \times 100\ \mu m^2$, on the object's surface just below the deposited thin strip of Pt. As the ions impacted the surface, individual atoms of the sample were ablated away. As each layer was removed, a new cross-sectional area, normal to the original surface, was revealed. After several layers have been removed, approximately 10 to 100 μm, a microstructural map of the surface and subsurface can be obtained. The "hole" created by this procedure is made in a step-shaped fashion, i.e. deeper at the cross-section face and shallower at the opposite end, to minimize the material removed and to maximize the viewing potential. The material removed from the objects measures on average approximately 20 × 30 ×

30 μm^3 and this microscopic damage is not visible to the unaided eye, thus enabling the objects to maintain their artistic and historical value. This technique can reveal an image of both the Ag-plated surface and the underlying bulk metal, which can be analyzed *in situ* using EDS to determine elemental compositions of each unique microstructural feature and EBSD to identify the phases present, although EBSD was not performed in our study.

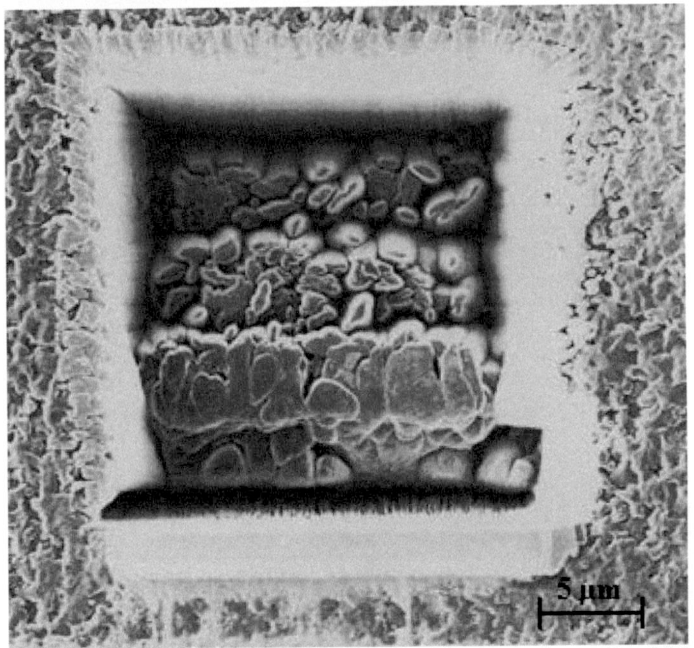

Figure 2. Ion-induced secondary electron (ISE) image of the support leg on *Candelabra* taken after milling procedure showing size and step-shaped structure of the "hole."

DESCRIPTION OF OBJECTS

Test Object

1939 New York World's Fair spoon

This spoon, which measures 2.223 × 12.7 × 3.175 cm^3, was designed in 1938 by Lillian V. M. Helander at the William Rogers Mfg. Co., a division of the International Silver Company.

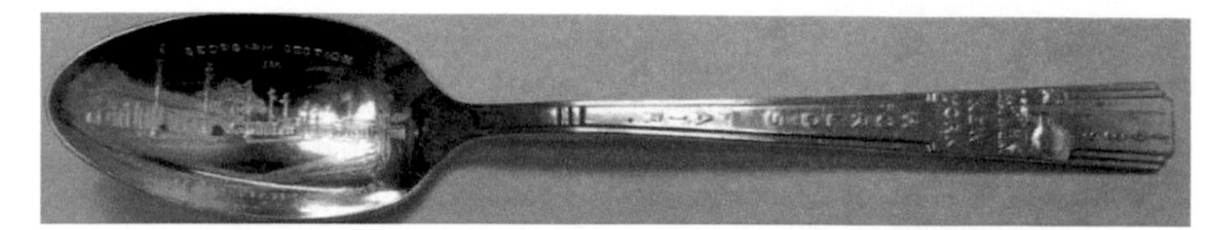

Figure 3. Photograph of a test object, 1939 New York World's Fair spoon, courtesy of Dr. Marcus L. Young.

Two objects from the DMA's Collection

"Century" spoon

This spoon was originally designed for the New York Central System's 20th-Century Limited, a train restyled in 1938 by Henry Dreyfuss. The flatware design is attributed to Frederick W. Stark of the International Silver Company. It was gifted to the Dallas Museum of Art by Kevin A. Tucker in honor of John R. Lane. It measures $1.905 \times 4.445 \times 15.24$ cm^3.

Candelabra

The Candelabra was designed circa 1935 by Helen Hughes Dulany Studio. Helen Hughes Dulany was an American artist who was active during the mid-1930s. This piece was examined courtesy of the Dallas Museum of Art from the Jewel Stern American Silver Collection. It measures $12.7 \times 22.9 \times 4.4$ cm^3.

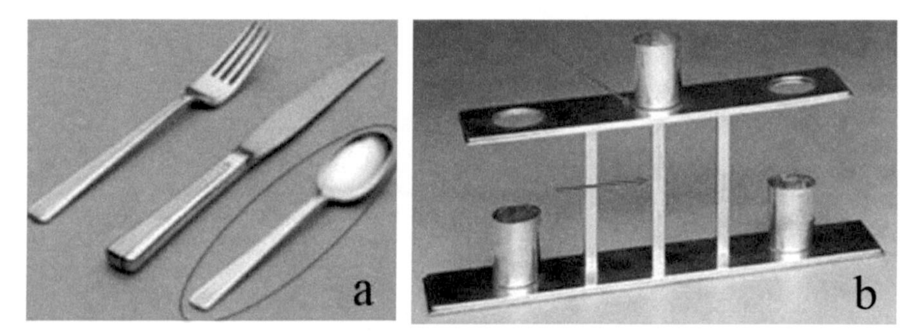

Figure 4. Photographs of a) "Century" flatware (for the New York Central System's 20th-Century Limited train) and b) Candelabra, circa 1935, Helen Hughes Dulany Studio. Arrows point to locations of SEM/FIB analysis. Both objects were analyzed with permission from the DMA.

RESULTS

1939 New York World's Fair Spoon

FIB milling of a 20 × 25 μm^2 area from the surface revealed a cross-section of both the silver-plating, Figure 5 a, as well as the underlying bulk material shown in Figure 5 b). The spoon was plated with a single layer of Ag with a thickness of about 10 μm. The spires visible around the cross-section are a combination of redeposition of the Ag and bulk material atoms and Ga^+ ions implanted on the surface from the ion beam. Spectra of the silver plated region and of the bulk material were collected for 120 seconds at 15 kV. As expected, the spectrum from the silver-plating consisted of predominantly Ag with small amounts of Cu, Ni, and Pt. The spectrum from the bulk, however, consisted of predominantly Cu and Ni with small amounts of Ag and Pt. In the silver-plated region, the small amount of Cu and Ni are most likely coming from the bulk metal rather than the silver-plating due to the large size of the EDS beam of 2-5 μm and from redeposition of the bulk during the milling process. Similarly, in the bulk metal region, the larger amount of Ag is most likely attributed to detection of redeposition from the surrounding silver-plating, and interaction of the EDS beam with the cross-section sidewalls. In both cases, the small Pt signal is coming from the Pt that was deposited before milling on the top edge of the surface. Furthermore, visual inspection of the cross-section shows that the Ag-plating consists of face-centered cubic grains with annealing twins on the order of a few microns similar to that observed in Metallography, Principles and Practice by Vander Voort [18], while the Cu-Ni-based bulk metal appears to have slightly larger grains, although EBSD is needed to clarify the precise structure of the bulk metal.

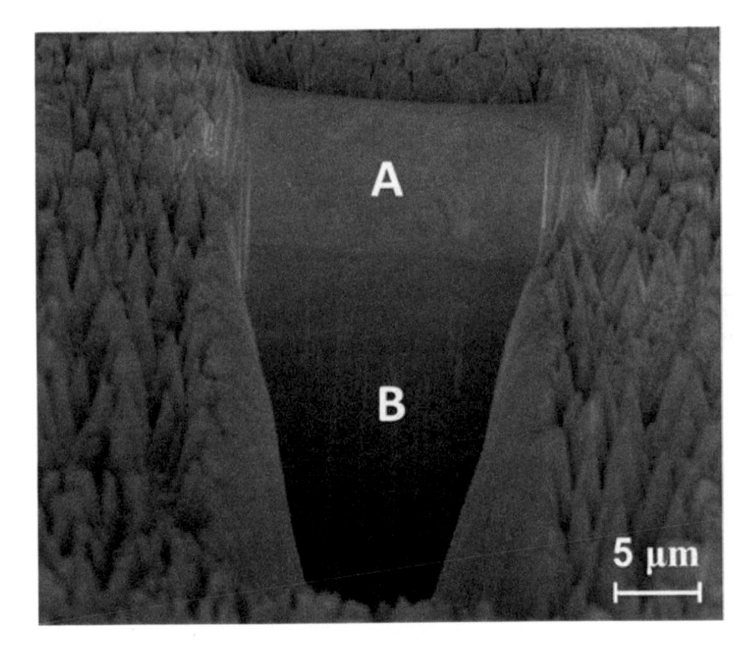

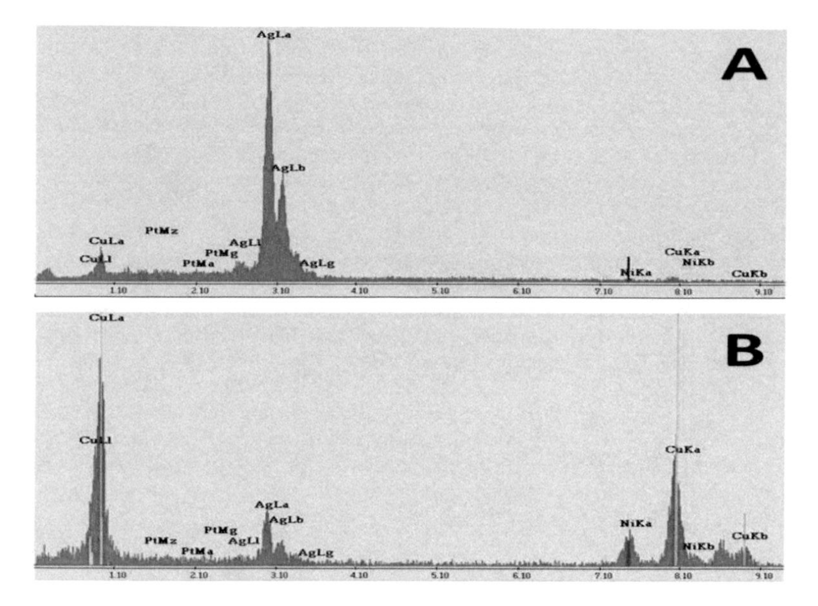

Figure 5. SEM image of a cross-sectional area of 1939 New York World's Fair spoon and corresponding EDS spectra. The "A" region is Ag-plate and "B" region is the underlying bulk metal which appears to be a Cu-Ni alloy or Cupronickel.

In order to determine the effects of overexposure to the ion beam and redeposition during milling, the test sample was milled well beyond the plating to about 50 μm below the plating. This overexposure resulted in a slight discoloration around the milled region which could be seen visually, as illustrated in Figure 6. The discolored area is approximately 150 μm in diameter. This brownish-red "halo" on the surface results from bulk metal redepositing onto the surface as the cross-section is being milled. In addition, it is a function of milling too deeply into the bulk material, and it is not visible when milling depths are below ~25 μm beyond the plating. No other objects in this study exhibited a visible remnant of the milling process.

Figure 6. Photograph in reflected light of the "halo", a slight discoloration due to overexposure to the ion beam and redeposition during milling, on the test object's reverse surface after milling. The "IS", which stands for International Silver, in the photograph is the silversmith's stamp not

damage. (New York World's Fair spoon).

"Century" spoon

FIB milling of a 20 × 35 μm^2 area from the surface revealed a cross-section for both the silver-plating (Figure 7 A) and the bulk metal (Figure 7 B). The "Century" flatware collection is said to be "triple plated". While this implies that three layers of Ag should be present, actually this term refers to the amount of Ag present, i.e. 3 times as much as the standard plating, in the electroplating bath. This was confirmed by comparison of the silver-plating from 1939 New York World's Fair spoon (~10 μm thick) to the "Century" spoon (~36 μm), which is approximately 3 times thicker . Spectra from the silver-plating (Figure 7 A area) and the bulk metal (Figure 7 B area) showed that the silver-plating consisted of predominately Ag with small amounts of Cu, Ga, and Pt, while the bulk metal region consisted of predominately Cu and Ag with small amounts Ga and Pt.

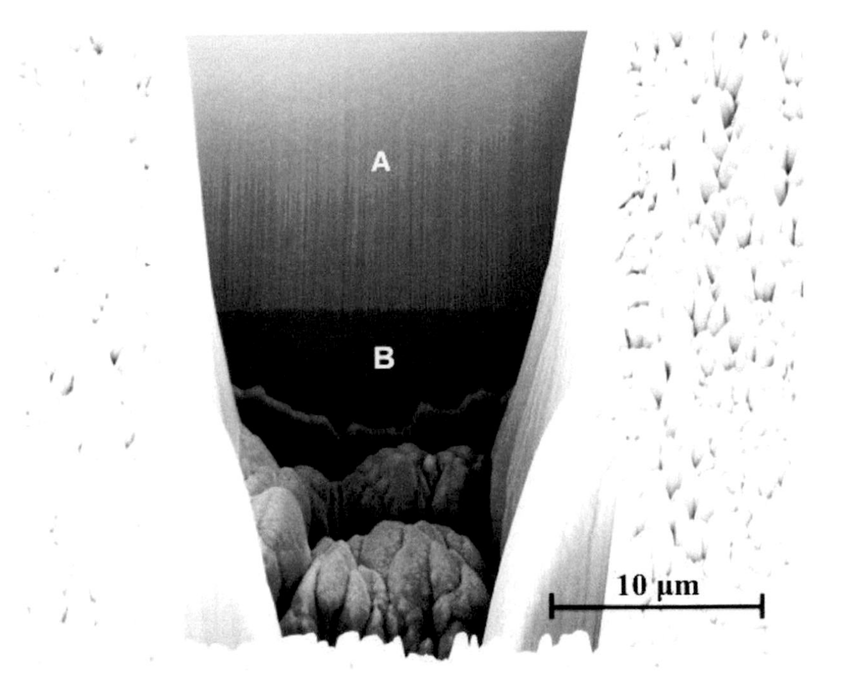

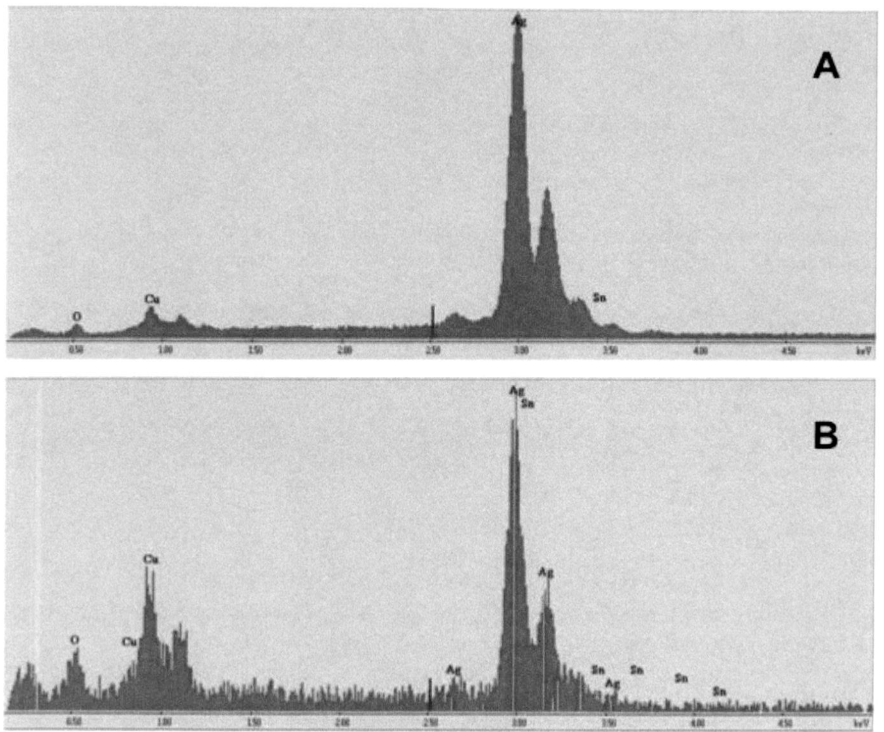

Figure 7. SEM image of a cross-sectional area of the "Century" spoon and corresponding EDS spectra. "A" region is Ag-plate and "B" region is the bulk material which shows mainly Ag due to the redeposited Ag atoms and the cross-section sidewalls being picked up by the EDS beam, as well as Cu.

Like the results from the sacrificial object, the Cu in the silver-plated region most likely comes from the bulk metal rather than the silver-plating. While the bulk metal shows mainly Ag, the spectra is most likely being skewed by redeposition of the Ag atoms, the large beam size detecting the Ag-plating on the sidewalls, the relatively small amount of bulk material able to be analyzed, and the significant depth of the "hole" which causes the secondary electrons to become trapped. Based on the increase in Cu in the spectrum from the "B" region and historical information, the bulk metal is most likely a Cu-based alloy, since Cu is much too soft to be used by itself as a bulk metal. While this is likely a Cu-Ni alloy (Cupronickel) or a Cu-Ni-Zn alloy (German silver), further research is needed to fully clarify the exact composition of the bulk metal. In both cases, the small amounts of Pt are from redeposition, while the Ga peaks result from FIB milling. Visual inspection of the spoon after analysis, showed no signs of visible damage despite having significantly thicker silver-plating than the previous New York World's

Fair spoon, thus confirming that when the redeposition of the underlying bulk metal is not extensive, no visual damage is observed.

Candelabra

The Candelabra was disassembled to fit individual components into the FIB chamber and analysis areas are illustrated by arrows in Figure 4b. Two components were analyzed, one in the middle of one of the middle base plates and the other on the middle surface of one of three supporting legs. In both cases, an approximate $20 \times 20\ \mu m^2$ area on the surface was milled to reveal cross-sectional areas of the Ag-plating and bulk material. The results were similar for both areas. In both cross-sections, the plating thickness is approximately 5 µm and shows porosity at the bulk-plating interface. This implies that there may have been adhesion problems during the electroplating process. The two plating spectra revealed high amounts of Ag with small amounts of Cu, while the bulk material once again revealed a Cu alloy, in this case Cu-Zn or brass. Since the plating was much thinner and the milled area was much cleaner, the spectra for the candelabra are not as skewed. Therefore a qualitative analysis of the composition of the bulk material was performed. This revealed an average composition of approximately 65 wt % Cu and 35 wt % Zn, a common brass composition known as yellow brass. A visual inspection of the objects after milling showed no visible damage.

a
A
AgL
AgL
CuL
CuL
AgL
AgL
CuK
CuK
keV
B
CuL
CuL
CuK
ZnK
CuK
ZnK
keV
b
A
B
5 µm

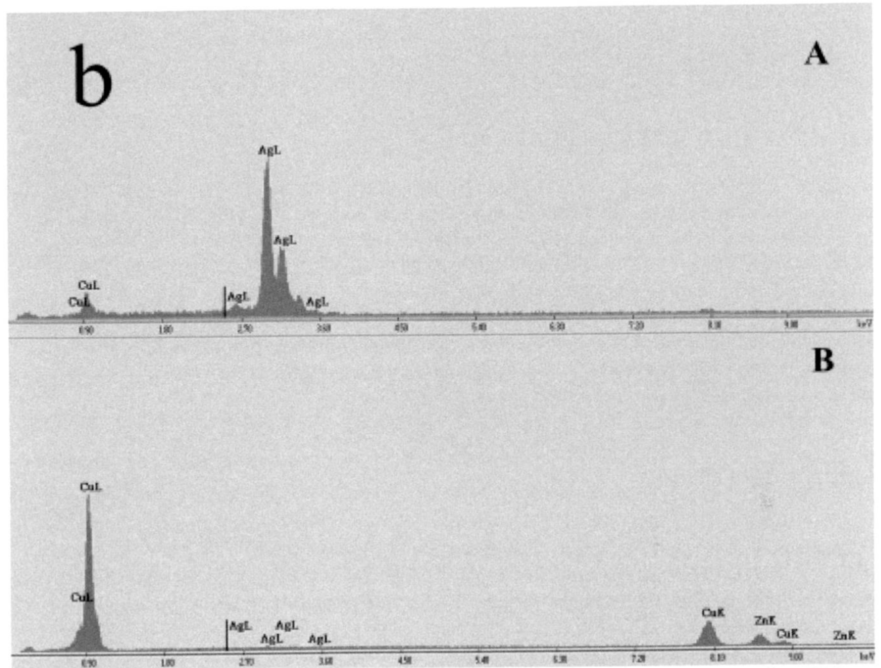

Figure 8. SEM image of revealed cross-sections from Candelabra and corresponding EDS spectra. a) is from the middle base plate and b) is from one of the three supporting legs. In both cases the "A" region is Ag-plate (lighter gray) and "B" region is the bulk material (darker gray).

DISCUSSION

Of the three objects tested, the thickness of the Ag-plating varied greatly from 5 – 36 μm. In addition, the quality of the Ag-plating on the candelabra was inferior to those of the New York World's Fair spoon and the "Century" spoon, as seen in the porosity at the interface. As the plating thickness grows, there is less bulk material present for analysis and more interference with the EDS detector which causes the elemental data gathered to become far less quantifiable but still allows for a qualitative analysis of the bulk material, as seen by the first two objects. However in the case of the candelabra (i.e. thinner plating), the data gathered from the EDS spectra yields more accurate nominal compositions of the bulk material. In order to gather this type of data on objects with thicker plating, greater care in milling the surface needs to be taken or a larger amount of material needs to be removed thereby allowing for addition bulk material to analyze. However the latter, creates a greater risk for potential visual damage to the objects and therefore caution must be undertaken. The visual damage experienced in the test New York World's Fair spoon was limited, but still significant enough to warrant attentiveness when using this method for analysis. Although experts have seen and assured us that this slight discoloration can easily be removed by a simple polishing, it is best to not have any visual change in the object. This goal can easily be achieved by careful observation during the milling process.

Limitations

Mounting the Objects and Tilting for Analysis

Milling the surface of the object to reveal a cross-section perpendicular to the surface requires that the sample stage and object be aligned perpendicular to the ion beam, thus the sample, assuming it is flat, is rotated approximately 52° from the horizontal position. Therefore, the object must be securely mounted to the sample stage in order to use the milling capabilities of the FIB and to prevent the object from slipping. Here, carbon tape, which is easily removed with acetone, can be used to secure the objects to the universal sample stage. Furthermore, the object must also be secured as to avoid all other detectors, pole pieces, (e.g. SEM, FIB, EBSD, EDS, etc…) and apparatuses in the chamber during rotating and positioning. Thus, an intrinsic size and shape limitation, associated with the chamber size, must be determined to safely test objects using this method.

Size and shape limitations

The chamber of the FIB is ~ 40 × 40 × 40 cm^3 with a sample working space of a maximum clearance height of 5.5 cm with x- & y-stage displacements of ~ 5 × 5 cm^2. This size limitation, as well as the need to rotate the sampled object 52° degrees from the horizontal, limits the accessibility to large or oddly shaped objects. Another size-limiting factor is the distance between the sample stage and the SEM/FIB dual beam detectors. Hence, any object placed in the chamber must fit in the chamber and on the platform without hitting the detectors when flat or rotated for the milling process. This challenge is exacerbated when the object has a curved surface because the sampling area must be normal relative to the ion beam column during milling. Many decorative hollowares have curved or irregular surfaces requiring different tilt angles, either greater than or less than 52° relative to normal of the platform, in order to achieve the necessary alignment with the ion beam column for milling. Furthermore, the problem of slipping is also more pronounced with curved or oddly shaped objects. Another challenge encountered with curved objects is their tendency to block or obstruct the secondary electron beam detectors and block or limit the amount of electrons collected during EDS testing; however, many of these problems could be alleviated with a larger chamber that would allow for a better collection of scattered electrons as well as allowing for larger objects to be tested,.

Time per experiment

Since the sample must be prepared for analysis by milling, a period of time required to complete the process. This time involves setup, Pt deposition, milling, cleaning, and then finally analysis of the sample using EDS. By far, the milling and cleaning process takes the majority of the time. Although this time, obviously, varies with plating thickness, the average pieces takes approximately 3-4 hours for analysis. This time can be reduced by increasing the ion beam current, 20nA is max for the FIB system used in this paper, or by milling smaller volumes if the plating depth can be properly estimated. At a current of 20 nA, a 20 × 25 × 25 μm^3 volume takes approximate one hour to complete. Then the cross-section must be cleaned since the face will be distorted by redepostion. This process can take upwards of 45 minutes to get a clean image and if the base metal is still not visible, then further milling will be needed which will also require another cleaning phase. However, other FIB systems can achieve higher beam currents that allow faster bulk milling and substantial reduce the time required for analysis. However, higher currents often cause more scattering and deposition of ions from different layers and this increases contamination.

Surface redeposition

As discussed earlier, during the milling process, the material (Ag, Cu, Ni, and Zn in the cases presented in this paper) is lifted off of the surface and bulk of the object and deposited along the surface surrounding the milling site. If the material being lifted out is the same as the material it is being deposited on, then there is no issue, but there can be a discoloration of the surface area, as seen in the New York World's Fair spoon. In addition this redeposition adds time to the cleaning process and it can also interfere with the EDS analysis, making it difficult to obtain quantitative data for the bulk material.

Milling depth limitations

The cross-section size determines the maximum depth possible and this size must be small enough (less than $\sim 100 \times 100\ \mu m^2$) so that it is not visually observable. Furthermore, once the milled "hole" is too deep, the sputtered atoms do not have enough energy to escape and will be re-deposited back into the "hole" or on the sidewalls, and thus obstructing the view of the cross-section. Since a deeper hole requires a greater viewing width to see the bottom, the milling area in the y-direction must also be expanded to create any benefit of the deeper cross-section. Eventually, the "hole" created will be visible to the unaided human eye, of about 100 μm.

Advantages

Despite the challenges and complications mention above, this technique offers unique benefits that traditional cultural heritage analytical techniques cannot provide. For instance, an undistorted picture of both the plating and the bulk material can be obtained without cutting of the sample. In addition, images of the grain structure within each layer can be seen without traditional etching of the sample due to the nature of ion beam. These images can be used for texture analysis, thus providing insight into the processing techniques employed during manufacturing. Furthermore, the elemental composition of all layers can be quantified from *in situ* EDS when the plating is not too thick (~100 μm maximum without incurring visible damage) and more importantly can be compared sacrificial test objects, which are either sectioned or milled. This technique also allows for large bodies of work, such as the DMA's Ag and Ag-plated collection, to be studied rapidly. Future dual beam SEM/FIB experiments involve using *in situ* EBSD to determine the grain texture and phase information as well as quantitative elemental information from EDS.

CONCLUSIONS

In this study, a dual beam SEM/FIB system with EDS and EBSD capabilities is used, for the first time, to analyze non-destructively Ag-plated cultural heritage objects from the Dallas Museum of Art. Compared to other characterization techniques, this technique can be used to characterize both the surface and bulk materials of large bodies of objects, due to its ease of availability and rapidity of use. With many analytical techniques, the surface composition can be misleading due to surface corrosion and alteration; however, by applying this technique with EDS to the DMA's silver collection, it is possible to characterize the composition of both the underlying bulk material and the Ag-plating.

Here, we examined three cultural heritage objects:

- 1939 New York World's Fair spoon, which was found to have a 10 μm Ag-plating on a Cu-Ni-based bulk metal. This test object was used to show that while the FIB technique is technically destructive, it is shown that the resulting damage is so small (on the order of a few 100 μm^2 for the worst case scenario) as to be insignificant and considered non-destructive.
- "Century" spoon, which was found to have a much thicker 36 μm Ag-plating on a Cu alloy base metal. This object was chosen from the DMA's collection to illustrate the depth capability of the novel technique.
- Candelabra, which was found to have a thin 5 μm Ag-plating on a yellow brass bulk metal. In addition, porosity was found at the plating-bulk interface suggesting poor adhesion during the electroplating process during manufacturing. For this object, the technique was performed on two different components to show that the method can yield repeatable and quantifiable results and can document variability.

Although the technique is only currently being applied to Ag-plated objects, it is easy to imagine applying it in other cultural heritage studies, especially since these SEM/FIB systems are becoming more readily available in both university and industrial settings.

ACKNOWLEDGMENTS

The authors would like to thank John Dennis, Mark Leonard, and Kevin Tucker from the Dallas Museum of Art for assistance with the objects and for their contributions to the project. The authors gratefully acknowledge the use of experimental facilities at the Center for Advanced Research and Technology (CART) at the University of North Texas and Dr. D. Jaeger for assistance with the FIB.

REFERENCES

[1] Ciliberto E, Spoto G. Modern Analytical Methods in Art and Archaeology. New York: John Wiley & Sons; 2000.
[2] Scott DA. Metallography and Microstructure in Ancient and Historic Metals. Singapore: J. Paul Getty Museum; 1991.
[3] Drakaki E, Kandyla M, Chatzzitheodoridis E, Zergioti I, Serafetinides AA, Terlixi A, et al. Laser Studies of Metallic Artworks. Appl Phys a-Mater 2010;101:349-55.

[4] Tate J. Some problems in analysing museum material by nondestructive surface sensitive techniques. Nuclear Instruments and Methods in Physics Research Section B: Beam Interactions with Materials and Atoms 1986;14:20-3.
[5] Beck L, Alloin E, Berthier C, Reveillon S, Costa V. Silver surface enrichment controlled by simultaneous RBS for reliable PIXE analysis of ancient coins. Nuclear Instruments and Methods in Physics Research Section B: Beam Interactions with Materials and Atoms 2008;266:2320-4.
[6] Mass JL, Matsen CR. Understanding silver hollow wares of the eighteenth and nineteenth centuries: Is there a role for X-ray fluorescence? The Decorative: Conservation and the Applied Arts 2012 IIC Congress. Vienna: The International Institute for Conservation of Historic and Artistic Works; 2012. p. S191-S8.
[7] Vittiglio G, Janssens K, Vekemans B, Adams F, Oost A. A compact small-beam XRF instrument for in-situ analysis of objects of historical and/or artistic value. Spectrochimica Acta Part B: Atomic Spectroscopy 1999;54:1697-710.
[8] Hanson VF. The Curator's Dream Instrument. In: Young WJ, editor. Application of Science in the Examination of Works of Art - proceedings of the seminar: June 15-19, 1970, conducted by the research laboratory. Boston, Massachussetts: Museum of Fine Arts; 1973. p. 18-30.
[9] Lahanier C, Preusser FD, Van Zelst L. Study and Conservation of Museum Objects: Use of Classical Analytical Techniques. Nuclear Instruments & Methods in Phys Res B 1986;14:1-9.
[10] Beck L, Bosonnet S, Reveillion S, Eliot D, Pilon F. Silver surface enrichment of silver-copper alloys: a limitation for the analysis of ancient silver coins by surface techniques. Nuclear Instruments & Methods in Phys Res B 2004;226:153-62.
[11] Quimby IMG, Johnson D, Museum HFdPW. American silver at Winterthur: Henry Francis du Pont Winterthur Museum; 1995.
[12] Dran J, Salomon J, Calligaro T, Walter P. Ion beam analysis of art works: 14 years of use in the Louvre. Nuclear Instruments & Methods in Phys Res B 2004;219-220:7-15.
[13] Grassi N, Giuntini L, Mando PA, Massi M. Advantages of scanning-mode ion beam analysis for the study of Cultural Heritage. Nuclear Instruments & Methods in Phys Res B 2007;256:712-8.
[14] Harbottle G, Gordon BM, Jones KW. Use of synchrotron radiation in archaeometry. Nuclear Instruments & Methods in Phys Res B 1986;14:116-22.
[15] Young ML. Archaeometallurgy using synchrotron radiation: a review. Reports on Progress in Physics 2012;75:036504.
[16] Stern J. Modernism in American Silver: 20th-Century Design. New Haven, CT: Yale University Press; 2005.
[17] Sciau P, Salles P, Roucau C, Mehta A, Benassayag G. Applications of focused ion beam for preparation of specimens of ancient ceramic for elctron microscopy and synchrotron X-ray studies. Micron 2009;40:597-604.
[18] Vander Voort GF. Metallography, Principles and Practice: ASM International; 1984.

Mater. Res. Soc. Symp. Proc. Vol. 1656 © 2014 Materials Research Society
DOI: 10.1557/opl.2014.708

The Potential of Low Frequency EPR Spectroscopy in Studying Pottery Artifacts and Pigments.

William J. Ryan,[1] Nicholas Zumbulyadis,[2] Joseph P. Hornak[1]

[1]RIT Magnetic Resonance Laboratory, RIT Rochester, NY, 14623
[2]Independent Researcher, Rochester, NY

ABSTRACT

Non-destructive investigation, chemically fingerprinting, and authentication of ceramic cultural artifacts is a challenging analytical problem. Electron paramagnetic resonance (EPR) spectroscopy is capable of distinguishing between clays based on the paramagnetic metals present, and firing temperature (T_F) based on the complexes of these metals formed at different T_F values. Unfortunately, the 9 GHz frequency of conventional X-band EPR restricts sample size to a few mm and limits its applicability to small fragments. Low frequency EPR (LFEPR) is based on an EPR spectrometer operating at a few hundred MHz. LFEPR can utilize larger samples on the order of a few cm, but has a lower sensitivity due to the smaller Boltzmann ratio. Additionally, LFEPR may not be capable of detecting a spectral transition if the LFEPR operating frequency is less then the zero-field splitting of the paramagnetic metal complex. We utilized an LFEPR operating at 300 MHz which scans the applied magnetic field between the local Earth's magnetic field and 26 mT to determine the feasibility of detecting EPR signals from clays, pigments, and glazes. Various clay samples were studied at $100 < T_F < 1200$ °C. Spectral differences were seen as a function of both clay type and T_F. Differences in the LFEPR spectra of Han, Egyptian, and Ultramarine blue support the ability to distinguish among pigments. Paramagnetic impurities in glass may allow distinction between glaze spectra. We have also explored the utility of LFESR by the use of a radio frequency surface coil rather than an enclosed resonator. Although the active volume of the surface coil is ~1 cm^3, objects as large as 20 cm in diameter might be easily characterized with our spectrometer.

INTRODUCTION

The main objective of the work reported in this paper is to explore the application of low frequency EPR spectroscopy (LFEPR) to the study of cultural heritage objects. Such objects can be works of art or artifacts of archaeological, ethnographic and anthropological significance. Cultural heritage science employs a diverse set of analytical methods, all which share the following attributes, at least to some extent. Ideally they should be (a) non-destructive, (b) fast, (c) simple, (d) provide relevant information, (e) mobile and (f) relatively inexpensive to implement.

The two most commonly used non-destructive techniques are X-ray fluorescence (XRF) [1] and vibrational spectroscopy using fiber optic setups, most notably mobile Raman spectroscopy. [2] These techniques have been successfully used in solving many problems in authentication and object conservation and have provided significant insights for the understanding of early produc-tion technologies, trade roots, etc. Nevertheless, both analytical approaches have their limitations. XRF is typically used for qualitative elemental analysis; accurate quantitation of XRF is very difficult. Raman spectroscopy gives information about functional groups, but may not be sensi-tive to subtle changes in short and intermediate range order and is overwhelmed by interference from fluorescence, by the chemical species of interest or the matrix in which the material of interest is embedded. X-ray microdiffractometry requires a very small amount of sample and can be viewed as a minimally invasive technique, but is applicable only to crystalline materials.

NMR and EPR provide more structural detail, unfortunately they are also significantly invasive requiring around 100 mg of sample. In the context of cultural heritage science they are appropriate for the evaluation of replication experiments, the characterization of model systems, and whenever large quantities of sacrificial samples are available. The applications of NMR spectroscopy to art and archaeology have been reviewed in the recent literature. [3]

There have also been numerous archaeometric applications of EPR spectroscopy. These include the characterization of lithic objects, [4] identification of different marble quarries used in antiquity, [5] comparison of medieval and early modern paper samples, [6] and the estimation of the T_F of archaeological clay objects. [7-9] All of these studies were carried out using conventional X-band EPR and required the intrusive removal of up to 100 mg of sample.

Our ultimate objective is to develop EPR methodology that is completely non-destructive and allows the examination of entire objects from as-found shards (≈1 cm) to vessels (≈30 cm). We believe that the most promising and realistic approach for attaining this objective is to use low-frequency EPR [10-12] together with surface coils. [13-19] In this paper we report preliminary results on clay, pigments, and glazes. We believe that it will also be possible to study marble and flint using LFEPR with surface coils.

The T_F of clay objects and the associated changes in the clay structure hold significant clues to the evolution of ceramic technology from the first artifacts made by hunter gatherer groups to the creations of the artisans of early modern Europe. Several analytical techniques including EPR, microscopy, and thermoluminescence have been used for the determination of T_F. Of those EPR techniques appear to be the most reliable. However, all of these methods require the removal of up to 100 mg of material from the objects and are thus significantly invasive. We will show that similar measurements can be performed in a non-destructive manner using LFEPR.

We have also examined the Cu^{+2} LFEPR spectra of Egyptian and Han blue. Egyptian blue, calcium copper silicate ($CaCuSi_40_{10}$), is the first synthetic pigment ever invented by mankind. [20] Han blue ($BaCuSi_40_{10}$) was discovered independently in China. Even though Han blue is isostructural to the Egyptian blue, the two pigments show subtle differences in their photophysical behavior. [21] These pigments are encountered mixed with other copper silicate phases. Distinguishing the various phases in a non-destructive way would be archaeologically very significant. Previously only the X-band EPR spectrum of Egyptian blue has been published. [22] Lapis lazuli, the pigment, was the earliest known naturally occurring blue inorganic pigment was widely used in paintings, illuminated manuscripts, and also for the decoration of European and Islamic medieval pottery. [23-24] The blue color stems from polysulfide radical anions such as S_3^- and S_2^- The correlation between EPR, Raman and colorimetric characteristics of synthetic ultramarine pigments has been reported. [25]

THEORY

EPR is used to study materials with unpaired electrons such as free radicals, transition metal ions, and electron holes. [26] When placed in an applied magnetic field (B_o), the magnetic moment of an unpaired electron can align either with or opposed to the field. This creates two energy states, probable by the absorption of a photon of frequency ν. B_o is related to ν by the nuclear magneton (β) and the g-factor, a unique constant for each electron spin environment.

$$h\nu = g\,\beta\,B_o$$

The position of an absorption peak in an EPR spectrum is identified by its g-factor. In many materials an anisotropic g-factor causes broad, powder pattern peaks. The spectral line shape can be further complicated by the first-derivative presentation of the EPR signal from the B_o field modulation and phase-sensitive detection at the modulation frequency. In summary, the qualitative parameters are g-factor and peak-to-peak linewidth (Γ_{PP}), and quantitative parameters are peak-to-peak signal height (S_{PP}) and absorption peak area.

The relative population of the two spin states (n^+/n^-) and the EPR signal is determined by Boltzmann statistics.

$$(n^+/n^-) = e^{-h\nu/kT} = e^{-g\beta B_o/kT}$$

The greater the energy difference between two levels, the greater the signal. LFEPR will inherently have a lesser signal then X-band EPR. The quantum mechanics of the spin system predict a zero-field splitting (E_o) in some energy levels diagrams. When $E_o > h\nu$, no EPR signal will be observed. Owing to the latter two limitations of LFEPR, we undertook this study to determine the potential of LFEPR to study pottery.

EXPERIMENTAL

The LFEPR was similar to that described previously [10] with the following modifications shown in Figure 1. The hybrid tee, magnet power supply, and lock-in amplifier were replaced, respectively, with a three-port circulator (UTE Microwave, CT1375-N); a 36 V, 30 A power supply (Kepco, ATE 36-30M); and a digital lock-in amplifier (Stanford Research Systems, SR830 DSP). The magnet was still swept by a 0 to 1 V control voltage. This voltage was produced by two 16 bit D/A converters resident in the lock-in amplifier and a voltage adder. This system allowed more flexibility and finer sweep steps than the original spectrometer. A tuning circuit was incorporated into the bridge which was disengaged during spectral acquisition. A mag-net insert and second DC power supply (Kepco, MPS 620M) were added to boost the magnetic field by 7.5 mT when larger field sweeps were needed. Magnetic field sweeps were calibrated with a hall probe gaussmeter (FW Bell, 5180).

The spectrometer was controlled through LabView code over a IEE-488.2 bus. Two, glass free, sample coils were used. The original single turn solenoid (STS) [27] was used with samples in 15 mm diameter styrene tubes. The second coil was a 7-turn, two cm diameter, spiral surface coil [28] with the previously described surrounding modulation coils. [29]

Samples were fired in a temperature-controlled, muffle furnace (Barnsteed-Thermolyne F47915) at the desired temperature for two hours, then slowly cooled to 20 ºC before studied by EPR. The clay sample was composed of the following weight percents of commercial pottery clay or components: Red Art (35), XX Sagger (17), Hawthorn Bond (17), Talc (7), and Grog (21). Samples were rolled into 1 cm diameter, 65 cm long rods and dried at 100 °C for two hours before firing. Other samples included a 1927, 50 pf, red stoneware coin from Saxony (Notgeld) ; a red terracotta clay flower pot; Han and Egyptian Blue (Kremer Pigments, NY, NY); Ultramarine Blue (Natural Pigments, Willits, CA); 5 mm glass beads (Quackenbush, Crystal Lake, IL); and the EPR line-shape and g-factor standard 2,2-diphenyl-1-picrylhydrazyl (DPPH) (Fluka Chemical).

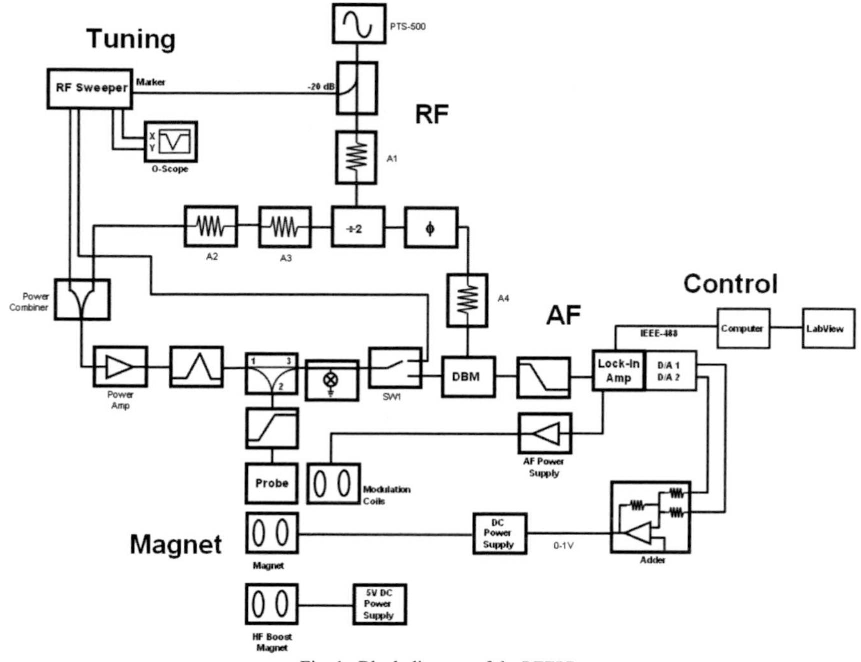

Fig. 1. Block diagram of the LFEPR.

RESULTS AND DISCUSSION

The LFEPR spectra of our clay samples fired at temperatures between 100 and 1200 °C showed three unique spectral peaks: a broad g≈4, a narrow g≈2, and a broad g≈2. The original sample has the narrow g≈2 and broad g≈4 peaks. The narrow g≈2 peak rapidly disappears on heating, while the g≈4 peak is more persistent. At 800 °C the broad, strong g≈2 peak starts to grow, reaching its maximum S_{PP} at 1000 °C. This broad g≈2 peak diminishes to approximately 5% of its 1000 °C size by 1200 °C. Fig. 2 summarizes the change in S_{PP} of these three g-factor absorptions with temperature. With these three temperature dependent spectral peaks it is possible to estimate the T_F. For example the ratio of S_{PP} for the g≈2 to g≈4 peaks can be used as a temperature marker between 100 and 500 °C. The presence of only the g≈4 peak indicates a temperature between 500 and 800 °C. The ratio of the broad g≈2 to g≈4 peaks can be used between 900 and 1200 °C.

The three blue pigments studied possessed unique LFEPR spectra. The copper paramagnetic center in Egyptian blue possesses a broader Γ_{PP} than Han blue (Fig. 3). Since the two pigments are isostructural except for the presence of Ca or Ba in the lattice, the Γ_{PP} difference must be related to these cations. Ultramarine blue is a modern day synthetic version of the natural mineral lapis lazuli. This pigment contains a S_3^- radical anion which is responsible for the EPR signal. Its Γ_{PP} value is narrower than the other two blue pigments and more asymmetric. The uniqueness of the three spectral peaks should allow discrimination of these three blue pigments from other related phases in a ceramic artifact.

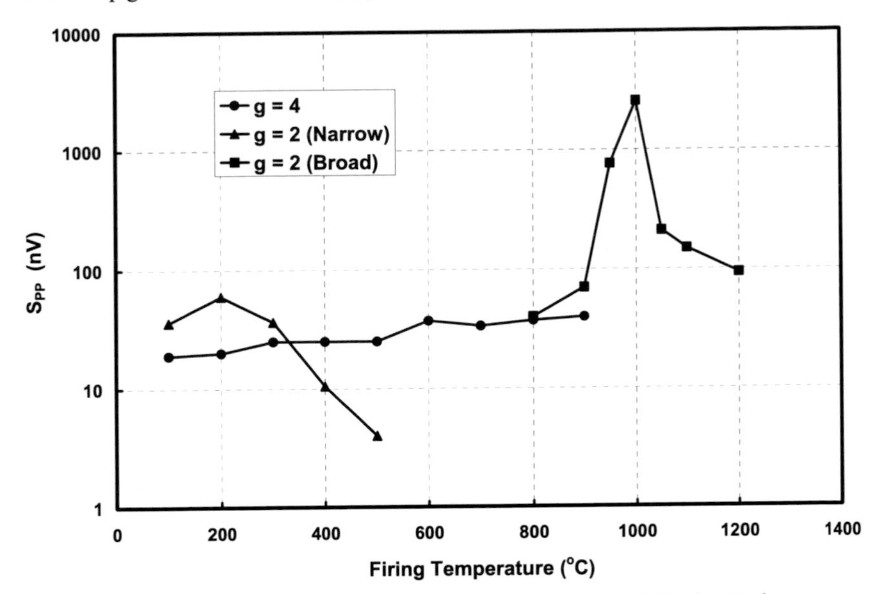

Fig. 2. Plots of S_{PP} vs. firing temperature for the three g-factors present in the clay samples.

Glazes are silicates and may possess an LFEPR signal due to paramagnetic impurities, paramagnetic pigments, or f-centers from radiation exposure. The glass bead sample displayed a large $g \approx 4$ peak from an Fe^{2+} impurity (Fig. 4).

Very few intact pottery samples could fit into the 15 mm diameter STS sample coil. Expanding the STS volume is possible [27]. However, the maximum size will be limited by the available power from the RF power amplifier. As an RF power amplifier is a significant cost factor in the construction of an LFEPR, it makes sense to explore alternatives to volume coils, namely surface coils. [15] This configuration will be limited by the size of the modulation coils. It also allows for localized EPR spectroscopy on large objects.

As an initial test of our prototype surface coil system, the spectrum of a DPPH sample in epoxy was measured. The spectrum displays a symmetrical first derivative shape indicating proper phasing of the RF and modulation field. Two fired clay samples were examined with this surface coil. The first was a common red terracotta clay flower pot. Terracotta is an earthen ware typically fired to 926–1150°C today, but to lower temperatures in early times. This sample gave an intense, broad, $g \approx 2$ iron signal (Fig. 5). The Notgeld coin also produced a broad peak, smaller in amplitude, and shifted to larger g-factor. The coin Γ_{PP} value was greater than that of the flower pot.

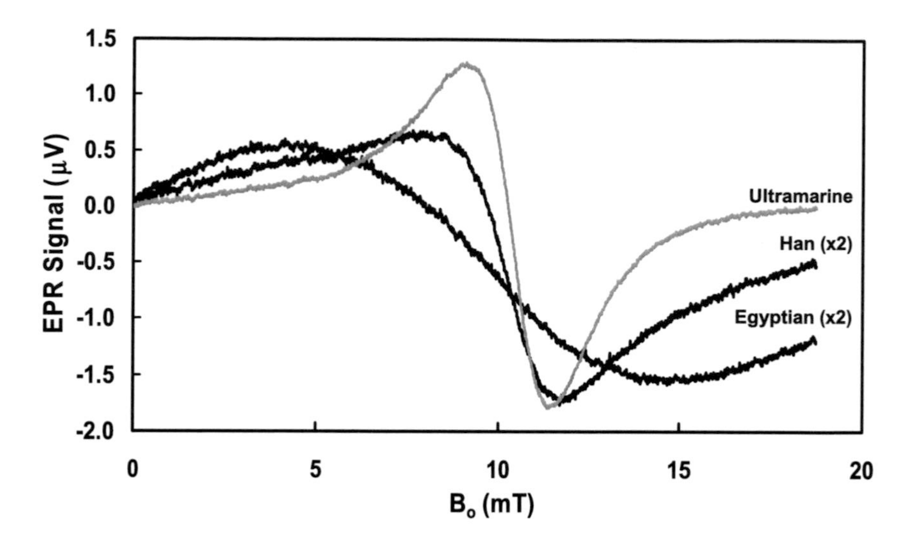

Fig. 3. LFEPR spectra of Ultramarine, Egyptian, and Han blue pigments.

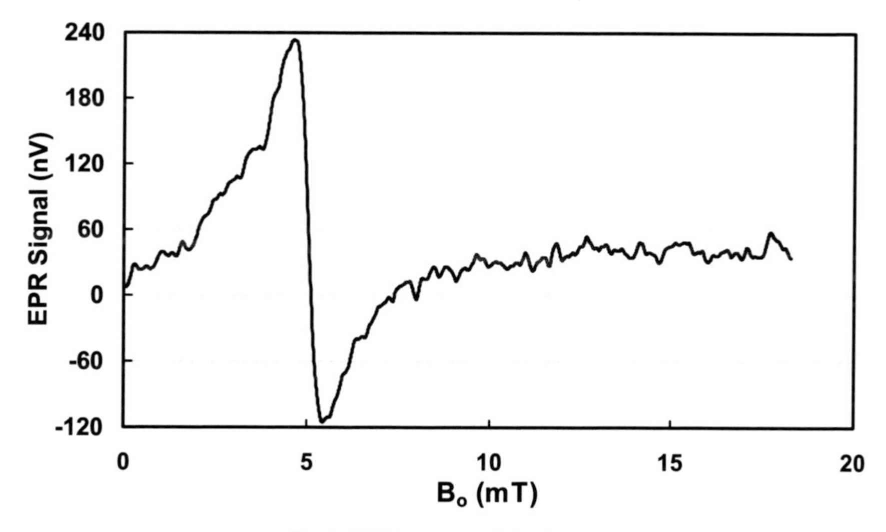

Fig. 4. LFEPR spectrum of glass beads.

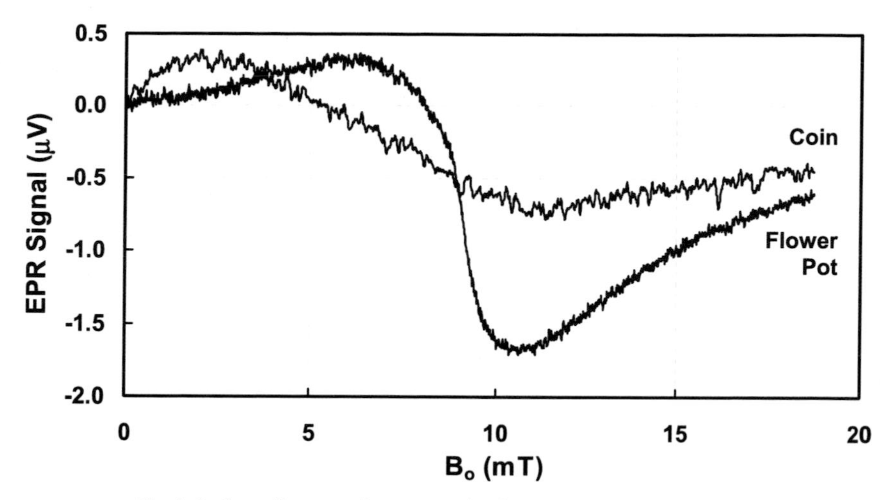

Fig. 5. Surface coil spectra of a terracotta clay flower pot and 1921 Notgeld coin.

CONCLUSIONS AND FUTURE DIRECTIONS

LFEPR is capable of detecting the signal from paramagnetic iron found in clay and thus serves distinguish between clay types. LFEPR can follow changes in the signal which occur as a consequence of the firing process, opening the unique possibility of T_F reporting. LFEPR can detect differences in three different blue pigments (two Cu^{2+} based and a S_3^- radical anion) found in pottery. Glazes with paramagnetic impurities, such as Fe^{2+}, or pigments will also yield a detectable LFEPR signal.

More development work is needed in two areas. First, clays with other paramagnetic metals such as Mn, Co, and Cr, need to be studied to determine whether $h\nu > E_o$ and a signal can be observed. Differences between the EPR signal of these metals will advance the ability of LFEPR to distinguish between clay types. Second, the sensitivity should be improved to allow detection of smaller concentrations of paramagnetic species. This will facilitate detection of differences between glazes and pigments found in thin layers on pottery surface.

As a result of this research, we envision that future EPR of cultural heritage objects to be carried out in a non-destructive and non-invasive manner. The LFEPR concept may also allow future spectrometers to be more compact, mobile, and less expensive.

ACKNOWLEDGEMENTS

We thank Jane Shellenbarger for the clay samples and helpful discussions.

REFERENCES

1. A.N. Shugar, J. L. Mass (eds.), Handheld XRF for Art and Archaeology (Studies in Archaeological Sciences), Leuven University Press, 2013.

2. P. Vandenabeele, J. Tate, L. Moens, Non-destructive Analysis of Museum Objects by Fibre-optic Raman Spectroscopy, Anal. Bioanal. Chem., 387, 813 (2007).
3. D. Capitani, V. Di Tullio, N. Proietti, Nuclear Magnetic Resonance to Characterize and Monitor Cultural Heritage, Progress Nucl. Magn. Reson., 64, 29 (2012).
4. G.V. Robins, N. J. Seeley, D. A. C. McNeil and M. C. R. Symons, Identification of Ancient Heat Treatment in Flint Artefacts by ESR Spectroscopy, Nature, 276, 703 (1978).
5. D. Cordischi, D. Monna and A. L. Segre, ESR Analysis of Marble Samples from Mediterranean Quarries of Archaeological Interest, Archaeometry, 25, 68 (1983).
6. D. Attanasio, D. Capitani, C. Federici, A. L. Segre, Electron Spin Resonance Study of Paper Samples Dating from the Fifteenth to the Eighteenth Century, Archaeometry, 37, 377 (1995).
7. T. Warashina, T. Higashimura, Y. Maeda, Determinationof the Firing Temperature of Ancient Pottery by Means of ESR Spectroscopy, British Museum Occasional Papers, 19, 117 (1981).
8. Y. Bensimon, B. Deroide, S. Clavel, J.V. Zanchetta, Electron Spin Resonance and Dilatometric Studies of Ancient Ceramics Applied to Determination of Firing Temperature, Jpn. J. Appl. Phys. 37, 4367 (1998).
9. F. Presciutti, D. Capitani, A. Sgamellotti, B.G. Brunetti, F. Costantino, S. Viel, A. Segre, Electron Paramagnetic Resonance, Scanning Electron Microscopy with Energy Dispersion X-ray Spectrometry, X-ray Powder Diffraction, and NMR Characterization of Iron-Rich Fired Clays, J. Phys. Chem. B, 109, 22147(2005).
10. J.P. Hornak, M. Spacher, R.G. Bryant, A Modular Low Frequency ESR Spectrometer, Meas. Sci. Technol. 2, 520 (1991).
11. G.A. Rinard, R.W. Quine, G.R. Eaton, S.S. Eaton, E.D. Barth, C.A. Pelizzari, H.J. Halpern, Magnet and Gradient Coil System for Low-Field EPR Imaging, Concepts Magn. Reson. (Magnetic Resonance Engineering), 15, 51(2002).
12. R.W. Quine, G.A. Rinard, S.S. Eaton, G.R. Eaton, A Pulsed and Continuous Wave 250 MHz Electron Paramagnetic Resonance Spectrometer, Concepts Magn. Reson. 15, 84(2002).
13. H. Nishikawa, H. Fuji, L.J. Berliner, Helices and Surface Coils for Low-Field *in vivo* ESR and EPR Imaging Applications. J. Magn. Reson. 62, 79 (1969).
14. A. Sotgiu, H Fuji, G Gualtieri, Toroidal Surface Coil for Topical ESR Spectroscopy. J. Phys. E: Sci. Instrum. 20, 1428 (1987).
15. M.R. Bendall, Surface Coil Technology. In Magnetic Resonance Imaging, ed. by C.L. Partain, R.R. Price, J.A. Patton, M.V. Kulkarni, A.E. James, Saunders, Philadelphia, 1988.
16. M. Ono, K. Ito, N. Kawamura, K C Hsieh, H Hirata, N Tsuchihashi, H. Kamada, A Surface-Coil-type Resonator for *in vivo* ESR Measurements. J. Magn. Reson., 104B, 180 (1994).
17. Y. Lin, H. Yokoyama, S.-I. Ishida, N. Tsuchihashi, T. Ogata, *In vivo* Electron Spin Resonance Analysis of Nitroxide Radicals Injected into a Rat by a Flexible Surface-Coil-Type Resonator as an Endoscope- or a Stethoscope-Like Device. MAGMA 5, 99 (1997).
18. M. Tada, H. Yokoyama, Y. Toyoda, H. Ohya, T. Ito, T. Ogata, Surface-Coil-Type Resonators for in Vivo Temporal ESR Measurements in Different Organs of Nitroxide-Treated Rats. Appl. Magn. Reson. 18, 575 (2000).
19. H. Yokoyama, M. Tada, T. Sato, H. Ohya, T. Akatsuka, Modified Surface-Coil-Type Resonators for EPR: Measurements of a Thin Membrane Like Sample. Appl Magn Reson 24, 233 (2003).
20. H. Berke, The Invention of Blue and Purple Pigments in Ancient Times, Chem Soc Rev 36, 15 (2007).

21. G. Pozza, D. Ajo, G. Chiari, F. De Zuane, M. Favaro, Photoluminescence of the Inorganic Pigments Egyptian Blue, Han Blue, and Han Purple, J. Cult. Herit., 1, 393 (2000).
22. P. Mirti, L. Appolonia, A. Casoli, Spectrochemical and Structural Studies on a Roman Sample of Egyptian Blue, Spectrochimica Acta (A), 51A, 437 (1995).
23. R.J.H. Clark, M.L. Curri, C. Laganara, Raman Microscopy: the Identification of Lapis Lazuli on Medieval Pottery Fragments from the South of Italy, Spectrochim. Acta (A) 53, 597 (1997).
24. P. Colomban, Lapis Lazuli an Unexpected Blue Pigment in Iranian Lajvardina Ceramics, J. Raman Spectr., 34, 420 (2003).
25. N. Gobeltz, A. Demortier, J.P. Lelieur, C. Duhayon, Correlation between EPR, Raman, and Colorimetric Characteristics of the Blue Ultramarine Pigments. J. Chem. Soc., Faraday Trans., 94, 677 (1998).
26. J.E. Wertz, J.R. Bolton, Electron Spin Resonance: Elementary Theory and Practical Applications. Chapman and Hall, NY, 1972.
27. J.P. Hornak, J. Szumowski, R.G. Bryant, Elementary Single Turn Solenoids Used as the Transmitter and Receiver in Magnetic Resonance Imaging, Magn. Res. Imag. 5, 233 (1987).
28. T. Munsat, W. M. Hooke, S. P. Bozeman, S. Washburnd, Two New Planar Coil Designs for a High Pressure Radio Frequency Plasma Source. Appl. Phys. Lett. 66, 2180 (1995).
29. E. Szczepaniak, J.P. Hornak, ESR Imaging Based on the Modulation Field Phase, J. Magn. Reson. 104A, 315 (1993).

Mater. Res. Soc. Symp. Proc. Vol. 1656 © 2017 Materials Research Society
DOI: 10.1557/opl.2017.3

RETRACTION

Ceramics at the Emergence of the Silk Road: A Case from Southeastern Kazakhstan - **RETRACTION**

MaryFran Heinsch, Pamela B. Vandiver, Kyra Lyublyanovics, Alice M. Choyke, Perry Tourtellotte, Claudia Chang

doi: 10.1557/opl.2014.430, Published by Materials Research Society, 09 May 2014.

An un-reviewed manuscript [1] was published online in error and it is therefore necessary to retract it. The correct and final version of record [2] is available at: https://doi.org/10.1557/opl.2015.841. Readers should cite this version.

The publisher and editors apologize to the author and readers for this mistake.

REFERENCES

1. Heinsch, M., Vandiver, P. B., Lyublyanovics, K., Choyke, A. M., Tourtellotte, P., Chang, C. (2014). RETRACTED–Ceramics at the Emergence of the Silk Road: A Case from Southeastern Kazakhstan. *MRS Proceedings, 1656*. doi: 10.1557/opl.2014.430
2. Heinsch, M., Vandiver, P.B., Lyublyanovics, K., Choyke, A. M., Reedy, C., Tourtellotte, P., Chang, C. (2015). Ceramics at the Emergence of the Silk Road: A Case of Village Potters from Southeastern Kazakhstan during the Late Iron Age. *MRS Proceedings, 1656*. doi: 10.1557/opl.2015.841.

Mater. Res. Soc. Symp. Proc. Vol. 1656 © 2018 Materials Research Society
DOI: 10.1557/opl.2018.2

CORRIGENDUM

Methods of Faience Manufacture in Antiquity: Investigation of Colorants and Technological Processes - **CORRIGENDUM**

Lesley Frame, Donna Bright DeSorda, Yuan-Chi Chang, and Pamela Vandiver

doi: 10.1557/opl.2011.927, Published by Materials Research Society, 20 June 2011.

In the original publication of Frame et al.[1], the numerical values in Figs. 1 and 2 on pp. 48-49 were not legible. The authors apologize to the readers and provide updated figures below.

REFERENCE

1. Frame, L., DeSorda, D.B., Chang, Y.-C., and Vandiver, P. (2014). Methods of Faience Manufacture in Antiquity: Investigation of Colorants and Technological Processes. *MRS Proceedings, 1319*. doi: 10.1557/opl.2011.297.

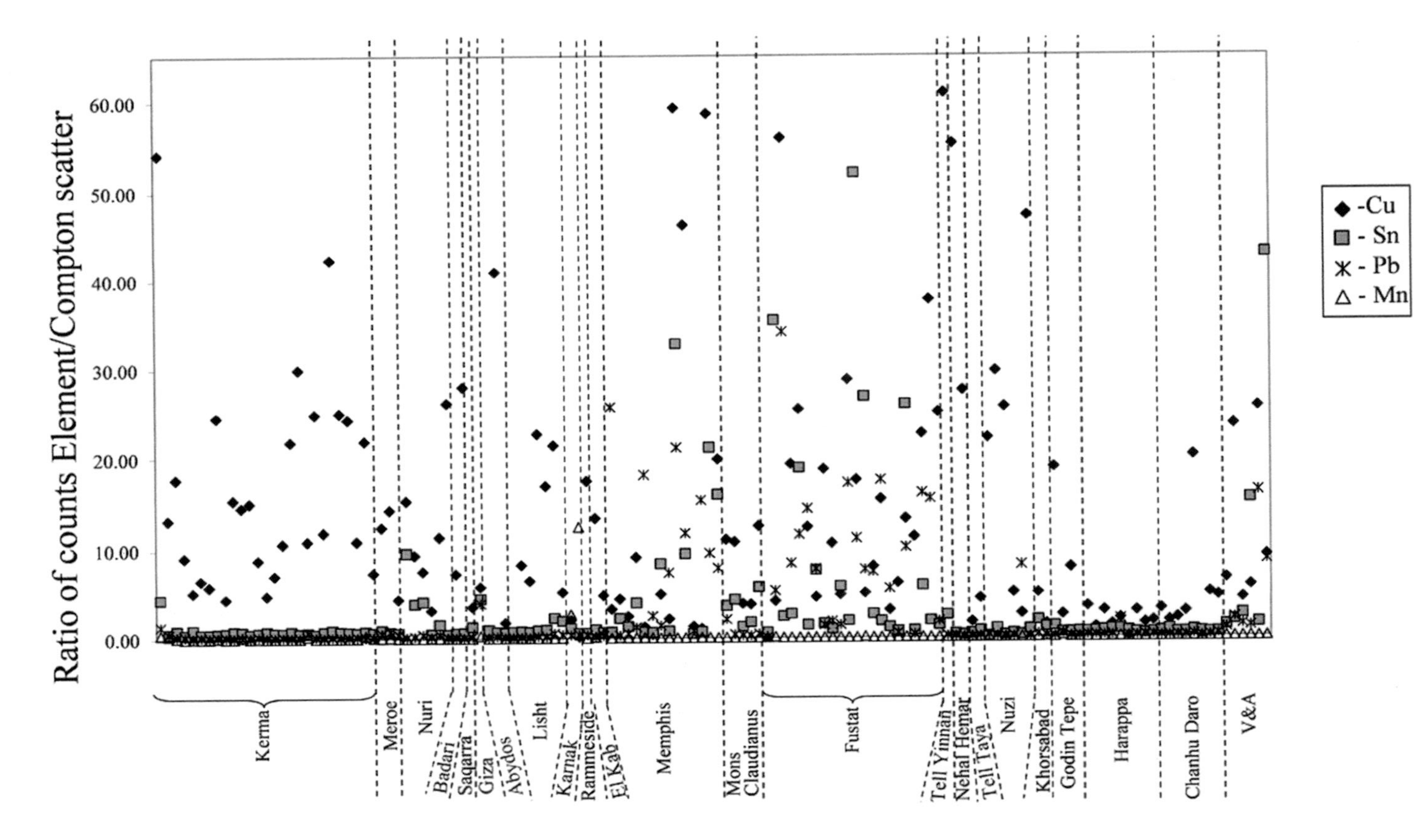

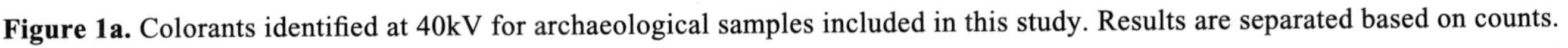
Figure 1a. Colorants identified at 40kV for archaeological samples included in this study. Results are separated based on counts.

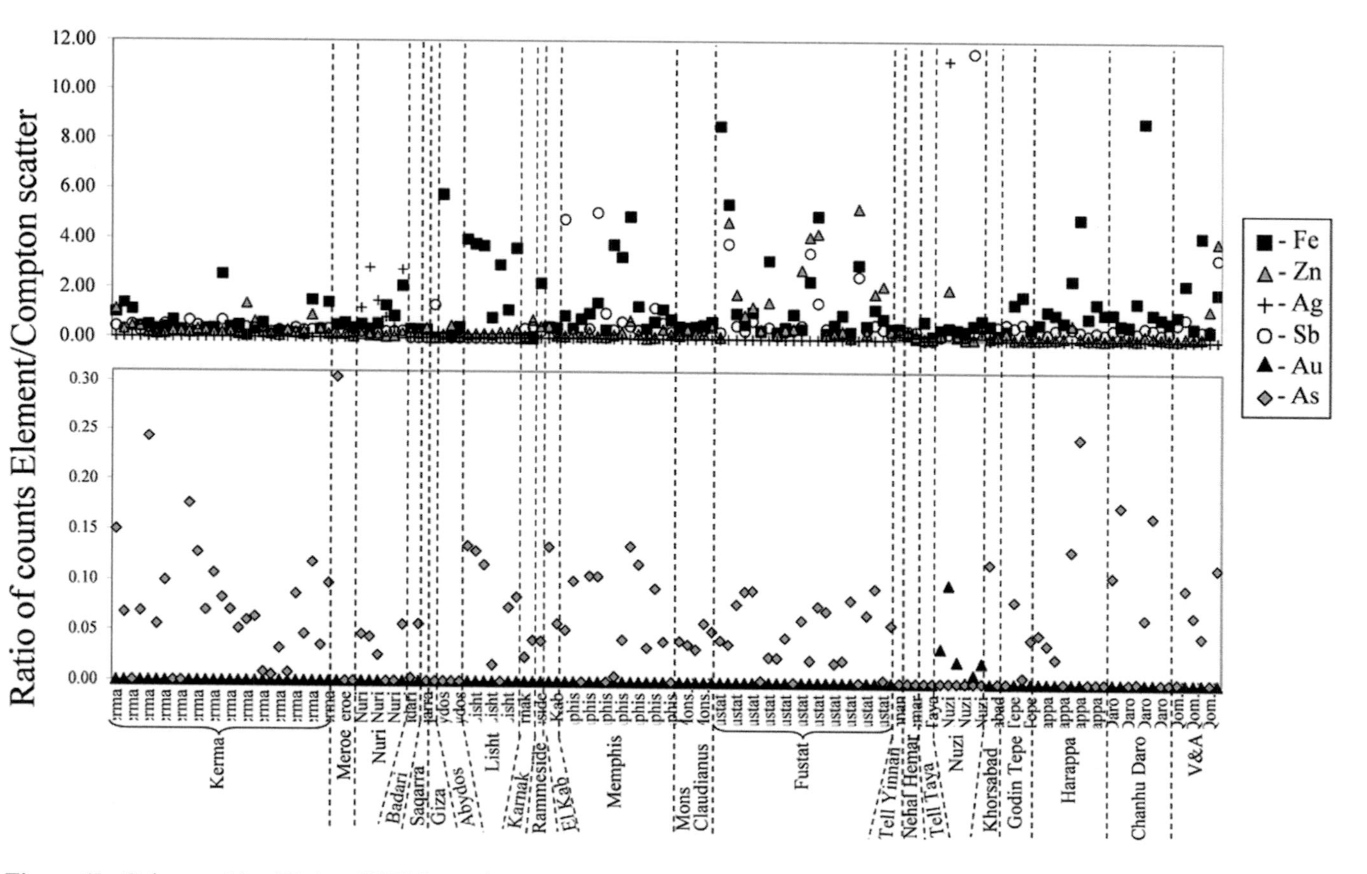

Figure 1b. Colorants identified at 40kV for archaeological samples included in this study. Results are separated based on counts.

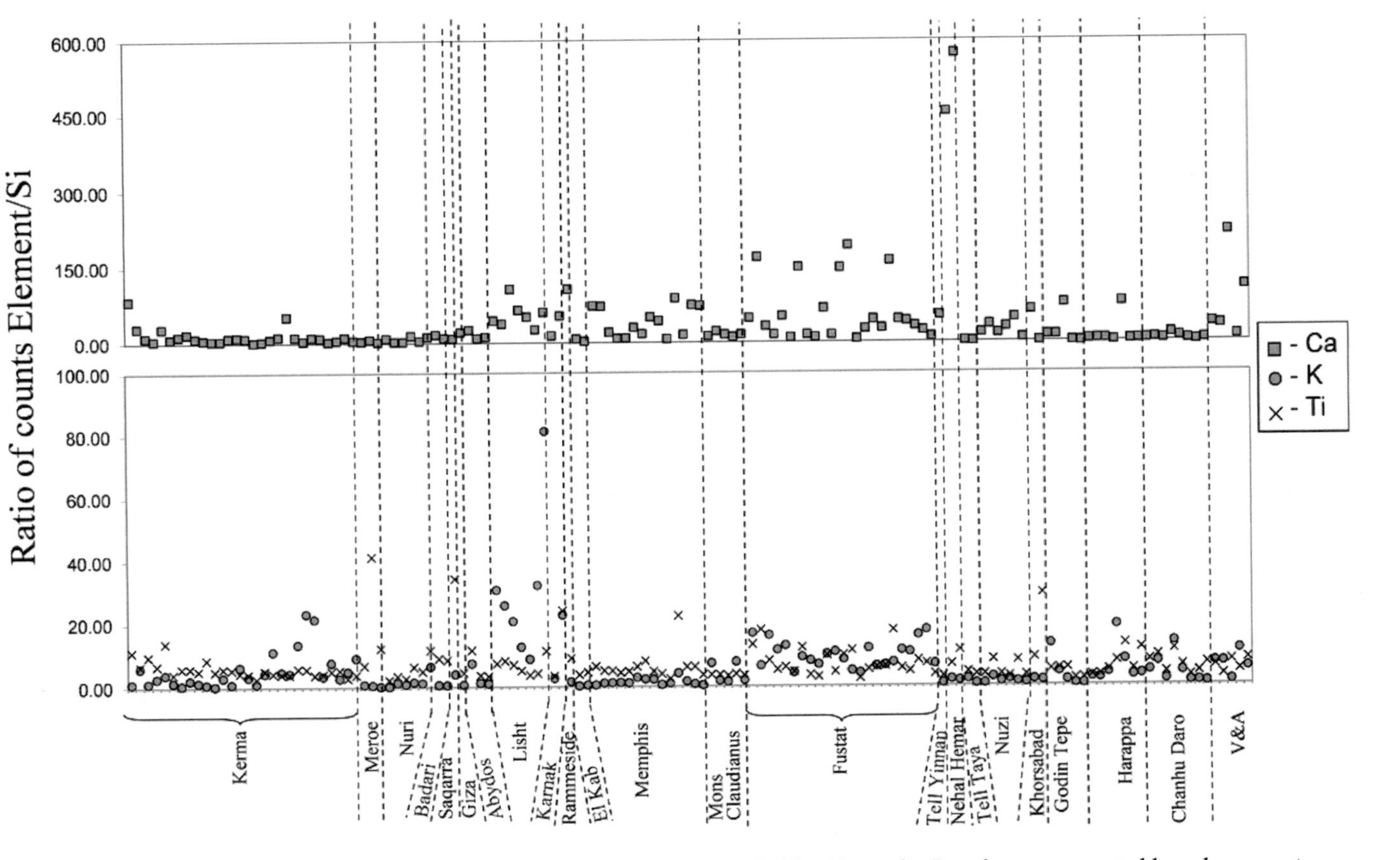

Figure 2a. Colorants identified at 15kV for archaeological samples included in this study. Results are separated based on counts.

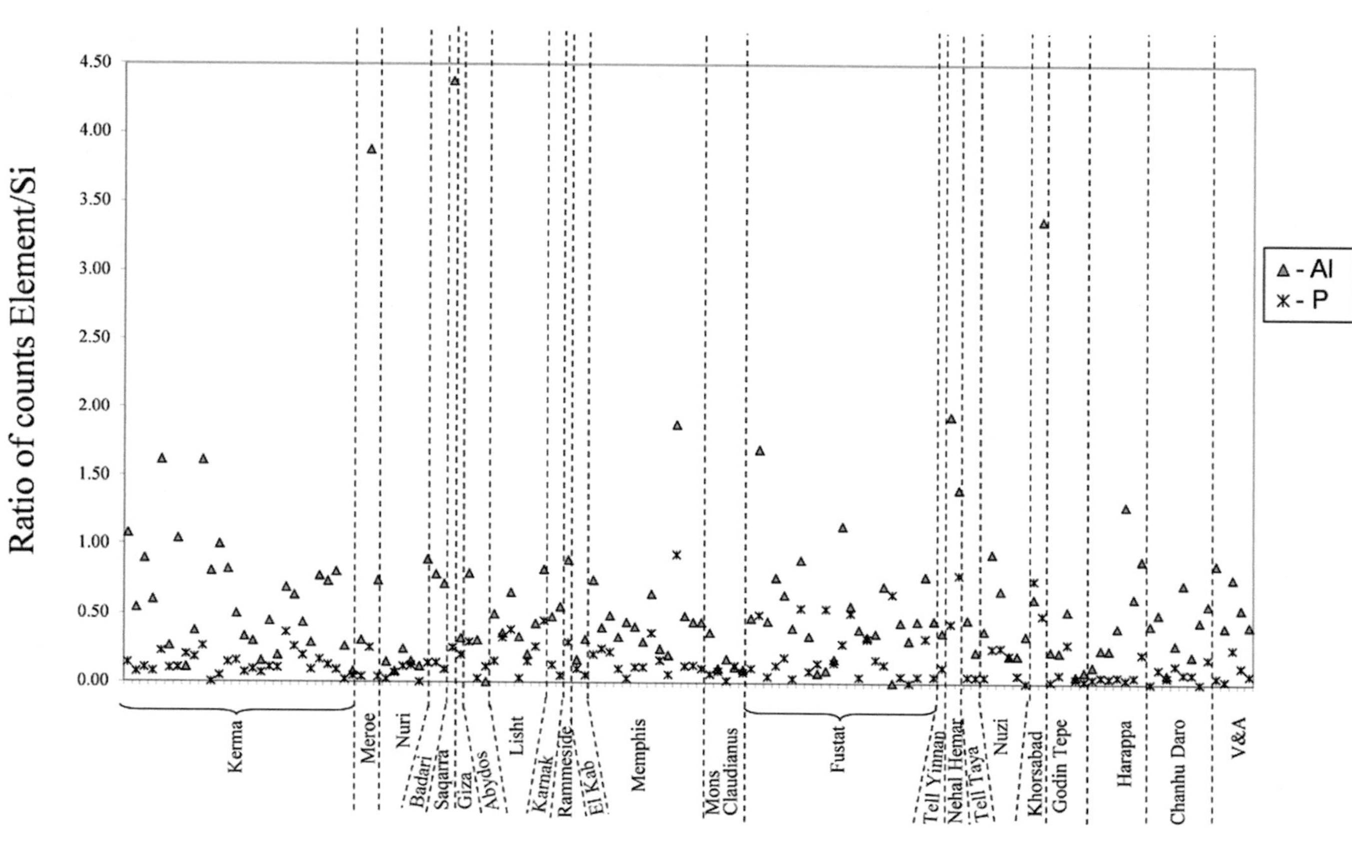

Figure 2b. Colorants identified at 15kV for archaeological samples included in this study. Results are separated based on counts.

AUTHOR INDEX

SUBJECT INDEX